D0847869

COMPLETE BUILDING CONSTRUCTION

edited by John Phelps
revised by Tom Philbin

Macmillan Publishing Company
New York

Collier Macmillan Publishers
London

SECOND EDITION

Copyright © 1978 by Howard W. Sams & Co., Inc.
Copyright © 1983 by The Bobbs-Merrill Co., Inc.
Copyright © 1986 by Macmillan Publishing Company, a division of
Macmillan, Inc.

All rights reserved. No part of this book may be reproduced or
transmitted in any form or by any means, electronic or mechanical,
including photocopying, recording or by any information storage
and retrieval system, without permission in writing from
the Publisher.

While every precaution has been taken in the preparation of this
book, the Publisher assumes no responsibility for errors or omissions.
Neither is any liability assumed for damages resulting from the use of
the information contained herein.

Macmillan Publishing Company
866 Third Avenue, New York, N.Y. 10022
Collier Macmillan Canada, Inc.

Library of Congress Cataloging-in-Publication Data

Phelps, John, 1932-
   Complete building construction.

   Previously published: Indianapolis : T. Audel, c1983.
   Includes index.
   1. Building—Handbooks, manuals, etc.   I. Philbin,
Tom, 1934-   . II. Title.
TH151.P49   1986   690   86-8700
ISBN 0-672-23377-0

Macmillan books are available at special discounts for bulk purchases
for sales promotions, premiums, fund-raising, or educational use.
For details, contact:

   Special Sales Director
   Macmillan Publishing Company
   866 Third Avenue
   New York, N.Y. 10022

10 9 8 7 6 5 4

Printed in the United States of America

Foreword

The ways and means of constructing a building have evolved into a fairly standardized procedure, and this procedure is presented in *Complete Building Construction*. This book assumes that you have a reasonable familiarity with the basic tools of carpentry and masonry and that you have a measure of mechanical aptitude. If you don't have these things, but are determined to construct a building, you will find that Audel's *Carpenters and Builders Library* and *Masons and Builders Library* will take you through the basics of each trade.

Plumbing and electrical wiring are not included because they are not strictly a part of building construction, and many buildings require neither one nor the other. In addition, many builders employ specialists to complete these phases, rather than attempt to do the job themselves.

Complete Building Construction will guide you through the tough parts with advice from accomplished craftsmen in carpentry and masonry, but once your building stands finished, it will be a testimonial to *your* workmanship.

One purpose of this book is to help you be justifiably pleased with that workmanship.

Contents

CHAPTER 4

CHAPTER 5

CHAPTER 6

CHAPTER 7

CHAPTER 8

CHAPTER 9

CHAPTER 10

CHAPTER 11

CHAPTER 12

CHAPTER 13

CHAPTER 14

CHAPTER 15

CHAPTER 16

CHAPTER 17

CHAPTER 18

CHAPTER 19

CHAPTER 20

CHAPTER 21

CHAPTER 22

CHAPTER 23

CHAPTER 24

CHAPTER 25

CHAPTER 26

CHAPTER 27

CHAPTER 28

CHAPTER 29

CHAPTER 30

CHAPTER 31

CHAPTER 32

CHAPTER 33

CHAPTER 1

Laying Out

The term *laying out* here means the process of locating and fixing reference lines, which define the position of the foundation and outside walls of a building to be erected, (Fig. 1-1).

SELECTION OF SITE

Preliminary to laying out (sometimes called *staking out*), it is important that the exact location of the building on the lot be properly selected. In this examination it may be wise to dig a number of small, deep holes at various points, extending to a depth a little below the bottom of the basement.

The *ground water*, which is sometimes present near the surface of the earth, will appear in the bottom of the holes if the holes extend down to its level. This water nearly always stands at the same level in all the holes.

Fig. 1-1. The first step in laying out a building is to measure overall property size. A long tape is invaluable in this and other operations.

If possible in selecting the site for the house, it should be located so that the bottom of the basement is above the level of the ground water. This may mean locating the building at some elevated part of the lot or reducing the depth of excavation. The availability of storm and sanitary sewers, and their depth, should have been previously investigated. The distance of the building from the curb is usually stipulated in city building ordinances.

STAKING OUT

After the approximate location has been selected, the next step is to *lay out the building lines.* The position of all corners of the building must be marked in some way so that when the excavation is begun, the workmen may know the exact boundaries of the basement walls. There are two main methods of laying out these lines:

1. With surveyor's instrument.
2. By method of diagonals.

The second method will do for most jobs, but the efficient carpenter or contractor will be equipped with a level or transit, with which the lines may be laid out with precision and more convenience than by the first method.

The Lines

Several lines must be located at some time during construction, and they should be carefully distinguished. They are:

1. The *line of excavation*, which is the outside line.
2. The *face line* of the basement wall inside the excavation line.
3. The *ashlar line*, which indicates the outside of the brick or stone walls.

In the case of a wooden structure, only the two outside lines need to be located, and often only the line of the excavation is determined at the outset.

Laying Out with Transit Instruments

A transit may be used (Fig. 1-2), and as mentioned is an instrument of precision, and the work of laying out is more accurate than when other methods are employed. In Fig. 1-3, let *ABCD* be a building already erected, and at a distance from this (at right angle) building *GHJK* will be erected. Level up the instrument at point *E*, making *A* and *E* the distance the new building will be from points *A* and *B*. Make points *B* and *F* the same length as points *A* and *E*. At this point, drive a stake in the ground at point *G*, making points *F* and *G* the required distance between the two buildings. Point *H* will be on the same line as point *G*, making the distance between the two points as required.

Place the transit over point *G*, and level it up. Focus the transit telescope on point *E* or *F* and lock into position. Turn the horizontal circle on the transit until one of the zeros exactly coincides with the vernier zero. Loosen the clamp screw and turn the telescope

15

and vernier 90 degrees. This will locate point *K*, which will be at the desired distance from point *G*. For detailed operation of the transit, see the manufacturer's information. The level may be used in setting floor timbers, aligning shafting, and locating drains.

Method of Diagonals

All that is needed in this method is a line, stakes, and a steel tape measure. Here, the right angle between the lines at the corners of a rectangular building is found by calculating the length of the diagonal which forms the hypotenuse of a right-angle triangle. By applying the following rule, the length of the diagonal (hypotenuse) is found.

Rule—The length of the hypotenuse of a right-angle triangle is equal to the square root of the sum of the squares of each leg.

Fig. 1-2. A Berger builders transit, used by builders, contractors, and agricultural planners for setting grades, batter boards, and various earth excavations.

Thus, in a right-angle triangle ABC, of which AC is the hypotenuse,

$$AC = \sqrt{AB^2 + BC^2} \quad\text{...} (1)$$

Suppose, in Fig. 1-4, $ABCD$ represents the sides of a building to be constructed, and it is necessary to lay out these lines to the dimensions given. Substitute the values given in equation (1), thus,

$$AC = \sqrt{30^2 + 40^2} = \sqrt{900 + 1600} = \sqrt{2500} = 50$$

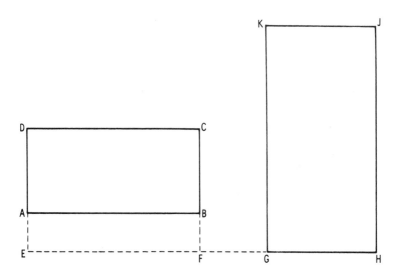

Fig. 1-3. Method of laying out with transit instrument.

To lay out the rectangle of Fig. 1-4, first locate the 40-ft. line AB with stake pins. Attach the line for the second side to B, and measure off on this line the distance BC (30 feet), point C being indicated by a knot. This distance must be accurately measured with the line at the *same tension* as in A and B.

With end of a steel tape fastened to stake pin A, adjust the position of the tape and line BC until the 50-foot division on the tape coincides with point C on the line. ABC will then be a right angle and point C will be properly located.

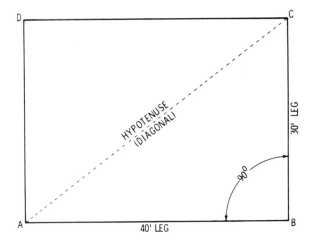

Fig. 1-4. How to find the length of the diagonal in laying out lines of a rectangular building by the method of diagonals.

The lines for the other two sides of the rectangle are laid out in a similar manner. After thus obtaining the positions for the corner stake pins, erect batter boards and permanent lines as shown in Fig. 1-5. A simple procedure may be used in laying out the foundations for a small rectangular building. Be sure that the opposite sides are equal, and then measure *both* diagonals. No matter what this distance may be, they will be equal if the building is square. No calculations are necessary, and the method is precise.

POINTS ON LAYING OUT

For ordinary residence work, a surveyor or the city engineer is employed to locate the lot lines. Once these lines are established, the builder is able to locate the building lines by measurement.

A properly prepared set of plans will show both the present countour of the ground on which the building is to be erected and the new grade line to be established after the building is completed. The most convenient method of determining old grade lines and establishing new ones is by means of a transit, or with a Y level and a rod. Both instruments work on the same principle in

18

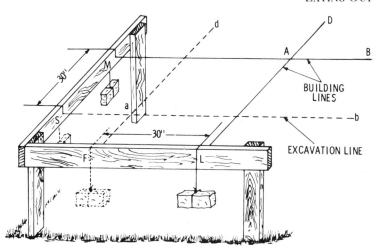

Fig. 1-5. Permanent location of layout lines made by cutting in batter boards (boards marked *S*, *M*, *F*, *L*). Slits *L* and *M* locate the building lines. Approximately 30 in. away are lines *F* and *S*, which are excavation lines.

grade work. As a rule, a masonry contractor has his own Y level and uses it freely as the wall is constructed, especially where levels are to be maintained as the courses of material are placed.

In locating the earth grade about a building, stakes are driven into the ground at frequent intervals and the amount of "fill" indicated by the heights of these stakes. Grade levels are usually established after the builders have finished, except that the mason will have the grade indicated for him where the wall above the grade is to be finished differently than below grade. When a Y level is not available, a 12- or 14-ft. straightedge and a common carpenter's level may be used, with stakes driven to define the level.

19

CHAPTER 2

Concrete Slabs, Walks, and Driveways

Successful placing of concrete on flat ground requires considerable preparation in advance. This preparation includes:

1. Determining the slope of the land and correcting it if necessary.
2. Removing the earth to the depth required for the thickness of the concrete.
3. Firming the subgrade to prevent subsequent sinking.
4. Placing edge forms or screeds.
5. Placing joints to allow for expansion.
6. Correct pouring of the concrete.
7. Leveling the concrete preparatory to finishing.

USING A TRANSIT OR LEVEL

Large sites, such as home development areas, must be staked out and graded to provide level individual homesites for pouring horizontal concrete slabs. Before bulldozers can be ordered in, grade stakes must be placed. This can be done with a builders level or transit.

Establishing Elevations With a Builders Level

Set up the instrument where locations for which elevations are to be determined may be seen through the telescope. Level up the instrument and take a reading on the measuring rod by means of the horizontal cross hair in the telescope. The rod is then moved to the second point to be established. Then the rod is raised or lowered until the reading is the same as the original. The bottom of the rod is then at the same elevation as the original point.

Measuring Difference in Elevation

To obtain the difference in elevation between two points, such as A and B in Fig. 2-1, set up and level the instrument at an intermediate point C. With the measuring rod on point A, note the reading where the horizontal cross hair in the telescope crosses the graduation marks on the rod. Then, with the rod held on point B, sight on the rod and note where the horizontal cross hair cuts the graduations on the rod. The difference between the reading at point A (5 ft.) and the reading at point B (5½ ft.) is the difference in elevation between points A and B. Thus point B is ½ ft. lower than point A.

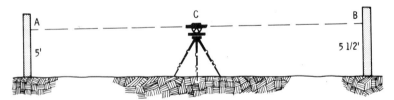

Fig. 2-1. The builder's level and measuring rod.

When, for any reason such as irregularity of the ground or a large difference in elevation, the two points whose difference in elevation is to be determined cannot be sighted from a single point, intermediate points must be used for setting up the instrument, as shown in Fig. 2-2.

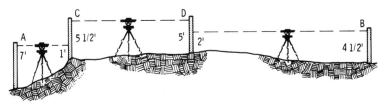

Fig. 2-2. Measuring elevation differences—several sightings are necessary.

Establishing Points on a Line with a Transit

Level the instrument and center it accurately over a point on the line by means of a plumb bob. Then sight the telescope on the most distant visible known point of that line. Lock the horizontal motion clamp screw to keep the telescope on line and place the vertical cross hair exactly on the distant point with the tangent screw. Then, by rotating the telescope in the vertical plane, the exact location of any number of stakes on that same line may be determined (Fig. 2-3).

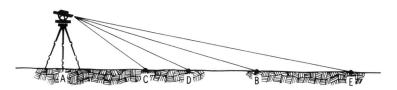

Fig. 2-3. A transit being used to establish points along a given line.

PREPARING THE SITE

Before placing any concrete, the forms or screeds must be set to proper grade. The grade is sometimes determined by using a builder's level. The level is sighted on a measuring rod or rule that is set on an established grade (Fig. 2-4A) and a reading is taken. Then the rod or rule is moved to where the form or screed will be placed (Fig. 2-4B). With a stake in the ground at this point, the rod or rule is raised or lowered until the desired grade is read by the level. A mark or nail is placed on the stake at the bottom of the rule

23

(Fig. 2-4C). This operation is continued until enough stakes are set and marked with the proper grade and alignment.

A string line is then strung tightly from stake to stake at the marked position (Fig. 2-4D). This string line will be at the top of the form or screed. The distance from the string line to the subgrade is checked to make sure there is enough depth to place the form or screed (Fig. 2-4E). If not, the subgrade is dug out until there is enough clearance. The forms or screeds are then set to proper line and grade by following the string line (Fig. 2-4F).

The forms or screeds are then well staked and braced. The stakes must be driven straight to ensure that the forms or screeds will be true and plumb. When the grade of a narrow walk is being established, it is sometimes more convenient to set one edge form and then use a spirit level, as shown in Fig. 2-5, to set the other edge form.

All sod and vegetable matter must be removed from the construction site (Fig. 2-6), and any soft or mucky places must be dug out, filled with at least 2 in. of granular material such as sand, gravel, or crushed stone, and thoroughly tamped (Fig. 2-7). Exceptionally hard compact spots must be loosened and then tamped to provide the same uniform support for the slab as the remainder of the subgrade.

SAND BASE

If the soil is porous and has good drainage, concrete may be poured directly on the earth if the earth is well tamped. If the soil has a lot of clay and drainage is poor, it would be well to put down a thin layer of sand or gravel on which to pour the concrete (Fig. 2-8).

A short time before pouring, give the soil a light water sprinkling (Fig. 2-9). Avoid developing puddles; when the earth is clear of excess water you can begin pouring concrete. When additional fills are required under walks, driveways, or floors to bring them to the proper grade, the fills should also be of a granular material thoroughly compacted in a maximum of 4-in. layers. It is best to extend the top of all fills at least 1 ft. beyond the edges of walks and drives, and to make the slope of the fill flat enough to prevent undercutting during rains.

The top 6 in. of the subgrade should be sand, gravel, or crushed stone where subgrades are water soaked most of the time. These granular subgrades must be drained to prevent the collection of water. Well-compacted, well-drained subgrades do not require such special granular treatment.

EXPANSION JOINTS

Where a new concrete slab abuts an existing walk, driveway, building, curb, lighting standard, fireplug, or other rigid object, a premolded material, usually ½ in. thick, should be placed at the joint. These joints are commonly called *expansion joints*. They are placed on all four sides of the square formed by the intersection of two walks. When the sidewalk fills the space between the curb and a building or wall, an expansion joint should be placed between the sidewalk and the curb and between the sidewalk and the building or wall. Expansion joints are not required at regular intervals in the sidewalk.

PREPARING THE MIX

The concrete should contain only enough water to produce a concrete that has a relatively stiff consistency, works readily, and does not separate. Concrete should have a slump of about 3 in. when tested with a standard slump cone. The adding of more mixing water to produce a higher slump than specified lessens the durability and reduces the strength of the concrete.

In northern climates where flat concrete surfaces are subjected to freezing and thawing, air-entrained concrete is necessary. Air-entrained concrete is made by using an air-entraining portland cement or by adding an air-entrained agent during mixing. Before the concrete is placed, the subgrade should be thoroughly dampened so that it is moist throughout, but without puddles of water.

Concrete should be placed between the forms or screeds as near to its final position as practicable. Precautions should be taken not

(A) Sighting transit on measuring rod.

(B) Sighting measuring rod for placing concrete forms

(C) Marking proper height for concrete form.

Fig. 2-4. Steps in using a transit to

(D) Placing string line at proper height for form.

(E) Checking string line for proper subgrade depth.

(F) Placing concrete forms using string lines as a guide.

set forms in a laterally straight line.

Fig. 2-5. With one concrete form in place, the other side is constructed by using a spirit level.

to overwork the concrete while it is still plastic because an excess of water and fine material will be brought to the surface, which may lead to scaling or dusting later on. The concrete should be thoroughly spaded along the forms or screeds to eliminate voids or honeycombs at the edges.

FORMS

Edge forms may be of wood or metal. Contractors would do well to purchase metal forms that are designed for concrete work and that may be used again and again, indefinitely. Wood forms are usually 2″ × 4″, 2″ × 6″, or wider, depending on the thickness of the concrete.

Forms must be placed carefully, as the tops of the forms become screed guides for leveling the concrete. Distances apart must be measured accurately and a spirit level used to assure that they are horizontal. However, if the forms are used on an inclined driveway, they must follow the incline. Forms for curved walks or driveways may be made from ½ -in. redwood. When using redwood, soak it with water for about 20 minutes, then it may be easily bent without splitting.

Stakes are placed at intervals along the outside of the forms and driven into the earth, then nailed to the forms to hold them securely in place. The tops of the stakes must be slightly below the edge of the forms so they will not interfere with the use of the strike-off board for screeding later.

Fig. 2-6. Removing sod or plant growth from construction sight.

Fig. 2-7. Tamping soil firmly.

DRIVEWAYS

Driveway construction varies, depending on the owner's preference. While most driveways are flat surfaces with a width great enough for a single car, or two cars, some are crowned, some depressed, and some consist of concrete runways laid parallel, which allows for grass in between. A few driveways will be below grade level, with curbs at the edges. Fig. 2-10 shows cross sections of representative driveways other than the flat slab. The crowned and center-depressed driveway will require making a strike-off board with the proper shape. Fig. 2-11 shows the details of a driveway with a 1-in. crown, and the strike-off board used for floating. The crown in the center of the driveway permits the run-off of water, especially important as ice or snow melts off the driveway.

Fig. 2-12 shows a cross section of a level driveway, which is often preferred, especially since a level slab is easier to screed and finish. Concrete highways are built perfectly level except on curves. It is important on high-speed roads that there be no pull to the side on the car as a result of a tilt in the road. This reduces driver fatigue on straight runs.

Fig. 2-8. Spreading a thin layer of sand or aggregate for a concrete base.

SIDEWALKS

Sidewalk construction is like that of driveways, except no curbs are used of the type shown and described. Figs. 2-13 and 2-14 show the construction of sidewalks, which include expansion joints.

31

WHERE TO PUT JOINTS

Concrete expands and contracts slightly due to temperature differences. It may also shrink as it hardens. Joints should be put in the concrete to control expansion, contraction, and shrinkage.

It is desirable to prevent slabs-on-ground, either inside or outside, from bonding to the building walls. Thus, the slab will be free to move with the earth. To prevent bonding, a continuous rigid waterproof insulation strip, building paper, polyethylene, or similar material is placed next to the wall. These materials are also used next to other existing improvements, such as curbs, driveways, and feeding floors. The continuous rigid waterproof insulation strip acts as an isolation joint.

Wide areas, such as floor slabs and feeding floors, should be paved in 10'- to 15'-wide alternate strips. A *construction joint*, also known as a *key joint*, is placed longitudinally along each side of the first strips paved. A construction joint is made by placing a beveled piece of wood on the side forms. This creates a groove in the slab edges. As the intermediate strips are paved, concrete fills

Fig. 2-9. Sprinkling before pouring concrete will settle dust and compact soil.

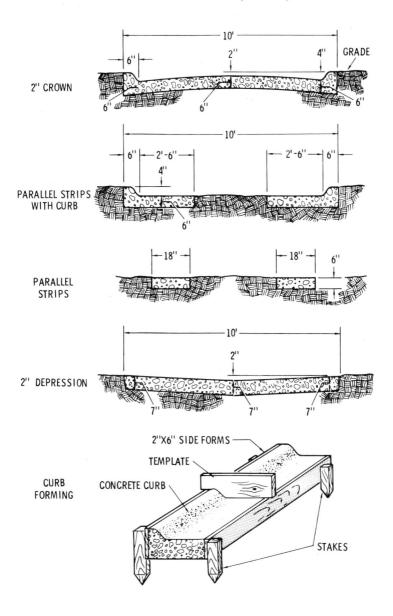

Fig. 2-10. A variety of driveway designs.

33

this groove, and the two slabs are keyed together. This type of joint keeps the slab surfaces even and transfers the load from one slab to the other when equipment is driven on the slabs.

Contraction joints, often called *dummy joints*, are cut across through the slab, but are cut to a depth of one-fifth to one-fourth the thickness of the slab, thus, making the slab weaker at this point. If the concrete cracks due to shrinkage or thermal contraction, the crack usually occurs at this weakened section (Fig. 2-15).

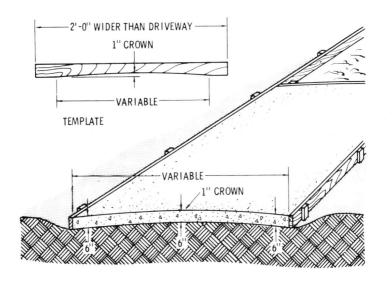

Fig. 2-11. Driveway with a 1-in. crown.

Contraction joints should be cut soon after the concrete has been placed in order to work the larger pieces of coarse aggregate away from the joint. A simple way to cut this type of joint is to lay a board across the fresh concrete and cut the joint to the proper depth with a spade, axe, or similar tool. A groover is then used to finish the joint. Contraction joints are generally placed 10 to 15 ft. apart on floor slabs, driveways, and feeding floors. They are placed 4 to 5 ft. apart on sidewalks. All open edges should be finished with an edger to round off the edge of the concrete slab to prevent spalling.

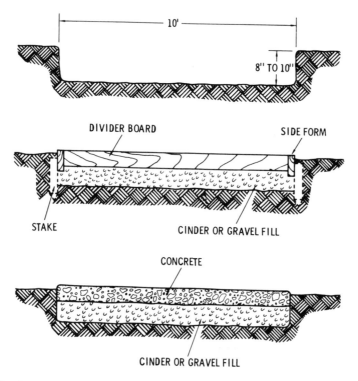

Fig. 2-12. Cross section of a typical single-car driveway.

POURING THE CONCRETE

Concrete should be poured within 45 minutes of the time it was mixed; otherwise, some curing begins to take place, and the concrete may become too thick to handle with ease. If you are pouring a large area, mix only as much as you can handle within the 45 minutes. In fact, the first batch should be a small one for trial purposes. After the first batch, you can determine whether the succeeding batches require more or less sand. *Remember*, do not vary the water for a thicker or thinner concrete—only the amount of sand. Once you have established the workability you like, stay with that formula.

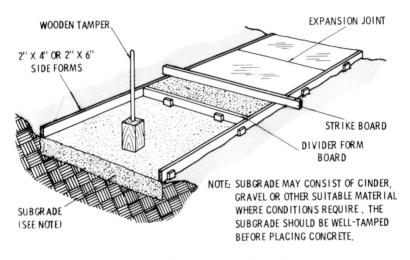

WOODEN TAMPER

2" X 4" OR 2" X 6" SIDE FORMS

EXPANSION JOINT

STRIKE BOARD

DIVIDER FORM BOARD

SUBGRADE (SEE NOTE)

NOTE: SUBGRADE MAY CONSIST OF CINDER, GRAVEL OR OTHER SUITABLE MATERIAL WHERE CONDITIONS REQUIRE. THE SUBGRADE SHOULD BE WELL-TAMPED BEFORE PLACING CONCRETE.

Fig. 2-13. Details in constructing a concrete sidewalk.

Fig. 2-14. This sidewalk width requires an expansion joint in the center.

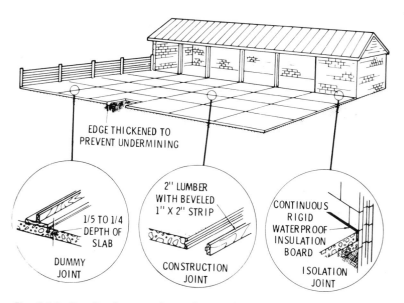

Fig. 2-15. Details of contraction and expansion joints.

Pour in the concrete up to the level of the form edges. Immediately after the first batch is poured into place, spade the concrete with an old garden rake or hoe to even it out approximately to the level of the form boards, and to make sure there are no air voids in the mix.

Level off the concrete with the striker board mentioned earlier in this chapter (Fig. 2-16). It frequently takes two men to do this—one at each end of the board. The object is to get a level and even top surface to the poured concrete using the board forms as the guide. Draw the striker board across the concrete while hugging the form edges; seesaw the board as you move it across. Now you can see why the stakes for the forms had to be below the edges of the form boards: they must not interfere with the movement of the strike board. The strike board takes off high spots and levels the concrete. If the board skips over pockets of low concrete, fill them in and go over it once more.

When the concrete is first poured and struck, there may be a sheen of water on the top, but not necessarily evenly across the top. You must wait until this sheen disappears before you do any

Fig. 2-16. After pouring, level concrete with a strike-off board.

other work on the concrete. This may be an hour or two, depending on temperature, humidity of the air, and the wind. During this time the concrete has begun to cure and harden slightly.

When pouring large areas of concrete, there is a right way and a wrong way. The right way is to provide the means for pouring the farthest point first, working back to the source of the concrete mix. This is illustrated in Fig. 2-17, showing boards laid across the forms or screeds. The boards are placed close enough together to make a runway for the wheelbarrow to take the concrete to the farthest point first. As the areas are filled, boards are removed for pouring the next load. Cross-board separators may be laid between the forms to make it possible to screed or level a section as a time. If a long area is poured at one time, the first concrete to be poured may begin to take a set and make leveling more difficult. As the strike-off board is drawn forward, excess concrete is allowed to fall into the next preceding section.

Fig. 2-18 shows the complete setup for pouring a concrete tennis court or the slab of a barn or house. Note how the material is set out for easy dumping into the mixer, in addition to the wheelbarrow runways for carrying the concrete to the farthest point first.

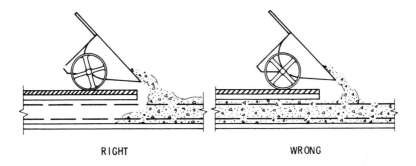

Fig. 2-17. The right and wrong way to pour concrete. Pour farthest point first.

When pouring concrete from a *ready-mix* truck, omit the edging form at the entrance for the truck. Provide means for the chute of the truck to reach the last section first. Ready-mix trucks are equipped with extension chutes which may be added to the fixed

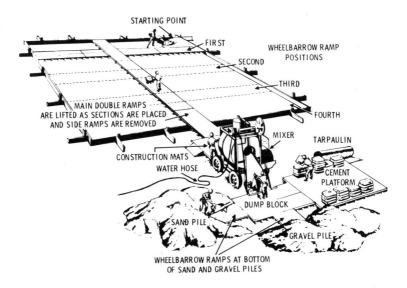

Fig. 2-18. Setup for pouring a large concrete slab.

39

chute on the truck. Chutes are on a swivel for laying the concrete exactly where needed. Since ready-mix trucks work fast, ample help must be on hand for leveling as the pouring takes place, or for wheelbarrows to haul to any position where truck chutes cannot reach.

CHAPTER 3

Reinforced Concrete

Concrete has tremendous compressive strength and durability. These are the reasons for its popularity in many construction projects. However, it has poor strength in tension. A beam made only of concrete and supported at its ends would not bear much weight at its center. Fig. 3-1 is a sketch of a concrete beam with stresses at the points shown with arrows. Cracks would develop at the small hash marks, and the concrete would soon give way.

By adding steel supports embedded into the concrete, its tensile strength is improved considerably. Steel reinforcement can add the equivalent of six times the added thickness of concrete for a given load. Fig. 3-2 shows a cross section of a concrete beam with reinforcing bars embedded near the bottom. The bars act in opposition to the stress at the bottom as a result of loads at the top.

Hollow concrete columns, frequently used to support upper floors and roofs, have a tendency to bulge outward with heavy top loading. This is illustrated in Fig. 3-3 for both square and round

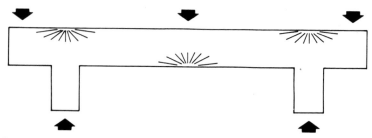

Fig. 3-1. A concrete beam showing stress points.

columns. Steel reinforcing supplies the lateral strength to overcome the effect of bulging.

Until concrete is fully cured, which may take up to a year, it has
a tendency to creep due to plastic flow. If subjected to loads after
the initial curing (typically 28 days), concrete may become slightly
deformed. Steel reinforcement reduces creep considerably. After

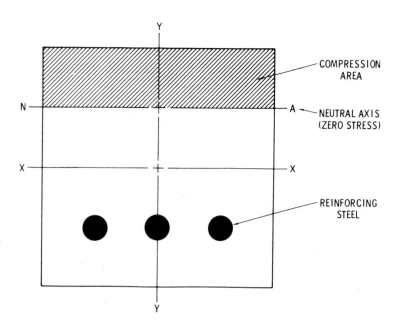

Fig. 3-2. Cross section of reinforcement in a concrete beam.

a period of time, perhaps one year, all plasticity in concrete is gone, and creeping no longer occurs.

Of importance in the successful reinforcement of concrete by steel is good bonding between the steel and the concrete. Reinforcing steel must be free of rust, scale, oil, or any surface material that could affect bonding. If a good bond does not exist between the steel and the concrete, all the beneficial effect of reinforcement is lost. Plain-surfaced bars have been replaced by bars with a rough design on the outside, called *deformed* bars. Fig. 3-4 shows three examples of the surface texture on deformed bars. Concrete

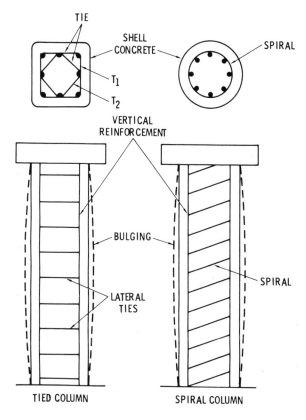

Fig. 3-3. The tendency of concrete columns to bulge laterally with a load is overcome by lateral steel reinforcement.

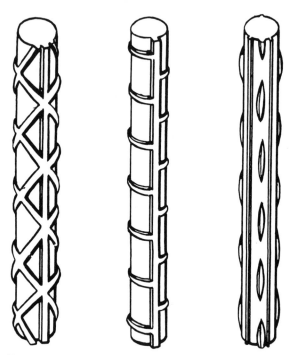

Fig. 3-4. Deformed bars are used for better bonding between concrete and steel bars.

flows into and around the deformations and, when cured, provides a perfect bond to the steel.

REINFORCING RODS AND SCREEN

Reinforcing rods or bars may be round or square. Smaller sizes are usually round, while those over 1 in. in diameter may be round or square. Table 3-1 shows the bar numbers and their sizes.

To support bars the proper distance above the bottom of concrete beams or slabs, *chairs* or *bolsters* are used. Fig. 3-5 shows four common types. Depending on the application, bar assemblies may contain hooks, trusses, and stirrups, as well as bolsters welded into place. Fig. 3-6 shows an example and names the parts. To reduce handling on the job, reinforcing steel trusses may be

44

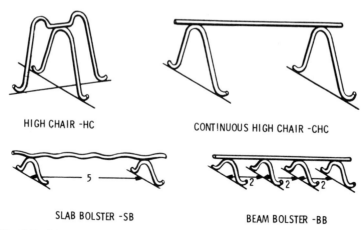

HIGH CHAIR -HC   CONTINUOUS HIGH CHAIR -CHC

SLAB BOLSTER -SB   BEAM BOLSTER -BB

Fig. 3-5. Chairs or bolsters used to support bars in concrete beams.

made to order for a specific need. Fig. 3-7 is one example of this. This one is called the *Kahn* trussed bar, a popular configuration over the years, but it has been replaced by made-to-order methods that better fit each job. Small jobs use straight stock or hooked bars

Table 1. Reinforcing Bar Numbers and Dimensions

Bar sizes*		Weight	Cross-sectional
Old (inches)	New numbers	(lbs. per. ft.)	area (sq. in.)
1/4	2	0.166	0.05
3/8	3	0.376	0.1105
1/2	4	0.668	0.1963
5/8	5	1.043	0.3068
3/4	6	1.502	0.4418
7/8	7	2.044	0.6013
1	8	2.670	0.7854
1/square	9	3.400	1.0000
1 1/8 square	10	4.303	1.2656
1 1/4 square	11	5.313	1.5625

*Note. The new bar numbers indicate the approximate number ot eighths of an inch included in the nominal diameter of the bar. Bars numbered 9, 10, and 11 are round bars and equivalent in weight and nominal cross-sectional area to the old type 1", 1 1/8" and 1 1/4" square bars.

that are laid on bolsters in a crosswise pattern. For thin slabs, screens are available in rolls, as shown in Fig. 3-8.

Steel reinforcement in columns may take a number of forms. Fig. 3-9 shows three of them. *A* shows the vertical rods held together with tranverse steel wire. Cross-bracing from corner to corner is often added. *B* is a spiral coil of steel wire fastened to vertical rods for support. This is the most popular type for round columns. Heavy columns may include steel T-beams, as illustrated in *C*.

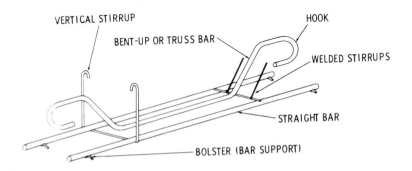

Fig. 3-6. Bar accessories may be welded in place.

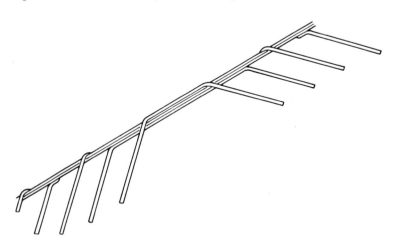

Fig. 3-7. Bars may be ordered with special configurations.

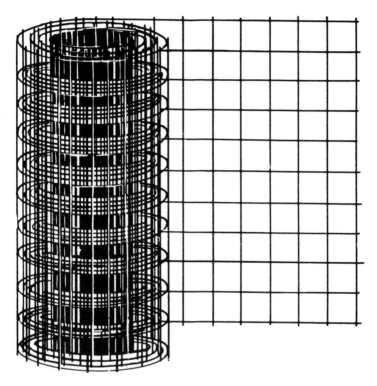

Fig. 3-8. Concrete slab reinforcement is usually in the form of screen mesh.

Another type of slab reinforcing steel (though seldom used) is the expanded steel sheet made in several patterns. Two of these are shown in Fig. 3-10. Expanded metal is made from flat steel stock punched with a pattern. When the pattern is pulled apart, it forms an open mesh.

HOW TO USE REINFORCEMENT

On industrial buildings, dams, and other heavy concrete construction projects, the architectural and engineering drawings for the structure also include details for the placement of reinforcement members in the concrete. Unless otherwise permitted by the

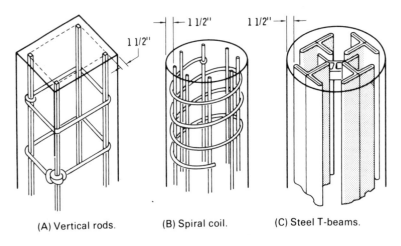

(A) Vertical rods.   (B) Spiral coil.   (C) Steel T-beams.

Fig. 3-9. Three forms of concrete column reinforcement.

plans, rods must never be bent after they are placed in the concrete. A few general rules which are followed by architects and engineers are listed below. These have to do with the thickness of the concrete adjacent to the reinforcement.

1. Where concrete is deposited against the ground *without* the use of forms, not less than 3 in. of concrete is used around the reinforcing steel bars.
2. Where concrete is poured on the ground, but *in* forms, not less than 2 in. of concrete is used around the reinforcing steel bars.
3. In slabs and walls not exposed to ground or weather, not less

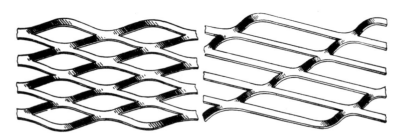

Fig. 3-10. Expanded-metal slab reinforcement.

than ¾ in. of concrete is used. In beams, columns, and girders not exposed to the ground or weather, not less than ½ in. of concrete is used.

4. In all cases, concrete thickness must be at least equal to the diameter of the bars.
5. Bars must be maintained clean of rust or other coatings to assure good bonding with the concrete.
6. Bars bent on the job must be cold bent.
7. Reinforcement must be placed according to the plans and supported with approved chairs, concrete blocks, or other metal spacers.
8. No splices shall be made unless the plans call for them.
9. All welding must follow the approved method outlined by the American Welding Society.

Reinforcement is applied near the area of greatest tension stress. This is usually near the surface opposite the load. This is shown in the cross-sectional view of Fig. 3-1. Large slabs poured on the ground, as for factory floors, barn floors, and house slabs (where this is no basement or crawl space) will be guided by architects drawings specifying the sizes and location of the reinforcing bars. The average distance from the ground is about 1½ in.

The bars are held above the ground by high chairs or bolsters, as previously described. Often it is acceptable to use other forms of raising the bars, such as precast concrete blocks, clean rocks of fairly uniform size, or anything that will become part of the poured concrete. Cross bars are also laid, and either wire-tied or welded to the perpendicular bars, depending on the mass of concrete to be poured.

On large jobs, bonding between the steel bars and the concrete must depend on the irregular surface of the bars alone. The ends of the bars must be bent to form hooks. The hooks make a firm grip on the cured concrete for greater strength and reduced shrinkage during hydration.

For large construction projects where a large number of reinforcing bars are ordered, their length may be specified and the hooks provided by the steel fabricator. On smaller jobs where stock bars are used, it becomes necessary to bend the hooks on the job. Hooks must be cold bent. Fig. 3-11 shows the recommended

RECOMMENDED SIZES - 180° HOOK

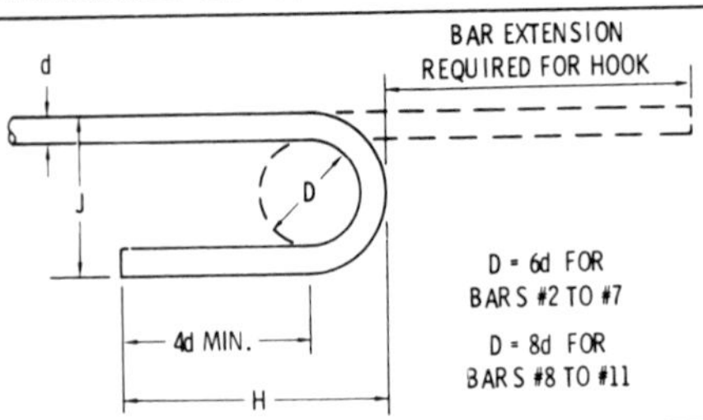
BAR EXTENSION
REQUIRED FOR HOOK
d
J
D
D = 6d FOR
BARS #2 TO #7
D = 8d FOR
BARS #8 TO #11
4d MIN.
H

J	BAR EXTEN.	APPROX. H	BAR SIZE D
2	4	3 1/2	#2
3	5	4	3
4	6	4 1/2	4
5	7	5	5
6	8	6	6
7	10	7	7
10	13	9	8
11 1/4	15	10 1/4	9
12 1/2	17	11 1/4	10
14	19	12 3/4	11

MIMIMUM SIZES - 180° HOOK

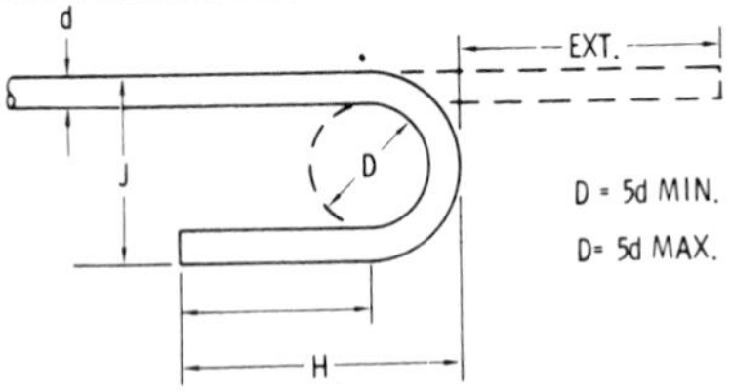
d
EXT.
J
D
D = 5d MIN.
D= 5d MAX.
H

NOTE: MINIMUM SIZE HOOKS TO BE USED ONLY FOR SPECIAL CONDITIONS; DO NOT USE FOR HARD-GRADE STEEL.

J	BAR EXTEN.	APPROX. H	BAR SIZE D
1 3/4	4	3 1/2	#2
2 3/4	5	4	3
3 1/2	5	4 1/4	4
4 1/4	6	4 3/4	5
5 1/4	7	5 3/4	6
6	9	6 1/2	7
7	10	7 1/2	8
8	11	8 1/2	9
9	13	9 1/2	10
10	14	10 1/2	11

RECOMMENDED MINIMUM SIZES - 90° HOOK

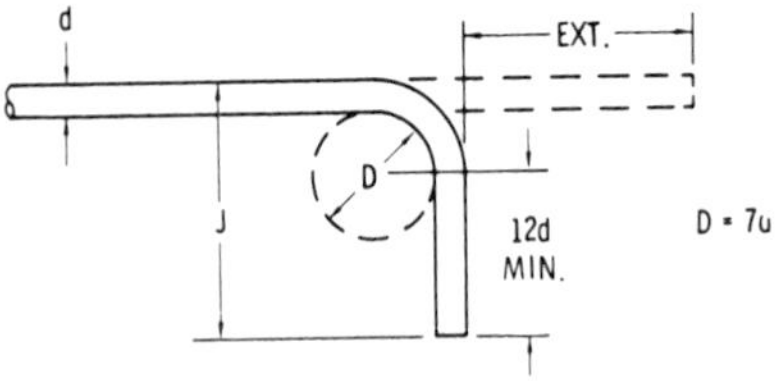
d
EXT.
D
J
12d
MIN.
D = 7u

BAR EXTEN.	APPROX. J	BAR SIZE D
3	3 1/2	#2
3	4	3
3	4 1/2	4
4	5	5
4	6	6
5	7	7
6	9	8
7	10	9
8	11 1/4	10
9	12 1/2	11

RECOMMENDED SIZES - 135° STIRRUP HOOK

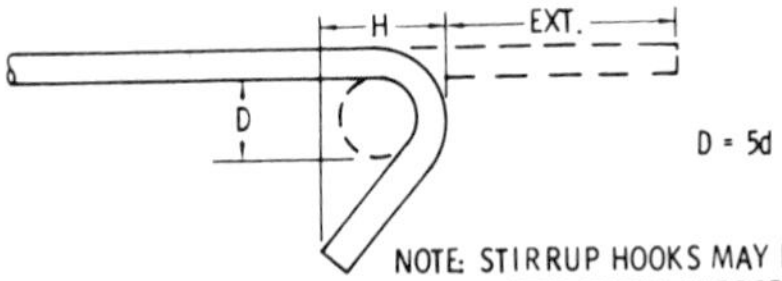
H
EXT.
D
D = 5d

NOTE: STIRRUP HOOKS MAY BE BENT TO THE DIAMETER OF THE SUPPORTING BARS.

BAR EXTEN.	H	BAR SIZE D
3 1/2	2	#2
4	2 1/4	3
4 1/2	2 1/2	4
5	2 3/4	5

Fig. 3-11. Hook dimensions for on-the-job bending of reinforcing bars.

radius of the bend and the length of stock to provide at the end. Fig. 3-12 shows the construction of a table for bending reinforcing bars. In emergencies, a hickey may be made from a $2'' \times 1\frac{1}{2}''$ pipe tee with a 3-ft. length of $1\frac{1}{2}''$ pipe screwed into the center junction. Saw off one arm of the tee to make sharper bends.

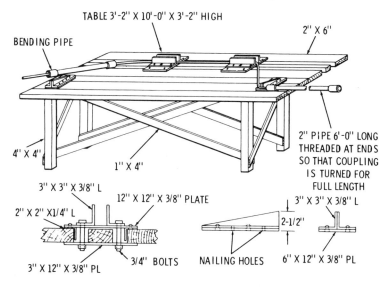

Fig. 3-12. Table for bending reinforcement bars on the job site.

Smaller slabs are reinforced with welded wire screen (available in rolls) as shown in Fig. 3-8. Pieces are cut off to size with a hacksaw or heavy-duty wire cutters. Small-size bolsters are used to support the material above the ground. Small rocks may also be used if they are first washed clean. For a section of 4-in.-thick driveway or walk, for example, the mesh only needs to be about 1 in. above the ground. Another method frequently used is to lay the screen on the ground, pour the concrete, then reach a hook down into the concrete and lift the screen up about 1 in. Remove the hook, then spade and level the concrete. The wire mesh will stay in the position to which it was lifted.

Where the width of the mesh roll is not wide enough to reach to the edges of the concrete, an overlap of screening is important to good strength. This is shown for two forms of screening in Fig. 3-13. Fig. 3-14 shows roll-type mesh for a concrete slab to be laid

51

for a home. This is a common type of home construction in the southwest. Wires are welded where they cross.

The need for reinforcement in footings will depend on the load to be supported. For example, footings for block walls around homes are sometimes poured into a ditch without reinforcement. While the concrete base may be capable of supporting a 5-ft.-high block wall, it could easily develop cracks due to subsoil settling. This settlement will cause cracks at the mortar lines in the block wall. Steel reinforcement rods should be laid in the ditch or form (if used), and supported just above the ground level. High and

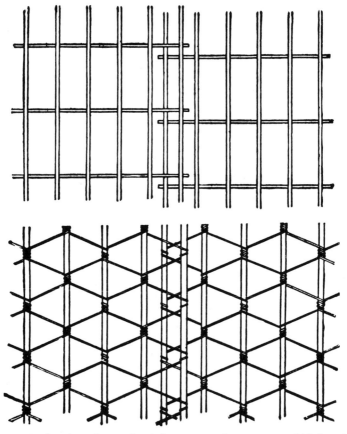

Fig. 3-13. Overlap wire mesh reinforcement where greater width is to be covered.

thick industrial walls, whether of concrete block or poured concrete, must be built on footings with steel reinforcement. Footings for columns, load-bearing walls, and beams must be reinforced.

Fig. 3-15 shows the bars laid in a typical footing for a column. The reinforcing steel bars are laid in place after the forms are in place. Crossed bars are tied together with wire and held above the subgrade by any of the means mentioned before.

Fig. 3-16 shows a method of installing reinforcement in a concrete poured wall. Individual bars, tied where they cross, or welded wire fabric may be used. A convenient method for holding the material in place, centered in the concrete, is to hang them from wood blocks fastened to one side of the form. As the concrete is poured and nears the top of the form, remove the wood blocks. Then complete the pouring to the top.

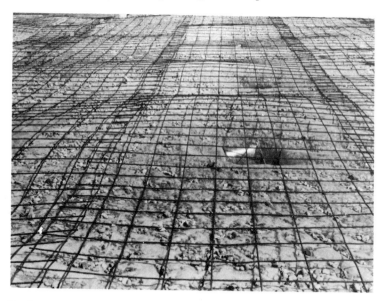

Fig. 3-14. Roll-type mesh reinforcement for a concrete slab.

Concrete beams must have steel reinforcement. The placing of the bars depends on the direction of stress, which in turn depends on the purpose of the beam. Fig. 3-17 is an example of a beam supported on a column, and on which the principal stress is upward in the center and downward at the ends.

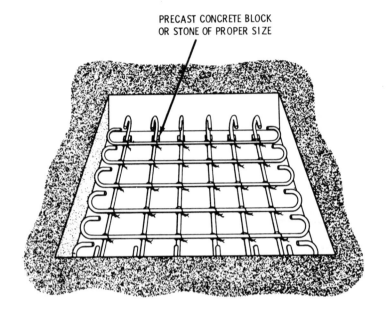

PRECAST CONCRETE BLOCK
OR STONE OF PROPER SIZE

Fig. 3-15. Reinforcement of a footing for a column.

PRESTRESSED CONCRETE

In long beams, as for use on bridges, increased strength and a lower volume of concrete and steel can be obtained by prestressing. Steel used in prestressed concrete is a special high-strength type capable of withstanding a pull of nearly 200,000 psi without excessive stretch. The steel rods used pull in on the ends of the beam and place the concrete under high compression to overcome the effect of any bend on the part of the concrete and to eliminate the effects of tensile pressures.

Two methods of prestressing are used—pre-tensioning and post-tensioning. In pretensioning, high-strength steel rods are stretched between abutments on a casting bed, then stretched. The concrete is poured in forms through which the rods (under tension) pass. After thorough curing, the rods are released from their abutments, and the cast concrete with the rods under tension

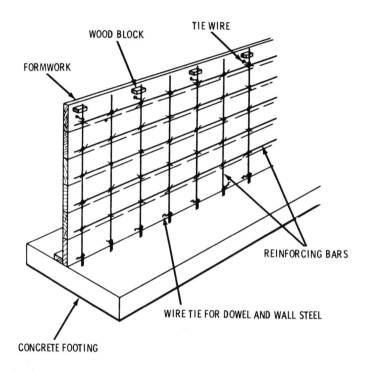

FORMWORK

WOOD BLOCK

TIE WIRE

REINFORCING BARS

WIRE TIE FOR DOWEL AND WALL STEEL

CONCRETE FOOTING

Fig. 3-16. Reinforcement bars in a poured concrete wall.

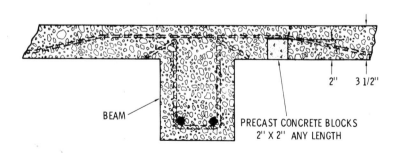

BEAM

PRECAST CONCRETE BLOCKS
2" X 2" ANY LENGTH

2"   3 1/2"

Fig. 3-17. In a concrete beam, reinforcement bars are placed to the far side
of the stress points.

55

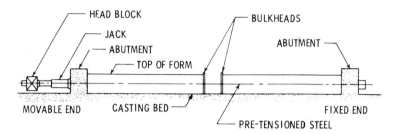

Fig. 3-18. In a pre-tensioned, prestressed steel reinforcement, the concrete is poured around rods under tension.

inside is lifted out of place. Fig. 3-18 illustrates this. The bond between the rods and concrete holds the concrete under tension.

Post-tensioning consists of pouring concrete in a form with rods or bars, not under tension, passing through. The rods are treated so there is no bond between them and the concrete. One end of each rod is fixed, with a heavy washer fastened in place. The other end has a heavy washer on it and is threaded for a large nut. After the concrete is thoroughly cured, pneumatic equipment connected to the free end of each rod will turn up on the nut, or other tightening device, thus putting the concrete under tension. The rod must be free to move within the concrete for post-tensioning.

CHAPTER 4

Concrete Forms

Since a concrete mixture is semifluid, it will take the shape of anything into which it is poured. Accordingly, molds or forms are necessary to hold the concrete to the required shape until it hardens. Such forms or molds may be made of metal, lumber, or plywood.

Form work may represent as much as one-third of the total cost of a concrete structure, so the importance of the design and construction of this phase of a project cannot be overemphasized. The character of the structure, availability of equipment and form materials, anticipated repeat use of the forms, and familiarity with methods of construction influence design and planning of the form work. Forms must be designed with a knowledge of the strength of the materials and the loads to be carried. The ultimate shape, dimensions, and surface finish must also be considered in the preliminary planning phase.

NEED FOR STRENGTH

Forms for concrete structures must be tight, rigid, and strong. If forms are not tight, there will be a loss of concrete, which may result in honeycombing, or a loss of water that causes sand streaking. The forms must be braced well enough to stay in alignment and strong enough to hold the concrete. Special care should be taken in bracing and tying down forms such as those for retaining walls in which the mass of concrete is large at the bottom and tapers toward the top. In this type of construction, and in other types, such as the first pour for walls and columns, the concrete tends to lift the form above its proper elevation. If the forms are to be used again, they must be designed so that they can easily be removed and re-erected without damage. Most forms are made of wood, but steel forms are commonly used for work involving large unbroken surfaces such as retaining walls, tunnels, pavements, curbs, and sidewalks, as shown in Fig. 4-1. Steel forms for sidewalks, curbs, and pavements are especially advantageous since they can be used many times.

Fig. 4-1. Steel forms in place for a concrete slab.

Any concrete laid below ground level for support purposes, such as foundations, must start below the freeze line. This will vary for different parts of the country, but is generally about 18 in. below ground level. The length of time necessary to leave the forms in place depends on the nature of the structure. For small construction work where the concrete bears external weight, the forms may be removed as soon as the concrete will bear its own weight, that is, between 12 and 48 hours after the concrete has been poured. Where the concrete must resist the pressure of the earth or water (as in retaining walls or dams), the forms should be left in place until the concrete has developed nearly its final strength; this may be as long as three or four weeks if the weather is cold or if anything else prevents quick curing.

BRACING

The bracing of concrete formwork falls into a number of categories. The braces that hold wall and column forms in position are usually 1-in.-thick boards or strips. Ordinarily such braces are not heavily stressed because the lateral pressure of the concrete is contained by wall ties and column clamps.

When it is not practical to use wall ties, the braces may be stressed depending on the height of the wall being poured. Braces of this kind are proportioned to support the wall forms against lateral forces. Deep beam and girder forms often require external braces to prevent the side forms from spreading.

The lateral bracing that supports slab forms is also important, not only in terms of safety, but also to prevent the distortions that can occur when the shores are knocked out of position. Lateral bracing should be left in place until the concrete is strong enough to support itself.

ECONOMY

One concern of the builder, particularly today when costs of materials are so high, is economy. Forms for a concrete building for example, can cost more than the concrete or reinforcing steel, or, in some cases, more than both together. Hence it is important to seek out every possible cost-cutting move.

Saving on costs starts with the designer of the building. He or she must keep in mind the forms required for the building's construction and look to build in every possible economy. For example, it may be possible to adjust the size of beams and columns so that they can be formed with a combination of standard lumber or plywood-panel sizes.

Specific example: In making a beam 10 or 12 in. wide, specially ripped form boards are needed. On the other hand, surfaced 1″ × 6″ (5½ in. actual size) will be just right for an 11-in. wide beam. Experience has shown that it is entirely practical to design structures around relatively standardized beam and column sizes for the sake of economy. When such procedures are followed, any diminution of strength can be made up by increasing the amount of reinforcing steel used—and you will still save money.

Another costly area to avoid is excessive design. It may just require a relatively small amount of concrete to add a certain look to a structure, but the addition can be very expensive in terms of overall cost.

Another economy may be found in the area of lumber lengths. Long lengths can often be used without trimming. Studs need not be cut off at the top or at a wall form, but can be used in random lengths to avoid waste. Random-length wales can also be used; and where a wall form is built in place, it does not harm if some boards extend beyond the length of the form.

Paradoxically, some very good finish carpenters are not good at form building because they spend a lot of time building the forms—neatness and exact lengths or widths are not required. Because they have overdone the form building, when it comes time to strip the form, there may be many nails to remove, and the job may be much more complicated.

FASTENING AND HARDWARE

All sorts of devices to simplify the building and stripping of forms are available. The simplest is the double-headed nail (Fig.

4-2). The chief advantage of these nails is that they can be pulled out easily because they are driven in only to the first head. There are also many different varieties of column clamps, adjustable shores, and screw jacks. Instead of wedges, screw jacks are especially suitable when solid shores are used.

Double-head Scaffold

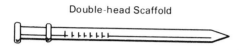

Fig. 4-2. Double-headed scaffold or framing nail.

A number of wedges are usually required in form building. Wedges are used to hold form panels in place, draw parts of forms into line, and adjust shores and braces. Usually the carpenter makes the wedges on the job. A simple jig can be rigged up on a table or a radial arm saw for cutting the wedges.

Form ties, available in many styles, are devices that support both sides of wall forms against the lateral pressure of concrete. Used properly, form ties practically eliminate external bracing and greatly simplify the erection of wall forms.

A simple tie is merely a wire that extends through the form, the ends of the wire doubled around a stud or wale on each side. Although low in cost, simple wire ties are not entirely satisfactory because, under pressure from the concrete, they cut into the wood members and cause irregularities in the wall. The most satisfactory ties can be partially or completely removed from the concrete after it has set and hardened completely.

SIZE AND SPACING

The size and spacing of ties is governed by the pressure of the concrete transmitted through the studs and wales to the ties. In other words, a wale acts as a beam and the ties as reactions. It might be assumed, therefore, that large wales and correspondingly large ties should be used, but in actuality size and spacing are limited by economics. The tie-spacing limit is generally considered to be 36 in. with 27 to 30 in. preferred. Many tie styles make it necessary to place spreaders in the forms to keep the two sides of the form from being drawn together. Generally, spreaders should be removed so that they are not buried in the concrete. Traditional spreaders are quite good and are readily removed. Fig. 4-3 shows how wood spreaders are placed and how they are removed.

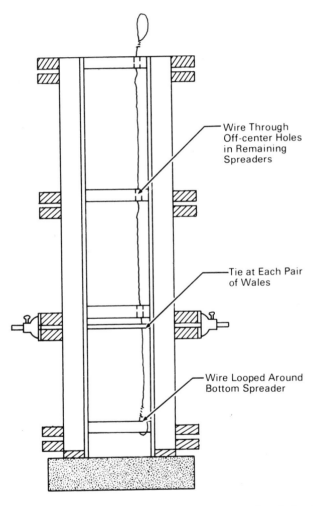

Wire Through Off-center Holes in Remaining Spreaders

Tie at Each Pair of Wales

Wire Looped Around Bottom Spreader

Courtesy National Forest Products Assn.

Fig. 4-3. How wood spreaders are placed and removed.

In more general use are several styles of spreaders combined with ties. Some of these are made of wire nicked or weakened in such a way that the tie may be broken off in the concrete within an inch or two from the face of the form. After you withdraw the ties, the small holes that remain are easily patched with mortar.

STRIPPING FORMS

Unlike most structures, concrete forms are temporary; the forms must later be removed or stripped. Sometimes, for example in building a single home, forms are used only once and then discarded. But in most cases economy dictates that a form be used and reused. Indeed, economical heavy construction depends on reusing forms.

Because forms are removable, the form designer has certain restrictions. He must not only consider erection but also stripping— the disassembly of the form. Thus, if a form is designed in such a way that final assembly nails are covered, it may be impossible to remove the form without tearing it apart and possibly damaging the partially cured concrete.

Another bad design may be that the form is built so that some of its members are encased in concrete and may be difficult to remove. This may result in defective work. It is often advisable to plan column forms so that they can be stripped without disturbing the forms for the beams and girders.

For easier stripping, forms can be coated with special oils. These are not always effective, however, and a number of coating compounds that work well have been developed over the years. These compounds reduce the damage to concrete when stripping is difficult or perhaps carelessly done. The use of these coatings reduces the importance of wetting formwork before placing concrete.

After stripping, forms should be carefully cleaned of all concrete before they are altered and oiled for reuse.

Stripping Forms for Arches

Forms for arches, culverts, and tunnels generally include hinges or loose pieces that, when removed, release the form (Fig. 4-4).

Shores are often set on screw jacks or wedges to simplify their removal. Screw jacks are preferable to wedges because forms held in jacks can be stripped with the least amount of hammering.

For small jobs, where jacks are not available, shores that are almost self-releasing can be made. Two 2 × 6s are fastened into a T-shaped section (Fig. 4-5) with double-headed nails. When wedged into position, this assembly is a stiff column. After the concrete hardens, the nails are drawn and the column becomes two 2 × 6s which are relatively easy to remove.

SPECIAL FORMS

Forms are required for building concrete piers, pedestals, and foundations for industrial machinery. The job basically still involves carpentry because such forms differ only in detail from building forms and usually do not have to withstand the pressures that are built up in deep wall or column forms.

Many piers are tapered upward from the footing. In these cases it is necessary to provide a resistance to uplift because the semi-liquid concrete tends to float such forms. Once the problem is recognized, it is easily solved. Two or more horizontal planks nailed or wired to spikes will hold down most forms. Sandbags placed on ledgers (Fig. 4-6) are usually enough for smaller forms.

PREFABRICATED FORMS

In addition to lumber and plywood forms, there is a tremendous variety of prefabricated forms. For example, there are prefab panels, and these panels can be made up in different widths and lengths, although the most common dimension is 2″ × 8″. There is a peg-and-wedge system to hold the panels together, and a variety of hinged forms are available, as well as circular ones made of metal and clamps.

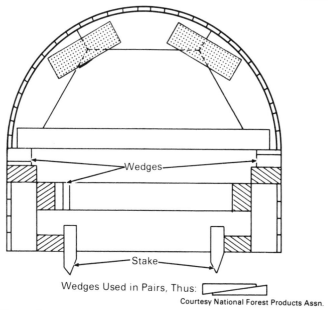

Wedges

Wedges

Stake

Wedges Used in Pairs, Thus:

Courtesy National Forest Products Assn.

Fig. 4-4. Forms for arches and culverts usually include loose sections or hinges; when removed, they release the form.

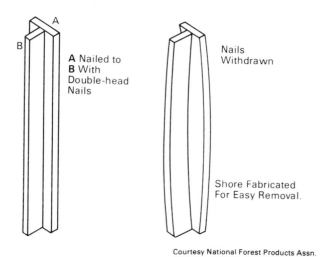

A
B

A Nailed to
B With
Double-head
Nails

Nails
Withdrawn

Shore Fabricated
For Easy Removal.

Courtesy National Forest Products Assn.

Fig. 4-5. Shores that are virtually self-releasing can be made with 2 × 6s and scaffold nails.

65

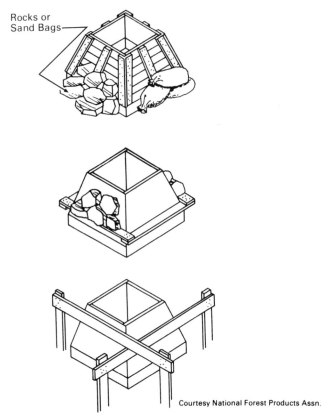

Rocks or
Sand Bags

Courtesy National Forest Products Assn.

Fig. 4-6. Various ways to anchor pier forms.

TYPES OF FORMS

Depending on application, forms may be very simple or very complex. The simpler the form, the easier the job, and the more economical. Bear in mind one thing—the form must be strong enough to support the concrete rigidly for a day or two until the concrete has taken a firm set.

The simplest form is one used for small column supports, like that of Fig. 4-7. An illustration of its use is shown in Fig. 4-8, in which it holds the concrete footing for a 4″ × 4″ post for a patio cover. If the form is made of redwood, it may be left in place as shown in the illustration. The earth itself may be the form if the sides are dug straight and smooth. Fig. 4-9 shows a trough dug as a footing for a concrete block wall. Long troughs are dug with a power shovel, but the sides must be finally smoothed by hand.

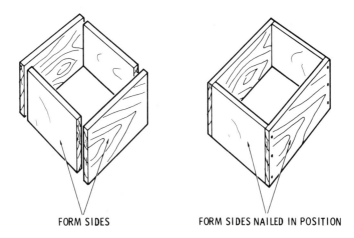

FORM SIDES   FORM SIDES NAILED IN POSITION

Fig. 4-7. A simple boxlike structure that will serve for small columns and posts.

Another example of the earth as a form is a simple clothesline pole installation. A pole digger will cut about a 6-in.-diameter hole. Set the pole in place in the hole, and pour concrete into the hole up to about 2 in. from the top of the soil. Cover the remainder with soil. The pole must be left undisturbed for several days for curing before use.

Slab concrete only requires simple 2″ × 4″ or 2″ × 6″ boards, as described previously, or metal rails as shown in Fig. 4-1. Since the pressure of a few inches of concrete is not great, the form boards may be staked to the earth.

FOOTINGS AND FOUNDATIONS

There is no substitute for an adequate foundation, a key part of every building. A concrete foundation represents only a small part of a typical building's cost and is one of the best investments in construction. An adequate footing provides a stable base and directly affects both the life and performance of the building. In addition, it gives protection against rats, mice, termites, water, and the elements.

A foundation consists of:

1. Its *bed*—the earth giving support.
2. Its *footing*—the widened part of the structure resting upon the bed.
3. Its *wall*—the structural part resting upon the footing.

Fig. 4-8. The form is being used to secure a 4″ × 4″ post in concrete.

The size of the footing depends on the load-carrying capacity of the soil and the weight of the building and its contents. Soils vary in their ability to support weight. The load-carrying capacity of several common soils is given in Table 4-1. Firm clay, for example, has a carrying capacity of 2 tons per sq. ft., meaning that under normal conditions, 1 sq. ft. of firm clay will support 4000 lbs.

Fig. 4-9. A concrete-block wall footing can be poured in the earth if the ditch is dug straight and smooth.

A common rule of thumb often used to dimension footings for lightly loaded buildings is that the footing shall be twice as wide as the foundation wall and as thick as the wall is wide. See Fig. 4-10. By this rule, the following footing sizes would be used:

69

Table 1. Load-Carrying Capacities of Soils

Type of soil	Tons per sq. ft.
Soft clay	1
Firm clay or fine sand	2
Compact fine or loose coarse sand	3
Loose gravel or compact coarse sand	4
Compact sand-gravel mixture	6

Courtesy Portland Cement Association.

For 8 in. thick foundation wall (or less), $8'' \times 16''$ footing.
For 10 in. thick foundation wall, $10'' \times 20''$ footing.
For 12 in. thick foundation wall, $12'' \times 24''$ footing.

If the structure is heavily loaded or the soil conditions are questionable, consult an engineer or architect. In areas subject to freezing, the bottom of the footing must be placed below the frost line to prevent frost-heave and resultant damage to the building. The trench bottom should be level and cut flat so the footing will bear evenly on undisturbed earth.

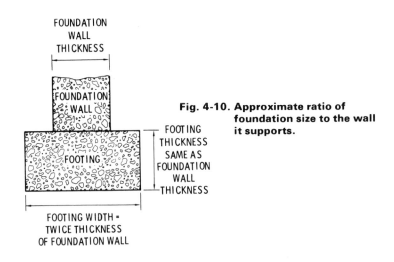

Fig. 4-10. Approximate ratio of foundation size to the wall it supports.

A wedge-shaped trough may be dug, like that shown in Fig. 4-11, in which case the sides of the trough may be used to support the form boards. This method and that of self-supporting side boards are shown in a cross-sectional view in Fig. 4-12. Fig. 4-13

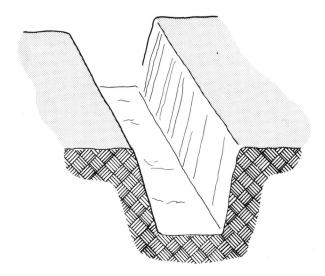

Fig. 4-11. If carefully made, a wedge-shaped trough can be used to support the side boards of a footing form.

SPREADER NAILED TO FORM SIDES

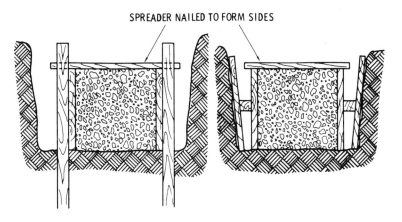

Fig. 4-12. Two methods used to support the side boards for a footing form.

71

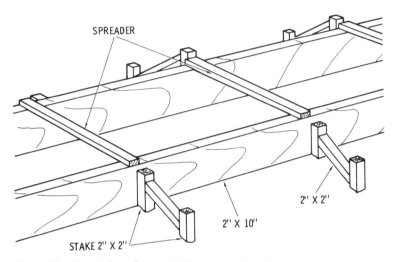

SPREADER

2" X 2"

2" X 10"

STAKE 2" X 2"

Fig. 4-13. Simple form for a typical concrete foundation.

Fig. 4-14. The complete house foundation using simple forms shown in Fig. 4-8.

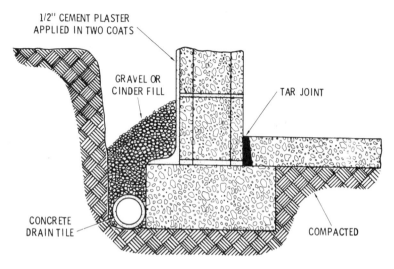

1/2" CEMENT PLASTER
APPLIED IN TWO COATS

GRAVEL OR
CINDER FILL

TAR JOINT

CONCRETE
DRAIN TILE

COMPACTED

Fig. 4-15. Drain tile should be included where foundations are made for basement wall support.

shows a simple form and lumber dimensions for a footing for an above-grade wall. Fig. 4-14 shows a completed concrete foundation using the simple forms in Fig. 4-13. Where footings are well below ground, as for a full-basement house, drainage must be provided.

For wet soil conditions, and in order to assure proper drainage, a drain tile should be laid around the footing as shown in Fig. 4-15. The drain tile should be laid with open joints and drained to a suitable outlet with a slope of 1 in. every 12 ft. In no case should the tile be lower than the footing. Joints between the tile should be covered with strips of tar paper, or roofing felt, to prevent sediment filling the tile during back filling. The tile line should be covered with not less than 18 in. of gravel or cinder fill. Filling of the dirt around the footing should be delayed until after the sub-floor is completed.

CHAPTER 5

Pouring, Finishing, and Curing

The placement of concrete has been discussed earlier, but a few reminders at this point are in order.

The concrete poured in forms should be spaded with a flat scraper or other thin-bladed tool or mechanically vibrated. This is done to eliminate a condition called "honeycombing," which occurs when coarse aggregate collects at the face of the wall. In inaccessible areas, the forms can be lightly tapped with a hammer to achieve the same result.

Spud vibrators are excellent tools to consolidate fresh concrete in walls and other formed work. The spud vibrator is a metal tube-like device that vibrates at several thousand cycles per minute. When inserted in the concrete for 5 to 15 seconds, the spud vibrator consolidates it and improves the surfaces next to the forms. When pouring flat slabs such as walks or driveways, make the necessary preparations to pour the farthest point first.

The supporting soil for all flat concrete work should be adequately compacted to prevent unequal settling. Prior to placing the concrete, the earth or granular sub-base should be dampened to prevent it from drawing water from the freshly placed concrete. Concrete should never be placed on frozen earth or earth that is flooded with water.

SCREEDING

To screed is to strike-off or level slab concrete after pouring. To do this, and go on with successful finishing, it is well to understand the nature of concrete.

Any craftsman or tradesman should understand the nature and properties of the materials with which he works. A cement mason should understand the nature of concrete and the different problems in working with concrete made with the various types of portland cement.

Generally, all the dry materials used in making quality concrete are heavier than water. Thus, shortly after placement, these materials will have a tendency to settle to the bottom and force any excess water to the surface. This reaction is commonly called "bleeding." This bleeding usually occurs with non-air-entrained concrete. It is of utmost importance that the first operations of placing, screeding, and darbying be performed before any bleeding takes place. The concrete should not be allowed to remain in wheelbarrows, buggies, or buckets any longer than is absolutely necessary. It should be dumped and spread as soon as possible and struck-off to the proper grade, then immediately screened, followed at once by darbying. These last two operations should be performed before any free water is bled to the surface. The concrete should not be spread over a large area before screeding— nor should a large area be screeded and allowed to remain before darbying. If any operation is performed on the surface while the bleed water is present, serious scaling, dusting, or crazing can result. This point cannot be overemphasized and is the basic rule for successful finishing of concrete surfaces.

The surface is struck off or rodded by moving a straightedge back and forth with a sawlike motion across the top of the forms or screeds. A small amount of concrete should always be kept ahead of the straightedge to fill in all the low spots and maintain a plane surface. For most slab work, screeding is usually a two-man job because of the size of the slab.

TAMPING OR JITTERBUGGING

On some flatwork construction jobs, the next operation is using the hand tamper or jitterbug. This tool should be used sparingly and in most cases not at all. If used, it should be used only on concrete having a low slump (1 in. or less) to compact the concrete into a dense mass. Jitterbugs are sometimes used on industrial floor construction because the concrete for this type of work usually has a very low slump, with the mix being quite stiff and perhaps difficult to work.

The hand tamper or jitterbug is used to force the large particles of coarse aggregate *slightly* below the surface in order to enable the cement mason to pass his darby over the surface without dislodging any large aggregate. After the concrete has been struck-off or rodded, and in some cases tamped, it is smoothed with a darby to level any raised spots and fill depressions. Long-handled floats of either wood or metal, called *bull floats*, are sometimes used instead of darbies to smooth and level the surface.

FINISHING

When the bleed water and water sheen have left the surface of the concrete, finishing may begin. Finishing may take one or more of several forms, depending on the type of surface desired.

Finishing operations must not be overdone, or water under the surface will be brought to the top. When this happens, a thin layer of cement is also brought up and later, after curing, it becomes a scale that will powder off with usage. Finishing can be done by hand or by rotating power-driven trowels or floats. The size of the job determines the choice, based on economy.

The type of tool used for finishing affects the smoothness of the concrete. A wood float puts a slightly rough surface on the concrete. A steel (or other metal) trowel or float produces a smooth finish. Extra-rough surfaces are given to the concrete by running a stiff-bristled broom across the top.

FLOATING

Most sidewalks and driveways are given a slightly roughened surface by finishing with a float. Floats may be small, handheld tools (Fig. 5-1), with the work done while kneeling on a board (Fig. 5-2), or they may be on long handles for working from the edge. Fig. 5-3 shows a workman using a long-handled float, and Fig. 5-4 shows the construction details for making a float.

When working from a kneeling board, the concrete must be stiff enough to support the board and the workman's weight without deforming. This will be within two to five hours from the time the surface water has left the concrete, depending on the type of concrete, any admixtures included, plus weather conditions. Experience and testing the condition of the concrete determines this.

Floating has other advantages. It also embeds large aggregate beneath the surface, removes slight imperfections such as bumps and voids, and consolidates mortar at the surface in preparation for smoother finishes, if desired.

Floating may be done before or after edging and grooving. If the line left by the edger and groover is to be removed, floating

Fig. 5-1. Typical wood float.

Fig. 5-2. Using the hand wood float from the edge of a slab. Often the worker will work out the slab on a kneeling board.

should follow the edging and grooving operation. If the lines are to be left for decorative purposes, edging and grooving will follow floating.

Troweling

Troweling, when used, follows floating. The purpose of troweling is to produce a smooth, hard surface. For the first troweling, whether by power or by hand, the trowel blade must be kept as flat against the surface as possible. If the trowel blade is tilted or pitched at too great an angle, an objectionable "washboard" or "chatter" surface will result. For first troweling, a new trowel is not recommended. An older trowel that has been "broken in" can be worked quite flat without the edges digging into the concrete. The

79

Fig. 5-3. The long-handled float.

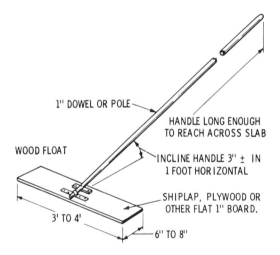

1" DOWEL OR POLE

HANDLE LONG ENOUGH
TO REACH ACROSS SLAB

WOOD FLOAT

INCLINE HANDLE 3" ± IN
1 FOOT HORIZONTAL

SHIPLAP, PLYWOOD OR
OTHER FLAT 1" BOARD.

3' TO 4'

6" TO 8"

Fig. 5-4. Construction details for making a long-handled float.

smoothness of the surface could be improved by timely additional trowelings. There should necessarily be a lapse of time between successive trowelings in order to permit the concrete to increase its set. As the surface stiffens, each successive troweling should be made by a smaller-sized trowel to enable the cement mason to use sufficient pressure for proper finishing.

Brooming

For a rough-textured surface, especially on driveways, brooming provides fine scored lines for a better grip for car tires. Brooming lines should always be at right angles to the direction of travel.

For severe scoring, use a wire brush or stiff-bristled push broom. This operation is done after floating. For a finer texture, such as might be used on a factory floor, use a finer-bristled broom. This operation is best done after trowelling to a smooth finish.

Brooming must be done in straight lines (Fig. 5-5), never in a circular motion. Draw the broom toward you, one stroke at a time, with a slight overlap at the edge of each stroke.

Fig. 5-5. A stiff-bristled broom puts parallel lines in the concrete for a better grip.

Grooving and Edging

In any cold climate there is a certain amount of freezing and thawing of the moist earth under the concrete. When water freezes, it expands. This causes heaving of the ground under the concrete, and this heaving can cause cracking of the concrete in random places.

Sometimes the soil base will settle because all air pockets were not tamped out, or because a leaky water pipe under the soil washed some of it away. A root of a nearby tree under that part of the soil can cause it to lift as the root grows. For all these reasons,

Fig. 5-6. Cutting a groove in a walk. If any cracking occurs, it will be in the groove, where it is less conspicuous.

the concrete can be subjected to stresses that can cause random cracking, even years later.

To avoid random cracking from occuring due to heaving, grooves are cut into the concrete at intervals. These grooves will become the weakest part of the concrete, and any cracking will occur in the grooves. Since, in many cases, heaving or settling cannot be avoided, it is better to have cracks occur in the least conspicuous place possible.

Run a groover across the walk, using a board as a guide to keep the line straight, as shown in Fig. 5-6. About a 1 in. deep groove will be cut, and at the same time, a narrow edge of smoothed concrete will be made by the flat part of the groover.

A rounded edge should be cut along all edges of concrete where it meets the forms, with an edging tool. Running it along the edge of the concrete, between the concrete and the forms, puts a slight round to the edge of the concrete, which helps prevent the edges from cracking off and also gives a smooth-surfaced border. See Fig. 5-7 for details.

In Fig. 5-8, masons are putting the finishing touches to a concrete sidewalk. One man is using an edger, while two men are floating the surface. In Fig. 5-9, a floor slab for a home has just been finished. It has a rough texture since the floor will be covered with carpeting when the house is up and ready for occupancy. Water and sewer lines were laid in position before the concrete was poured.

FINISHING AIR-ENTRAINED CONCRETE

Air entrainment gives concrete a somewhat altered consistency that requires a little change in finishing operations from those used with non-air-entrained concrete.

Air-entrained concrete contains microscopic air bubbles that tend to hold all the materials in the concrete (including water) in suspension. This type of concrete requires less mixing water than non-air-entrained concrete, and still has good workability with the same slump. Since there is less water, and it is held in suspension, little or no bleeding occurs. This is the reason for slightly different finishing procedures. With no bleeding, there is no waiting for the evaporation of free water from the surface before starting the floating and troweling operation. This means that general floating and troweling should be started sooner—before the surface becomes too dry or tacky. If floating is done by hand, the use of an aluminum or magnesium float is essential. A wood float drags and greatly increases the amount of work necessary to accomplish the same result. If floating is done by power, there is practically no difference between finishing procedures for air-entrained and non-air-entrained concrete, except that floating can start sooner on the air-entrained concrete.

Fig. 5-7. An edger being used to round off the edge of a driveway.

Practically all horizontal surface defects and failures are caused by finishing operations performed while bleed water or excess surface moisture is present. Better results are generally accomplished, therefore, with air-entrained concrete.

Fig. 5-8. Finishing a concrete sidewalk.

CURING

Two important factors affect the eventual strength of concrete.

1. The water/cement ratio must be held constant. This was discussed in detail in previous chapters.
2. Proper curing is important to eventual strength. Improperly cured concrete can have a final strength of only 50% of that of fully cured concrete.

It is hydration between the water and the cement that produces strong concrete. If hydration is stopped due to evaporation of the water, the concrete will become porous and never develop the compressive strength it is capable of producing.

Fig. 5-9. The finished slab of a home under construction. Water and sewer pipes were placed before concrete was poured.

The following relates various curing methods and times compared to the 28-day strength of concrete when moist-cured continuously at 70°F.

1. Completely moist-cured concrete will build to an eventual strength of over 130% of its 28-day strength.
2. Concrete moist-cured for 7 days, and allowed to air dry the remainder of the time, will have about 90% of the strength of example 1 at 28 days and only about 75% of eventual strength.
3. Concrete moist-cured for only 3 days will have about 80% of the 28-day strength of example 1 and remain that way throughout its life.
4. Concrete given no protection against evaporation will have about 52% of 28-day strength, and remain that way.

Curing, therefore, means applying some means of preventing evaporation of the moisture from the concrete. It may take the form of adding water, applying a covering to prevent evaporation, or both.

Curing Time

Hydration in concrete begins to take place immediately after the water and cement are mixed. It is rapid at first, then tapers off as time goes on. Theoretically, if no water ever evaporates, hydration goes on continuously. Practically, however, all water is lost through evaporation, and after about 28 days, hydration nearly ceases, although some continues for about a year.

Actually, curing time depends on the application, the temperature, and the humidity conditions. Lean mixtures and large massive structures, such as dams, may call for a curing period of a month or more. For slabs laid on the earth, with a temperature around 70°F and humid conditions with little wind, effective curing may be done in as little as 3 days. In most applications, curing is carried for 5 to 7 days.

Table 5-1 shows the relative strength of concrete between an ideal curing time and a practical time. The solid line is the relative strength of concrete kept from any evaporation. Note that its

Table 1. Relative Concrete Strength Versus Curing Method

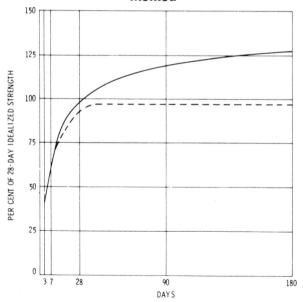

strength continues to increase, but at a rather slow rate with increase in time. The dotted line is the relative strength of concrete that has been cured for 7 days, then allowed to be exposed to free air after that. Strength continues to build until about 28 days, then levels off to a constant value after that. Curing methods should be applied immediately on concrete in forms, and immediately after finishing of flat slabs.

Curing Methods

On flat surfaces such as pavements, sidewalks, and floors, concrete can be cured by *ponding*. Earth or sand dikes around the perimeter of the concrete surface retain a pond of water within the enclosed area. Although ponding is an efficient method for preventing loss of moisture from the concrete, it is also effective for maintaining a uniform temperature in the concrete. Since ponding generally requires considerable labor and supervision, the method is often impractical except for small jobs. Ponding is undesirable if fresh concrete will be exposed to early freezing.

Fig. 5-10. Plastic sheeting is a popular covering for curing concrete slabs.

Continuous *sprinkling* with water is an excellent method of curing. If sprinkling is done at intervals, care must be taken to prevent the concrete from drying between applications of water. A fine spray of water applied continuously through a system of nozzles provides a constant supply of moisture. This prevents the possibility of "crazing" or cracking caused by alternate cycles of wetting and drying. A disadvantage of sprinkling may be its cost. The method requires an adequate supply of water and careful supervision.

Wet coverings such as burlap, cotton mats, or other moisture retaining fabrics are extensively used for curing. Treated burlaps that reflect light and are resistant to rot and fire are available.

Fig. 5-11. A concrete slab with curing compound sprayed on the surface.

Forms left in place provide satisfactory protection against loss of moisture if the top exposed concrete surfaces are kept wet. A soil-soaker hose is an excellent means of keeping concrete wet. Forms should be left on the concrete as long as practicable.

Wood forms left in place should be kept moist by sprinkling, especially during hot, dry weather. Unless wood forms are kept moist, they should be removed as soon as practicable and other methods of curing started without delay.

The application of *plastic sheets* or *waterproof paper* over slab concrete is one of the most popular methods of curing. To do this, sprinkle a layer of water over the slab and lay the sheets on top. Tack the edges of the sheets to the edge forms or screeds to keep the water from evaporating. If the sheets are not wide enough to cover the entire area with one piece, use a 12-in. overlap between sheets. Use white-pigmented plastic to reflect the rays of the sun, except in cold weather when you want to maintain a warm

Table 2. Curing Methods

Method	Advantage	Disadvantage
Sprinkling with water or covering with wet burlap	Excellent results if constantly kept wet.	Likelihood of drying between sprinklings. Difficult on vertical walls.
Straw	Insulator in winter.	Can dry out, blow away, or burn.
Curing compounds	Easy to apply. Inexpensive.	Sprayer needed; inadequate coverage allows drying out; film can be broken or tracked off before curing is completed; unless pigmented, can allow concrete to get too hot.
Moist earth	Cheap, but messy.	Stains concrete, can dry out, removal problem.
Waterproof paper	Excellent protection, prevents drying.	Heavy cost can be excessive. Must be kept in rolls; storage and handling problem.
Plastic film	Absolutely watertight, excellent protection. Light and easy to handle.	Should be pigmented for heat protection. Requires reasonable care and tears must be patched; must be weighed down to prevent blowing away.

temperature on the concrete. Waterproof paper is available for the same application. Keep the sheets in place during the entire curing period. Fig. 5-10 shows plastic sheeting being laid on a newly finished walk.

The use of a liquid *curing compound* that may be sprayed on the concrete is increasing in popularity. It is sprayed on from a hand spray or power spray. It forms a waterproof film on the concrete that prevents evaporation. Its disadvantage is that the film may be broken if the concrete bears the weight of a man or vehicle before the curing period is completed. An advantage is that it may be sprayed on the vertical portions of cast-in-place concrete after the forms are removed.

Fig. 5-11 shows a slab after applying a curing compound. The black appearance is the color of the curing compound which is sprayed on top of the concrete. A white-pigmented compound is better when the concrete is exposed to the hot sun. Table 5-2 lists various curing methods, their advantages and disadvantages.

CHAPTER 6

Concrete Block

Many people think of concrete blocks as those gray, unattractive blocks used for foundations and warehouse walls, but this is not true today. Modern concrete blocks come in a variety of shapes and colors and are used for many purposes, including partition walls (Fig. 6-1).

BLOCK SIZES

Concrete blocks are available in many sizes and shapes. Fig. 6-2 shows some of the sizes in common use. They are all sized on the basis of multiples of 4 in. The fractional dimensions shown allow for the mortar (Fig. 6-3). Some concrete blocks are poured concrete made of standard cement, sand, and aggregate. An 8″ × 8″ × 16″ block weighs about 40 to 50 lbs. Some use lighter natural aggregates, such as volcanic cinders or pumice; and some

(A) Using grille block in a partition wall.

(B) Using standard block and raked joints in a partition wall.

Fig. 6-1. Examples of concrete block used in home construction.

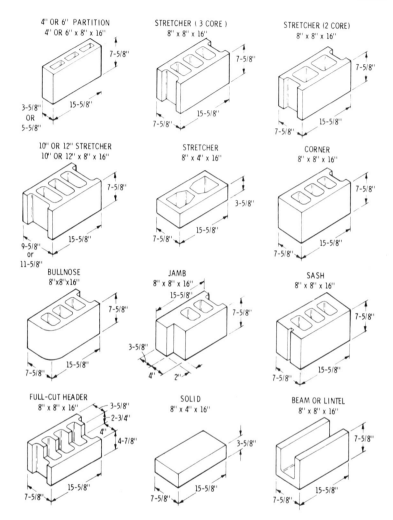

Fig. 6-2. Standard sizes and shapes of concrete blocks.

are manufactured aggregates such as slag, clay, or shale. These blocks weigh 25 to 35 lbs.

In addition to the hollow-core types shown, concrete blocks are available in solid forms. In some areas they are available in sizes other than those shown. Many of the same type have half the

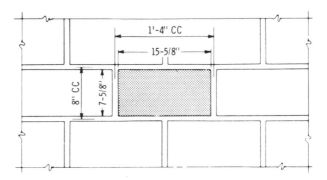

Fig. 6-3. Block size to allow for mortar joints.

height, normally 4 in., although actually 3⅝ in. to allow for mortar. The 8″ × 8″ × 16″ stretcher (center top illustration of Fig. 5-2) is most frequently used. It is the main block in building a yard wall or a building wall. Corner or bullnose blocks with flat finished ends are used at the corners of walls. Others have special detents for window sills, lintels, and door jambs.

Compressive strength is a function of the face thickness. Concrete blocks vary in thickness of the face, depending on whether they are to be used for non-load-bearing walls, such as yard walls, or load-bearing walls, such as for buildings.

DECORATIVE BLOCK

In addition to standard rectangular forms, concrete masonry blocks are made in unusual designs and with special cast-in colors and finishes to make them suitable architectural designs for both indoor and outdoor construction. A few decorative blocks are described and illustrated in the next few pages.

Split Block

Resembling natural stone, split block is made from standard 8-in.-thick pieces split by the processor into 4-in.-thick facing blocks. The rough side faces out. Split block is usually gray, but some have red, yellow, buff, or brown colors made as an integral part of the cast concrete.

Split block is especially handsome as a low fence. By laying it up in lattice-like fashion, it makes a handsome carport, keeping out the weather but allowing air and light to pass through (Fig. 6-4).

Slump Block

When the processor uses a mix that slumps slightly when the block is removed from the mold, it takes on an irregular appearance like old-fashioned hand molding. Slump block strongly resembles adobe or weathered stone, and it too is available with integral colors and is excellent for ranch-style homes, fireplaces, and garden walls (Fig. 6-5).

Fig. 6-4. Split block laid up in a latticelike pattern adds a nice touch.

97

Fig. 6-5. A wall of slump block adds rustic charm and is especially suitable for ranch-type homes.

Grille Blocks

Some of the most attractive of the new concrete blocks are grille blocks, which come in a wide variety of patterns, a few of which are shown in Fig. 6-6. In addition to its beauty, the grille block provides the practical protection of a concrete wall, yet allows some sun and light to enter (Fig. 6-7). They are especially useful in cutting the effects of heavy winds without blocking all circulation of air for ventilation. They are usually 4 to 6 in. thick and have faces that are 12 to 16 in. square.

Screen Block

Similar to grille block but lighter in weight and more open, screen block is being used more and more as a facing for large

98

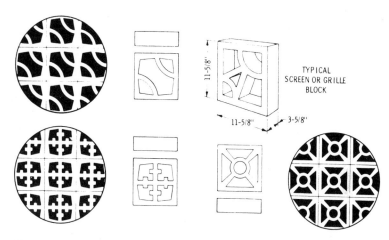

TYPICAL
SCREEN OR GRILLE
BLOCK

11-5/8"

11-5/8" 3-5/8"

Fig. 6-6. Examples of grille blocks. The actual patterns available vary with processors in different areas.

window areas. In this way, beauty and some temperature control is added. They protect the large window panes from icy winter blasts and strong summer sun, yet provide privacy and beauty to the home. Special designs are available, or they may be made up by laying single-core standard block on its side as shown in Fig. 6-8.

Patterned Block

Solid block may also be obtained with artistic patterns molded in for unusual effects both indoors and out. Some carry the trade names of *Shadowal* and *Hi-Lite*. The first has depressed diagonal recessed sections, and the second has raised half-pyramids. Either can be placed to form patterns of outstanding beauty.

Special Finishes

Concrete block is produced by some manufacturers with a special bonded-on facing to give it special finishes. Some are made with a thermosetting resinous binder and glass silica sand, which give a smooth-faced block. A marbleized finish is produced by another manufacturer; a vitreous glaze by still another. Some

Fig. 6-7. Outside and inside views of a grille block wall.

blocks may be obtained with a striated bark-like texture. Blocks with special aggregate can be found—some ground down smooth for a terrazo effect.

STANDARD CONCRETE BLOCK

Standard concrete blocks with hollow cores can make handsome walls, depending on how they are laid and on the sizes chosen. Fig. 6-9 shows a solid high wall of standard block which provides maximum privacy. A few are laid hollow core out for air circulation. Fig. 6-10 is standard block with most of the blocks laid sideways to expose the cores. Fig. 6-11 shows single-cored, thin-edged concrete blocks used to support a sloping ground level. It prevents earth runoff and adds an unusual touch of beauty.

In Fig. 6-12, precast concrete block of standard size is laid like brickwork for use as a patio floor. Level the earth and provide a gentle slope to allow for rain runoff. Put in about a 2-in. layer of sand, and lay the blocks in a two-block crisscross pattern. Leave a thin space between blocks and sweep sand into the spaces after the blocks are down. Fig. 6-13 shows a walk made of precast concrete blocks of extra large size. These are not standard, but they are available.

WALL THICKNESS

Garden walls under 4 ft. high can be as thin as 4 in., but it is best to make them 8 in. thick. Walls over 4 ft. high must be at least 8 in. thick to provide sufficient strength.

A wall up to 4 ft. high needs no reinforcement. Merely build up the fence from the foundation with block or brick, and mortar. Over 4 ft., however, reinforcement will be required, and the fence should be of block, not brick. As shown in the sketch of Fig. 6-14, set $1/2$-in.-diameter steel rods in the poured concrete foundation at 4-ft. centers. When you have laid the blocks (with mortar) up to the level of the top of the rods, pour concrete into the hollow cores around the rods. Then continue on up with the rest of the layers of blocks.

Fig. 6-8. A large expanse of glass in a home can be given some privacy without cutting off all light.

In areas subject to possible earthquake shocks or extra high winds, horizontal reinforcement bars should also be used in high walls. Use No. 2 (¼ in.) bars or special straps made for the purpose. Fig. 6-15 is a photograph of a block wall based on the sketch of Fig. 6-14. The foundation is concrete poured in a trench dug out of the ground. Horizontal reinforcement is in the concrete, with vertical members bent up at intervals. High column blocks are laid (16″ × 16″ × 8″) at the vertical rods. The columns are evident in the finished wall of Fig. 6-16.

Load-bearing walls are those used as exterior and interior walls in residential and industrial buildings. Not only must the wall support the roof structure, but it must bear its own weight. The greater the number of stories in the building, the greater the thickness the lower stories must be to support the weight of the concrete blocks above it, as well as roof structure.

FOUNDATIONS

Any concrete-block or brick wall requires a good foundation to support its weight and prevent any position shift that may produce cracks. Foundations or footings are concrete poured into forms or trenches in the earth. Chapter 4 describes forms and their construction for footings. For non-load-bearing walls, such as yard fences for example, an open trench with smooth sides is often satisfactory. Recommended footing depth is 18 in. below the grade level. In areas of hard freezes, the footing should start below the frost line.

Footings or foundations must be steel reinforced, with reinforc-

Fig. 6-9. An attractive garden wall with a symmetrical pattern.

ing rods just above the bed level. Reinforcing rods should be bent to come up vertically at regular intervals into the open cores of column sections of walls, where used, or regular sections of the wall if columns are not used. Columns of double thickness are recommended where the wall height exceeds 6 ft. Even lower height walls are better if double-thick columns are included. (See Fig. 6-17). The earth bed below the foundation must be well-tamped and include a layer of sand for drainage.

MORTAR

Mortar bonds the masonry units together to form a strong durable wall. The mortar must be chemically stable and resist rain penetration and damage by freezing and thawing Mortar must

Fig. 6-10. Standard double-core block laid with cores exposed for ventilation.

Fig. 6-11. Single-core corner blocks used in a garden slope.

Fig. 6-12. Solid concrete blocks used as a patio floor.

have sufficient strength to carry all loads applied to the wall for the life of the building with a minimum of maintenance.

Mortar is widely used in home construction for all types of masonry walls. Masonry cement eliminates the need to stockpile and handle extra material and reduces the chance of improper on-the-job proportioning. Consistent mortar color in successive batches is easy to obtain when using masonry cement.

Water is added until the mortar is plastic and handles well under a trowel. Mortar should be machine-mixed whenever practical. Masonry cement contains an air-entraining agent that causes the formation of tiny air bubbles in the mortar. These bubbles make the mortar more workable when plastic, slow the absorption of water by the masonry unit, and reduce the possibility of weather damage.

BUILDING WITH CONCRETE BLOCKS

Proper construction of concrete-block walls, whether for yard fencing or building structures, requires proper planning. Standard

Fig. 6-13. Extra-large precast concrete slabs used to make an attractive walk.

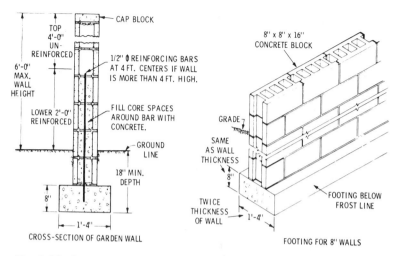

Fig. 6-14. Cross section and view of a simple block wall. Vertical reinforcement rods are placed in the hollow cores at various intervals.

concrete blocks are made in 4-in. modular sizes. Their size allows for a ⅜-in.-thick mortar joint. By keeping this in mind, the width of a wall and openings for windows and doors may be planned without the need for cutting any of the blocks to fit.

Fig. 6-18 shows the right and wrong way to plan for openings. The illustration on the top did not take into account the 4-in. modular concept, and a number of blocks must be cut to fit the window and door opening. The illustration on the bottom shows correct planning, and no blocks need to be cut for the openings.

Having established the length of a wall on the basis of the 4 in. modular concept (total length should be some multiple of 4-in.), actual construction begins with the corners. Stretcher blocks are then laid between the corners.

Laying Block at Corners

In laying up corners with concrete masonry blocks, place a taut line all the way around the foundation with the ends of the string tied together. It is customary to lay up the corner blocks, three or four courses high, and use them as guides in laying the walls.

A full width of mortar is placed on the footing, as shown in Fig.

6-19, and the first course is built two or three blocks long each way from the corner. The second course is half a block shorter each way than the first course; the third, half a block shorter than the second, etc. Thus, the corners are stepped off until only the corner block is laid. Use a line, and level frequently to see that the blocks are laid straight and that the corners are plumb. It is customary that such special units as corner blocks, door and window jamb blocks, fillers, and veneer blocks be provided prior to commencing the laying of the blocks.

Building the Wall Between Corners

In laying walls between corners, a line is stretched tightly from corner to corner to serve as a guide (Fig. 6-20). The line is fastened

Fig. 6-15. Vertical reinforcement rods through double-thick column blocks.

Fig. 6-16. A newly finished concrete-block wall. Note reinforcement columns at various intervals.

Fig. 6-17. A garden wall of concrete blocks. The columns are spaced 15 ft. apart.

to nails or wedges driven into the mortar joints so that, when stretched, it just touches the upper outer edges of the block laid in the corners. The blocks in the wall between corners are laid so that they will just touch the cord in the same manner. In this way, straight horizontal joints are secured. Prior to laying up the outside wall, the door and window frames should be on hand to set in place as guides for obtaining the correct opening.

109

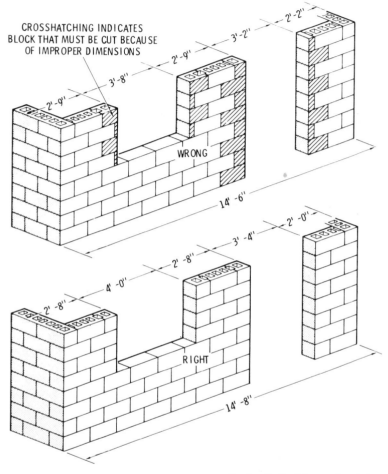

CROSSHATCHING INDICATES
BLOCK THAT MUST BE CUT BECAUSE
OF IMPROPER DIMENSIONS

Courtesy Portland Cement Association.

Fig. 6-18. The right and wrong way to plan door and window openings in block walls.

Applying Mortar to Blocks

The usual practice is to place the mortar in two separate strips, both for the horizontal or bed joints and for the vertical or end joints, as shown in Fig. 6-21. The mortar is applied only on the face shells of the block. This is known as *face-shell bedding*. The air

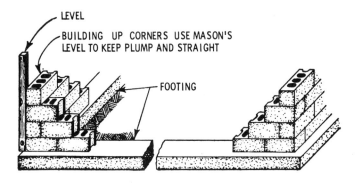

Fig. 6-19. Laying up corners when building with concrete masonry block units.

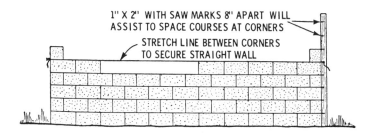

Fig. 6-20. Procedure in laying concrete block walls.

spaces thus formed between the inner and outer strips of mortar help produce a dry wall.

Masons often stand the block on end and apply mortar for the end joint as shown in Fig. 6-21. Sufficient mortar is put on to make sure that all joints will be well filled. Some masons apply mortar on the end of the block previously laid as well as on the end of the block to be laid next to it to make certain that the vertical joint will be completely filled.

Placing and Setting Blocks

In placing, the block that has mortar applied to one end is picked up, as shown in Fig. 6-22, and shoved firmly against the block previously placed. Note that mortar is already in place in the bed or horizontal joints.

111

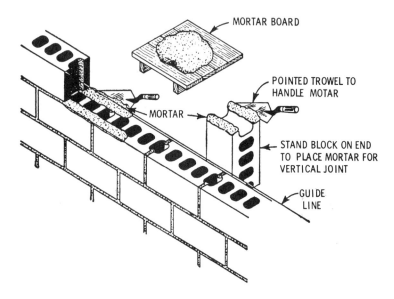

Fig. 6-21. The usual practice in applying mortar to concrete blocks.

Mortar squeezed out of the joints is carefully scraped off with the trowel and applied on the other end of the block or thrown back onto the mortar board for later use. The blocks are laid to touch the line and are tapped with the trowel to get them straight and level as shown in Fig. 6-23. In a well-constructed wall, mortar joints will average ⅜ in. thick. Fig. 6-24 shows a mason building up a concrete-block wall.

Building Around Door and Window Frames

There are several acceptable methods of building door and window frames in concrete masonry walls. One method used is to set the frames in the proper position in the wall. The frames are then plumbed and carefully braced, after which the walls are built up against them on both sides. Concrete sills may be poured later.

The frames are often fastened to the walls with anchor bolts passing through the frames and embedded in the mortar joints. Another method of building frames in concrete masonry walls is to build openings for them, using special jamb blocks as shown in

Fig. 6-25. The frames are inserted after the wall is built. The only advantage of this method is that the frames can be taken out without damaging the wall, should it ever become necessary.

Placing Sills and Lintels

Building codes require that concrete-block walls above openings shall be supported by arches or lintels of metal or masonry (plain or reinforced). Arches and lintels must extend into the walls not less than 4 in. on each side. Stone or other nonreinforced masonry lintels should not be used unless supplemented on the inside of the wall with iron or steel lintels. Fig. 6-26 illustrates typical methods of inserting concrete reinforced lintels to provide for door and window openings. These are usually prefabricated, but may be made up on the job if desired. Lintels are reinforced with steel bars placed 1 ½ in. from the lower side. The number and

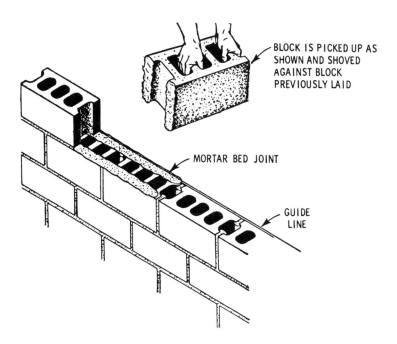

BLOCK IS PICKED UP AS SHOWN AND SHOVED AGAINST BLOCK PREVIOUSLY LAID

MORTAR BED JOINT

GUIDE LINE

Fig. 6-22. Common method used in picking up and setting concrete blocks.

size of reinforcing rods depend upon the width of the opening and the weight of the load to be carried.

Sills serve the purpose of providing watertight bases at the bottom of wall openings. Since they are made in one piece, there are no joints for possible leakage of water into walls below. They are sloped on the top face to drain water away quickly. They are usually made to project 1½ to 2 in. beyond the wall face, and are made with a groove along the lower outer edge to provide a drain so that water dripping off the sill will fall free and not flow over the face of the wall causing possible staining.

Slip sills are popular because they can be inserted after the wall proper has been built, and therefore require no protection during construction. Since there is an exposed joint at each end of the sill,

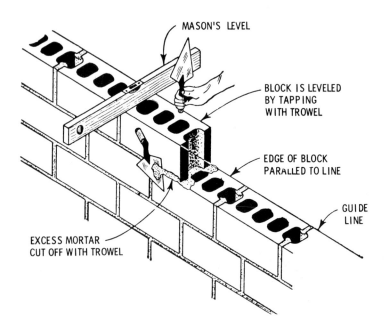

Fig. 6-23. A method of laying concrete blocks. Good workmanship requires straight courses with the face of the wall plumb and true.

(A) Several blocks are receiving mortar on the end.

(B) Blocks are tapped into position.

Fig. 6-24. Construction of a concrete-block wall.

115

(C) Excess mortar is removed and alignment checked .

Fig. 6-24. Construction of a concrete block wall.—Cont.

special care should be taken to see that it is completely filled with mortar and the joints packed tight.

Lug sills project into the masonry wall (usually 4 in. at each end.) The projecting parts are called *lugs*. There are no vertical mortar joints at the juncture of the sills and the jambs. Like the slip sill, lug sills are usually made to project from 1 ½ to 2 in. over the face of the wall. The sill is provided with a groove under the lower outer edge to form a drain. Frequently, they are made with washes at either end to divert water away from the juncture of the sills and the jambs. This is in addition to the outward slope on the sills.

At the time lug sills are set, only the portion projecting into the wall is bedded in mortar. The portion immediately below the wall opening is left free of contact with the wall below. This is done in case there is minor settlement or adjustments in the masonry work during construction, thus avoiding possible damage to the sill during the construction period.

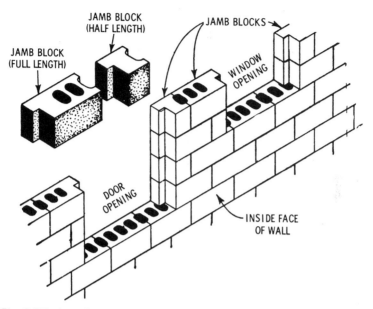

Fig. 6-25. A method of laying openings for doors and windows.

BASEMENT WALLS

Basement walls shall not be less in thickness than the walls immediately above them, and not less than 12 in. for unit masonry walls. Solid cast-in-place concrete walls are reinforced with at least one ⅜-in. deformed bar (spaced every 2 ft.) continuous from the footing to the top of the foundation wall. Basement walls with 8-in. hollow concrete blocks frequently prove very troublesome. All hollow block foundation walls should be capped with a 4-in. solid concrete block, or else the core should be filled with concrete.

BUILDING INTERIOR WALLS

Interior walls are built in the same manner as exterior walls. Load-bearing interior walls are usually made 8 in. thick; partition walls that are not load bearing are usually 4 in. thick. The recom-

117

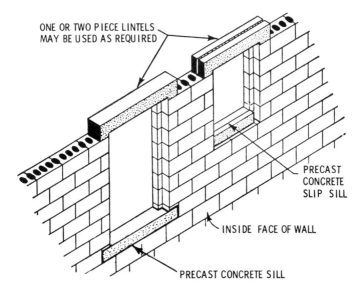

ONE OR TWO PIECE LINTELS
MAY BE USED AS REQUIRED

PRECAST
CONCRETE
SLIP SILL

INSIDE FACE OF WALL

PRECAST CONCRETE SILL

Fig. 6-26. A method of inserting precast concrete lintels and sills in concrete-block wall construction.

mended method of joining interior load-bearing walls to exterior walls is illustrated in Fig. 6-27.

BUILDING TECHNIQUES

Sills and plates are usually attached to concrete block walls by means of anchor bolts, as shown in Fig. 6-28. These bolts are placed in the cores of the blocks, and the cores filled with concrete. The bolts are spaced about 4 ft. apart under average conditions. Usually ½-in. bolts are used and should be long enough to go through two courses of blocks and project through the plate about 1 in. to permit use of a large washer and anchor bolt nut.

Installation of Heating and Ventilating Ducts

These are provided for as shown on the architect's plans. The placement of the heating ducts depends on the type of wall—whether it is load bearing or not. A typical example of placing the

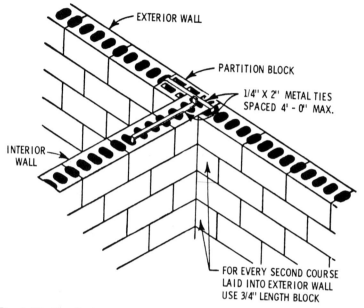

EXTERIOR WALL

PARTITION BLOCK

1/4" X 2" METAL TIES
SPACED 4' - 0" MAX.

INTERIOR
WALL

FOR EVERY SECOND COURSE
LAID INTO EXTERIOR WALL
USE 3/4" LENGTH BLOCK

Fig. 6-27. Detail of joining an interior and exterior wall in concrete-block construction.

heating or ventilating ducts in an interior concrete masonry wall is shown in Fig. 6-29.

Interior concrete-block walls which are not load bearing, and which are to be plastered on both sides, are frequently cut through to provide for the heating duct, the wall being flush with the ducts on either side. Metal lath is used over the ducts.

Electrical Outlets

These are provided for by inserting outlet boxes in the walls, as shown in Fig. 6-30. All wiring should be installed to conform with the requirements of the National Electrical Code and local codes in the area.

Fill Insulation

In masonry construction, insulation is provided by filling the cores of concrete block units in all outside walls with granulated

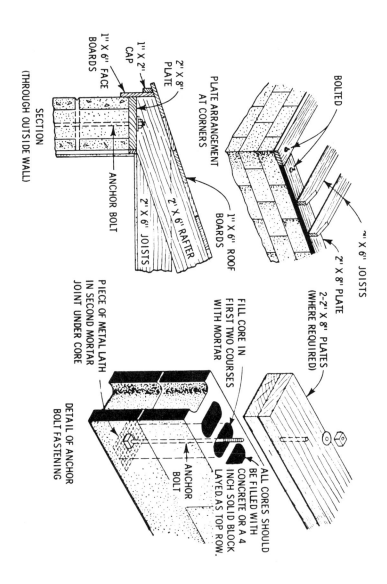

Fig. 6-28. Details of methods used to anchor sills and plates to concrete-block walls.

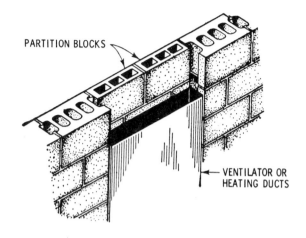

Fig. 6-29. A method of installing ventilating and heating ducts in concrete-block walls.

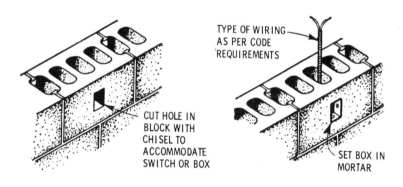

Fig. 6-30. A method of installing electrical switches and outlet boxes in concrete-block walls.

insulation or, the preferred way, inserting rigid foam insulation.

Flashing

Adequate flashing with rust- and corrosion-resisting material is of the utmost importance in masonry construction, since it pre-

121

vents water from getting through the wall at vulnerable points. Points requiring protection by flashing are:

1. Tops and sides of projecting trim under coping and sills.
2. At the intersection of a wall and the roof.
3. Under built-in gutters.
4. At the intersection of a chimney and the roof.
5. At all other points where moisture is likely to gain entrance.

Flashing material usually consists of No. 26 gauge (14-oz.) copper sheets or other approved noncorrodible material.

TYPES OF JOINTS

The concave and V joints are best for most areas. Fig. 6-31 shows four popular joints. While the raked and the extruded styles are recommended for interior walls only, they may be used outdoors in warm climates where rains and freezing weather are at a

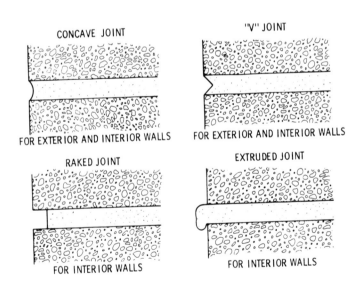

CONCAVE JOINT

"V" JOINT

FOR EXTERIOR AND INTERIOR WALLS

FOR EXTERIOR AND INTERIOR WALLS

RAKED JOINT

EXTRUDED JOINT

FOR INTERIOR WALLS

FOR INTERIOR WALLS

Fig. 6-31. Four joint styles popular in block wall construction.

minimum. In climates where freezes can take place, it is important that no joint permits water to collect.

In areas where the raked joint can be used, you may find it looks handsome with slump block. The sun casts dramatic shadows on this type of construction. Standard blocks with extruded joints have a rustic look, and make a good background for ivy and other climbing plants (Fig. 6-32).

Tooling the Joints

Tooling consists of compressing the squeezed-out mortar of the joints back tight into the joints and taking off the excess mortar. The tool should be wider than the joint itself (wider than ½ in.). You can make an excellent tooling device from ¾-in. copper tubing bent into an S-shape. By pressing the tool against the

Fig. 6-32. Brick wall with extruding joint construction.

Fig. 6-33. Block with V joints. Block can be painted—pad painter works well on it.

mortar, you will make a concave joint—a common joint but one of the best. Tooling not only affects appearance but it makes the joint watertight, which is the most important function. It helps to compact and fill voids in the mortar. Fig. 6-33 shows V-joint tooling of block.

CHAPTER 7

How to Lay Brick

The art of laying brick has not changed much in thousands of years, and happily, it is an art well within the doing of the careful do-it-yourselfer. Brick itself has not changed a great deal either, though there are some new forms. And, most important, brick-work still looks modern and still has a durability that few other materials can equal. What follows is a variety of techniques to learn how to lay brick.

ESTIMATING NEEDS

Table 7-1 shows the amount of brick and mortar needed for various areas of brick wall surface and at several common thicknesses. This table is based on ½-in.-thick mortar which, with the use of standard brick sizes, will result in the 4-in. modular dimensions. Since actual brick sizes will vary slightly from processor to processor, it will be necessary to adjust the mortar thickness to achieve the modular dimensions.

The dimensional volume of brick to mortar is about 7 to 1. A ⅜-in. mortar thickness would mean a 25% reduction in mortar volume and about a 3% increase in brick numbers for the same area coverage. A ⅝-in. mortar thickness calls for a 25% increase in mortar and 3% decrease in brick numbers. If face brick is used on the first thickness of wall and common brick on other tiers, adjust the number of different brick accordingly. Remember to allow for extra face brick depending on how many face-brick headers are to be laid.

Table 7-2 shows the amount of cement, lime, and sand, to use for batches of a little over 1 cu. yd. of mortar. The lower amounts of lime are to be used,when portland masonry cement is used. The higher sand ratios are permitted only if the sand is well graded. Higher volumes of poorly graded sand can only result in a weak mortar with poor bonding qualities.

SCAFFOLDING

Equally important to the handling of the trowel for reduction of worker fatigue and improved production are the placement of material and the amount of physical movement required to lay brick. The mortar board and supply of brick should be immediately behind the bricklayer so he has a minimum amount of steps to take for material. An apprentice should be kept busy supplying the mortar from a mixer and seeing that the brick pile is always adequate.

A brick wall up to about 4 ft. high can be erected with the brick mason standing at ground level. Above 4 ft., it is necessary that he and his material be raised to reduce the amount of reach. Scaffolding for raising the mason and his material takes on many forms, from simple to complex. One- or two-man operations on low walls will find a simple wood scaffold sufficient. Contractors with a variety of jobs will make a worthwhile investment in adjustable tubular-metal scaffolding.

Fig. 7-1 shows the dimensions for making a wood scaffold out of

Table 7-1. Ratio of Cement, Sand, and Lime for Various Mortar Mixes

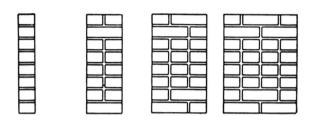

Area of wall in (sq. ft.)	4-inch wall		8-inch wall		12-inch wall		16-inch wall	
	Number of brick	Cubic feet of mortar	Number of brick	Cubic feet of mortar	Number of brick	Cubic feet of mortar	Number of brick	Cubic feet of mortar
1......	6.2	0.075	12.4	0.195	18.5	0.314	24.7	0.433
10......	62	1	124	2	185	3½	247	4½
20......	124	2	247	4	370	6½	493	9
30......	185	2½	370	6	555	9½	740	13
40......	247	3½	493	8	740	13	986	17½
50....	309	4	617	10	925	16	1,233	22
60......	370	5	740	12	1,109	19	1,479	26
70......	432	5½	863	14	1,294	22	1,725	31
80......	493	6½	986	16	1,479	25	1,972	35
90.....	555	7	1,109	18	1,664	28	2,218	39
100....	617	8	1,233	20	1,849	32	2,465	44
200....	1,233	15	2,465	39	3,697	63	4,929	87
300....	1,849	23	3,697	59	5,545	94	7,393	130
400....	2,465	30	4,929	78	7,393	126	9,857	173
500....	3,081	38	6,161	98	9,241	157	12,321	217
600....	3,697	46	7,393	117	11,089	189	14,786	260
700....	4,313	53	8,625	137	12,937	220	17,250	303
800....	4,929	61	9,857	156	14,786	251	19,714	347
900....	5,545	68	11,089	175	16,634	283	22,178	390
1,000..	6,161	76	12,321	195	18,482	314	24,642	433
2,000..	12,321	151	24,642	390	36,963	628	49,284	866
3,000..	18,482	227	36,963	584	55,444	942	73,926	1,299
4,000..	24,642	302	49,284	779	73,926	1,255	98,567	1,732
5,000..	30,803	377	61,605	973	92,407	1,568	123,209	2,165
6,000..	36,963	453	73,926	1,168	110,888	1,883	147,851	2,599
7,000..	43,124	528	86,247	1,363	129,370	2,197	172,493	3,032
8,000..	49,284	604	98,567	1,557	147,851	2,511	197,124	3,465
9,000..	55,444	679	110,888	1,752	166,332	2,825	221,776	3,898
10,000	61,605	755	123,209	1,947	184,813	3,139	246,418	4,331

Note: Mortar joints are ½'' thick

127

Table 7-2. Amount of Brick and Mortar Needed for Various Wall Sizes

Mix by volume, cement-lime-sand	Cement, sacks	Lime, lb.	Sand, cu. yd.
1—0.05—2	13.00	26	0.96
1—0.05—3	9.00	18	1.00
1—0.05—4	6.75	44	1.00
1—0.10—2	13.00	52	.96
1—0.10—3	9.00	36	1.00
1—0.10—4	6.75	27	1.00
1—0.25—2	12.70	127	.94
1—0.25—3	9.00	90	1.00
1—0.25—4	6.75	67	1.00
1—0.50—2	12.40	250	.92
1—0.50—3	8.80	175	.98
1—0.50—4	6.75	135	1.00
1—0.50—5	5.40	110	1.00
1—1—3	8.60	345	.95
1—1—4	6.60	270	.98
1—1—5	5.40	210	1.00
1—1—6	4.50	180	1.00
1—1.5—3	8.10	485	.90
1—1.5—4	6.35	380	.94
1—1.5—5	5.30	320	.98
1—1.5—6	4.50	270	1.00
1—1.5—7	3.85	230	1.00
1—1.5—8	3.40	205	1.00
1—2—4	6.10	490	.90
1—2—5	5.10	410	.94
1—2—6	4.40	350	.98
1—2—7	3.85	310	1.00
1—2—8	3.40	270	1.00
1—2—9	3.00	240	1.00

homemade or ready-made saw horses. It is important that the top planks be secured with nails to the trestles, to allow freedom of movement without worry about loose planks. Mortar board and a supply of brick are also kept on the scaffold. The scaffold structure should be no closer than 3 in. from the wall, to be sure it does not bear against the wall and push it out of alignment.

Fig. 7-2 shows a typical tubular-metal scaffold structure. It

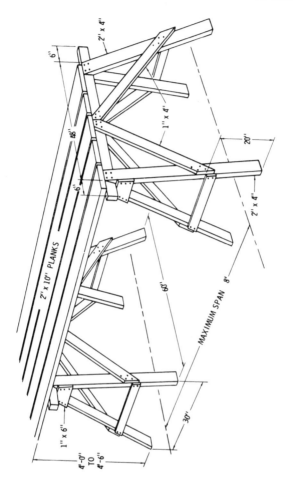

Fig. 7-1. How to make a simple wood scaffold for bricklaying.

allows for considerable adjustment of the working platform for a minimum amount of bending and reach. The mason's platform does not bear against the brick wall, as it appears to do in the illustration. It is counterbalanced by the weight of the brick and mortar and needs no front support. This type of scaffolding may be used on multistory structures if brick is laid from the inside. At the finish of the first story of the building, a rough floor is placed

129

STACK MATERIALS
HERE FIRST.

4'-6"

2'-0"

PLANK AND STACK
MATERIALS ON TOP.

MASON'S BRACKET
ON EXTENDER.

Fig. 7-2. An adjustable tubular-metal scaffold provides greater flexibility for material placement and height adjustment.

130

and the scaffolding is used on the floor for second-story brick work.

FOOTINGS

It is important that the weight of brick walls, especially those which carry a load, be supported by a base that will provide an even distribution of weight. Otherwise, any settling will result in cracks in the mortar.

Although some firm, well-packed earth foundations will be sufficient, the usual practice is to pour a concrete foundation in which reinforcing rods have been embedded. The concrete foundation or footing should be twice the width of the wall and have a depth below the frost line. This is particularly important in colder climates where there is considerable freezing and thawing, which can produce some heaving of the earth. The footing can be poured into a dug trench, with fairly straight sides, with or without forms.

The importance of a perfectly straight and horizontal footing cannot be overemphasized. The footing and the first course of brick will establish the accuracy of a horizontal wall rising course by course above it. After the concrete of the footing has set enough to support the weight of brick above it, usually about seven days, place a line of 1-in.-thick mortar along it and lay an entire course of brick along the length. Use a spirit level and tap the brick into place to make a perfectly horizontal first course. Thereafter, the taut line method of measuring, described later, will keep the wall going up with horizontal accuracy.

MIXING THE MORTAR

The structural strength of a brick wall depends more on the ingredients and mixing of the mortar, and the workmanship in applying it, than on the strength of the brick itself. Mortar must have the proper plasticity for easy handling, yet the correct mixture of ingredients for strength over a long period. This was partially covered in Chapter 6.

A large proportion of sand, while economical, will result in

131

weak mortar, unless the sand is well graded. Since well-graded sand costs more than ordinary sand, it may be just as economical to use less sand and more cement to keep the mortar quality good. Mortar must be constantly mixed fresh. If it is more than two or three hours old, depending on weather, it will begin to take a set and become harder to handle. If the mortar begins to stiffen before that time, it is probably due to water evaporation, and some water may be added (called retempering). If stiffening is due to hydration, retempering will only result in weak mortar.

WETTING THE BRICK

Always give brick the sprinkle test before using. It only takes a minute. Sprinkle a few drops of water on a brick. If it is absorbed by the brick within one minute, the brick has too much absorption and should be wetted before using. If left untreated, the brick will absorb water from the mortar leaving less in the mortar for complete hydration.

If wetting is called for, direct water from a hose on the entire pile of brick until the water runs down the sides. Let the water soak in and the surface water evaporate from the faces of the brick before using them.

PROPER USE OF THE TROWEL

Since the laying of brick is a hand operation, and the trowel is in the bricklayer's hand constantly, there is probably no more important operation in laying up a brick wall than the proper use of the trowel. How it is held in the hand and how it is turned as mortar is applied affects efficiency and fatigue.

The student of bricklaying must first learn how to handle the trowel in picking up, "throwing," and spreading mortar. The practical way for the apprentice to acquire this knowledge is to practice on the mortar board during lunch periods and before working hours. In this practice he should learn how to pick up a trowelful of mortar cleanly, and to spread sufficient mortar to lay at least three brick with one trowelful of mortar. In one operation.

some bricklayers throw enough mortar to lay four or five brick, depending on the thickness of the joint.

In order to use a trowel properly, it should be held firmly, yet loosely, with the full grasp of the right hand and applied with the play of the muscles of the arm, wrist, and fingers. Only actual practice can give the various necessary mechanical movements. Lifting a trowelful of mortar from the tub or mortar board, up to the courses of brick on the wall and throwing the mortar the length of three or more brick, is done with the muscles of the forearm.

In throwing the mortar, *the trowel is turned through an angle of 180°* (that is, turned upside down) while the trowel is being moved the length of three or more brick. In order to turn the trowel upside down, evidently it must be held as shown in Fig. 7-3, because the hand, unlike the owl's head, does not work on a pivot and 180° is about the limit it will turn without elevating the elbow. Fig. 7-4 shows how the mortar is picked up from the mortar board preliminary to throwing it on the brick.

In order to fully understand how the bricklayer throws the mortar, the operation is first shown by the diagrams in Fig. 7-5. Here, only the trowel is shown without hand or mortar so its various positions may be seen as it travels the length of the spread of a three-brick length, and back again to begin the spreading stroke.

APPLYING THE MORTAR

The placing of the mortar consists of four distinct operations each of which is completed in turn in order to form a strong, durable masonry joint:

1. Throwing.
2. Spreading.
3. Cutting off.
4. Buttering end joint.

The operation of throwing the mortar results in a rounded column of mortar along the central portion of the brick leaving the outside portions bare as shown in Fig. 7-6. In order that the brick shall have a full bed of mortar to lie on so that the load will be

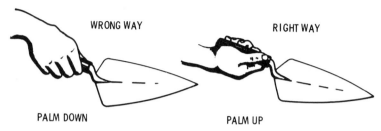

Fig. 7-3. The wrong and right way to hold a trowel. Note position of the thumb.

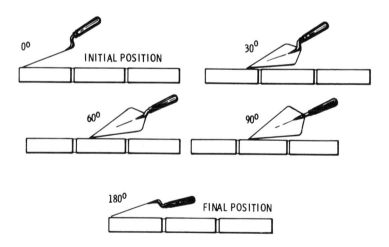

Fig. 7-4. How mortar is taken from the board. The trowel is always held with the thumb up for easier turning.

Fig. 7-5. How the trowel is turned during a single stroke of throwing a mortar line.

134

distributed over its entire face, the mortar, after being thrown, should be spread by going over it with the point of the trowel as shown in Fig. 7-7.

When the operation of spreading the mortar has been perfectly done, no cutting off is necessary. If, however, too much mortar was thrown, or too much pressure exerted on the trowel in spreading the mortar, some of it will hang over the side of the brick as shown in Fig. 7-8. In this case it must be cut off so that it will not at any point project over the side of the brick, as shown in Fig. 7-9. In addition to laying a bed of mortar for the brick to lie on, the end of each brick, when laid, must be covered or "buttered" with mortar so there will be a layer of mortar in the vertical joints as well as in the horizontal joints.

LAYING THE BRICK

In lifting a brick from the pile on the ground or scaffold, in order to place it on the bed of mortar, the bricklayer grasps the brick in his left hand, as shown in Fig. 7-10. He butters one end, and in "laying" the brick, first places it on top of the bed of mortar (previously spread) a little in advance (to the right) of its final position as shown in Fig. 7-11A. He presses the brick into the mortar with a downward slanting motion as indicated by positions M, S, in Fig. 7-11B, so as to press the mortar up into the end joint. During this operation the brick moves from its initial position M, shown in dotted lines M, (corresponding to the position shown in Fig. 7-11A) to some intermediate position S, as shown in Fig. 7-11B.

This is the shoving method of bricklaying, and if the mortar is not too stiff and is thrown into the space between the inner and outer courses of brick with some force, it will completely fill the upper part of the joints not filled by the shoving process. After the brick is shoved down and against the mortar in the end joint, it is forced home, or down until it aligns with the brick previously laid by tapping it either with the blade of the trowel, as shown at L (Fig. 7-11C), or with the handle butt of the trowel in position F, shown in dotted lines. During the operation just described, more or less mortar is squeezed out through the face and end joints as

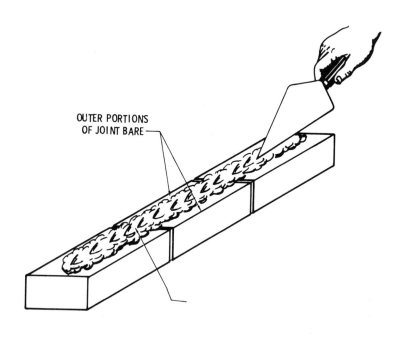

OUTER PORTIONS
OF JOINT BARE

Fig. 7-6. As the trowel is turned the mortar is spread over the center of a row of three to five brick.

shown in fig. 7-11D. For appearance and to save mortar it is cut off flush with the trowel. This mortar on the trowel thus cut off should be used for buttering the end of the next brick. It should never be thrown from the trowel back onto the mortar board. When thrown back onto the mortar board, a large portion may daub up the brick instead of landing on the board, and the operation results in an unnecessary motion each time.

Fig. 7-12 shows the laying of a course of rowlock header brick. In Fig. 7-12A the largest face of the brick has been buttered and is ready to be shoved into position. In Fig. 7-12B a closure is placed as the last brick in the course. Both faces are buttered and the brick is shoved down into place. If the measurement of total length and thickness of mortar has been correct, the closure brick should result in the two facing vertical mortar lines being the same thickness as the others in the wall.

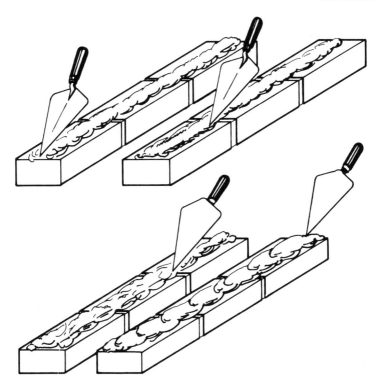

Fig. 7-7. The point of the trowel is used to spread the center furrow of mortar over most of the brick surface.

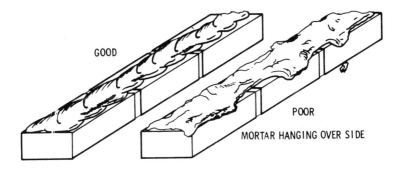

GOOD

POOR

MORTAR HANGING OVER SIDE

Fig. 7-8. Good practice is to spread the mortar over the brick surface without any excessive overhang.

137

USING THE TAUT LINE

Clearly, if no guide is provided and the brick is laid by eye, a true wall surface could not be obtained as some of the brick would be laid too far out and others too far in. In order to guide the bricklayer so that the brick will lay straight, a taut line secured by pins is used, or the equivalent. In order to have supports for the line, a corner of the wall is first built up several courses, and then a lead or support is placed at some point along the course.

The line is made fast around the end or corner, stretched taut and wound around a brick on the lead, as shown in Fig. 7-13. This is better than using a nail or pins because if the latter pulls loose, the nail may hit a bricklayer in the eye, resulting in injury to or loss of an eye.

The line should be placed $\frac{1}{32}$ in. outside the top edge of the brick and exactly level with it. In order to hold the line at $\frac{1}{32}$ in. distance outside the top edge, make two distance pieces out of cardboard, or preferably tin, shaped and attached to the line as shown in Fig. 7-14. The reason for this offsetting of the line is that

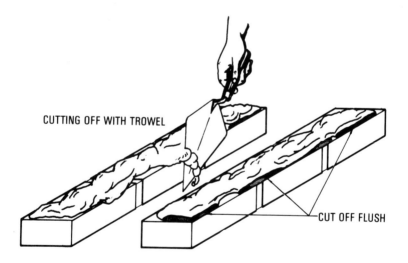

CUTTING OFF WITH TROWEL

CUT OFF FLUSH

Fig. 7-9. If mortar overhang does occur, it must be cut off as shown.

the brick should be laid without touching the line—the ¹⁄₃₂ in. marginal distance being gauged by the eye.

Obviously, if the brick are laid so that they touch the line, the latter would be shoved out of place resulting in irregularities in the wall. Hence, see that no brick touches the line. The tendency of inexperienced bricklayers is to "crowd the line" or as it is called laying brick *strong* on the line. This effort to work with precision does not accomplish the desired result for the reason just given.

The student who works with precision will not be satisfied with the instruction to set the line level with the top face of the brick. He will want to know whether the top or bottom of the line should be level with the top of the brick, especially if it is a thick line. Of course, bricklaying is not a machinist's job and one is not expected to work with machinist's precision, however, precision methods cannot be criticized when they can be used without any extra effort or loss of time. Fig. 7-15 shows the wrong and right ways to set the line when precision is considered.

A skillful bricklayer will never touch the line even in applying

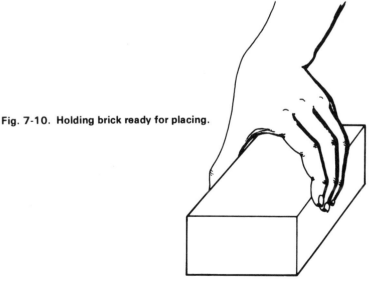

Fig. 7-10. Holding brick ready for placing.

the mortar or laying the brick. There are two ways of holding the brick, as shown in Fig. 7-16, so the line will not be disturbed. It should be understood that even the fingers must not touch the line—otherwise it will be pushed out of place while other work-

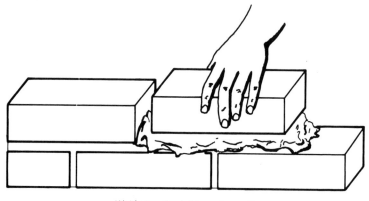

(A) Placing the brick on the wall.

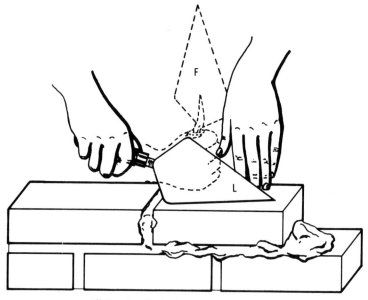

(C) Tapping the brick for proper alignment.

Fig. 7-11. Four steps in placing a brick in position.

140

men are using it as a guide. The method of laying the brick without touching the line is shown in Fig. 7-17. Of course, practice is necessary to do this successfully. The student should practice before laying to the line so that he will acquire the habit of bringing

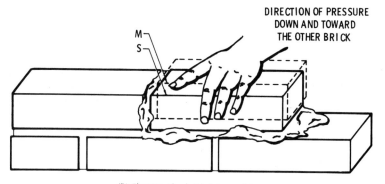

DIRECTION OF PRESSURE
DOWN AND TOWARD
THE OTHER BRICK

M
S

(B) Shoving the brick into position.

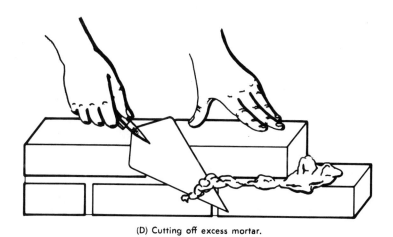

(D) Cutting off excess mortar.

141

(A) Buttering the wide part of the brick.

(B) Placing the closure or last brick in place.

Fig. 7-12. Rowlock leader brick placement.

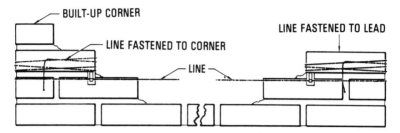

Fig. 7-13. With corners built up, a line is drawn taut to establish a level for each course of brick.

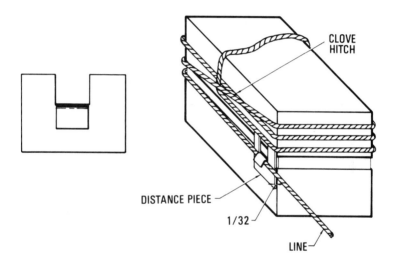

Fig. 7-14. A spacer made to hold the taut line 1/32 in. away from brick face.

his thumb and fingers up as the brick goes down near the line.

Fig. 7-18 shows the laying and cutting off of the front tier of a wall. In Fig. 7-18A, the end of a brick has been buttered, placed in position, and shoved toward the brick already in position, to the correct vertical mortar-joint thickness. If the mortar is plastic enough, the pressure of the hand on the brick will align it with the taut string. In Fig. 7-18B, a closure brick is placed and pressed down to the level of the line. Fig. 18C shows a brickmason cutting

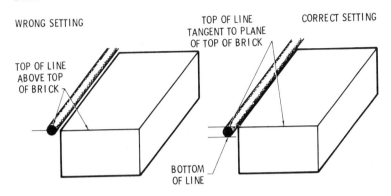

WRONG SETTING

TOP OF LINE
TANGENT TO PLANE
OF TOP OF BRICK

CORRECT SETTING

TOP OF LINE
ABOVE TOP
OF BRICK

BOTTOM
OF LINE

Fig. 7-15. Correctly set the line 1/32 in. out from the edge of the brick and the top surface even with the top of the brick.

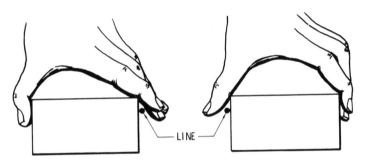

LINE

Fig. 7-16. Placing the brick in position without disturbing the line.

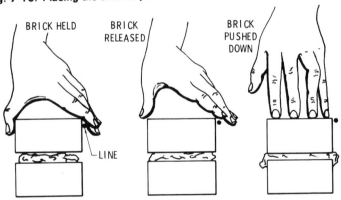

BRICK HELD

BRICK
RELEASED

BRICK
PUSHED
DOWN

LINE

Fig. 7-17. Steps in placing a brick without disturbing the line.

144

(A) The brick is placed and shoved to the right aligning it with the string.

(B) A closure brick put into place.

Fig. 7-18. Brick placement and mortar cutoff.

off the squeezed out mortar. With experience this can be done without touching the line.

In Fig. 7-19, a second layer of brick, with a shallow cavity between, is laid and mortar cut off. In Fig. 7-19A, the line of mortar is thrown and readied for a few bricks of a course. In Fig. 7-19B, the bricks are in place, properly aligned to the string. The only difference between the two series of illustrations is the final cutting off of excess mortar. With a shallow cavity, cutting off excess mortar upward is difficult because of the small amount of room. It is more economical to cut off downward and let the excess mortar fall into the cavity, as shown in Fig. 7-19C.

The three illustrations in Fig. 7-20 show a tier of bricks being laid up for the front of a frame home. Because only one tier is used, this is called brick-veneer construction. At point (A), the taut line is in place and a layer of mortar is being thrown across between two halves of a course, with a few brick in place near the center. Two

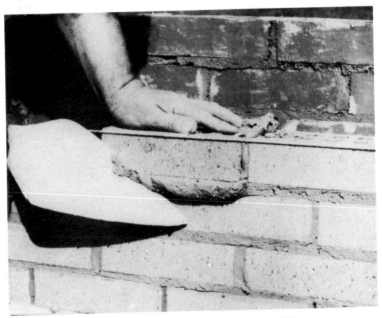

(C) Mortar cutoff which will be used to butter the next brick.

(A) Mortar is spread over several bricks.

(B) Bricks in place and aligned.

Fig. 7-19. Laying the rear tier of brick with a cavity between.

147

brickmasons will be working, one at each half of the course. In Fig. 7-20B, one mason is working on the short course around the corner. He is tapping a ½-bat, with header end showing, into place, with the constant use of a spirit level. Note the vertical board temporarily nailed into place. This board has been carefully aligned with a vertical level and serves as the guide for the corner. In Fig. 7-20C, the closure brick has been placed and excess mortar has been cut off. The line will now be moved up (⅔ of 4 in. for modular brick), checked for horizontal accuracy, and the next course laid.

TOOLING THE JOINTS

Tooling consists of compressing the squeezed out mortar of the joints back tight into the joints and taking off the excess mortar.

(C) Mortar cutoff is downward allowing excess to drop into cavity.

(A) With a line in place, mortar is thrown across several bricks.

(B) A ½-bat header is being tapped into place.

Fig. 7-20. Laying up a brick-veneer wall.

(C) A closure brick has been placed to complete one course.

The tool should be wider than the joint itself. Jointing tools are available in a number of sizes and shapes. They are generally made of pressed sheet steel or solid tool steel, S-shaped, and are convex, concave, or V-shaped. The convex side is pressed against the mortar.

By pressing the tool against the mortar you will make a concave joint—a common joint but one of the best. Tooling not only affects appearance but it makes the joint watertight which is the most important function. It helps to compact and fill voids in the mortar. Fig. 7-21 shows concave-joint tooling.

TYPES OF JOINTS

The concave and V joints are the best for most areas. Fig. 7-22 shows four popular joints. While the raked and the extruded styles are recommended for interior walls only, they may be used outdoors in warm climates where rains and freezing weather are at a

150

Fig. 7-21. Using a pointing tool to press mortar into the joints.

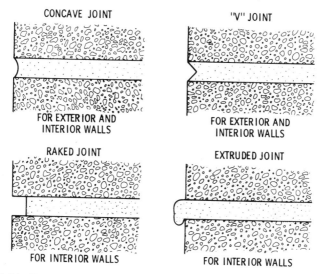

CONCAVE JOINT

FOR EXTERIOR AND
INTERIOR WALLS

"V" JOINT

FOR EXTERIOR AND
INTERIOR WALLS

RAKED JOINT

FOR INTERIOR WALLS

EXTRUDED JOINT

FOR INTERIOR WALLS

Fig. 7-22. Four popular mortar joints. The raked and extruded joints are not
recommended for exterior walls in cold climates.

151

minimum. In climates where freezing can take place, it is important that no joint permits water to collect.

In areas where the raked joint can be used, you may find it looks handsome with slump style brick. The sun casts dramatic shadows on this type of construction.

CHAPTER 8

Bonding

To bond means to bind or hold together. Bonding is important in brick construction to make a solid and secure structure. There may be three different meanings to the word *bond*, as it refers to masonry. These are:

Structural bond. The interlocking of masonry units by overlapping bricks or by metal ties.

Pattern bond. Interlocking and overlapping brick work following a fixed sequence. Pattern bonds for structural purposes have become standardized and are given names. Some pattern bonds are used for appearance purposes only, or combined to provide a special appearance plus structural bonding qualities.

Mortar bond. The adhesion of the mortar to the masonry or to steel reinforcement ties placed in the masonry. Mortar alone is not strong enough to provide sufficient bonding for secure structures.

In overlapping brick construction, most building codes require that no less than 4% of the wall surface consist of headers, with the distance between headers no less than 24 in. vertically or horizontally. Headers are bricks laid with their longest dimension perpendicular to the front or facing tier or brick, and overlap into the second row or tier behind, for a double-thick wall. Steel ties are finding wider use, with some codes requiring at least one tie for each 4 ½ sq. ft. of surface.

TERMS APPLIED TO BONDING BRICK

Common names have been applied to brick depending on their position in the structure. Fig. 8-1 is a sketch of their names and positions.

Most wall construction is with *stretchers*. Stretchers are brick with their largest surface horizontal and longest edge facing out. *Headers* have their smallest dimension facing out and are used to interlock with a wall of brick behind. A *soldier* is a brick standing

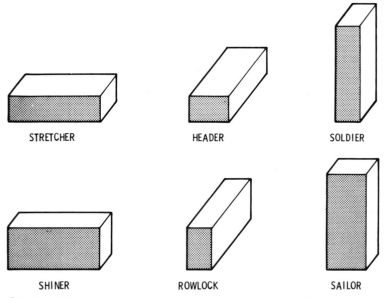

STRETCHER   HEADER   SOLDIER

SHINER   ROWLOCK   SAILOR

Fig. 8-1. Brick are given names, based on their use and position.

on its end, and is often used as the top course of a wall, forming a sill for the support of ceiling or roof joists (Fig. 8-2). When a soldier is placed with its largest face forward, it is often called a *sailor*. A *rowlock* is a header standing on edge. A rowlock with its largest surface facing out is sometimes called a *shiner*. Rowlock placement reduces the number of brick in a double-thick wall but it is not as strong.

Fig. 8-3 illustrates the methods used to bond the brick shown in Fig. 8-1. Each layer of brick is called a course. Stretchers are laid overlapping and provide the interlock between courses. Rowlocks are placed as stretchers or headers for interlocking. The dimensions are such as to leave a cavity between vertical tiers. *Wythe* (or *with*) is the term generally applied to the grout between brick faces. The grout used is a thinner mortar to permit its flowing completely into every crevice and irregularity of the brick surfaces. The word *wythe* also refers to the wall between flue cavities of a chimney.

Fig. 8-2. How soldiers are used for the top course as a sill. Generally its main purpose is decorative.

155

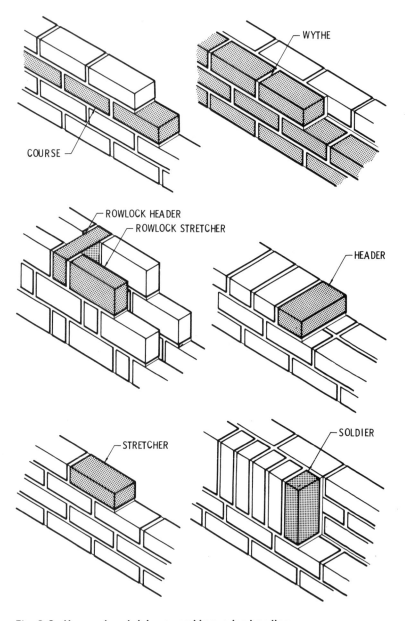

Fig. 8-3. How various brick are used in overlap bonding.

OVERLAPPING AND THE MODULE CONCEPT

As mentioned in an earlier chapter, nominal brick dimensions are based on the 4-in. module concept. The face and long edge are a multiple of 4 in., while the thickness is a simple fractional part of 4 in. (usually ⅓ of 8 in.). Actual dimensions are slightly less to allow for mortar thickness. Modular dimensions permit placing brick in walls with both stretchers and headers to result in a fixed pattern in appearance, without the need for cutting brick into small pieces, except in a few cases.

Fig. 8-4 shows how a stretcher is twice the width of a header and, conversely, headers are one-half the width of a stretcher.

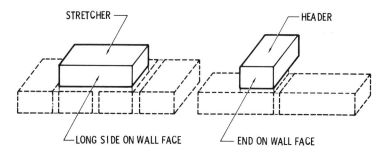

STRETCHER

HEADER

LONG SIDE ON WALL FACE

END ON WALL FACE

Fig. 8-4. The nominal length of a brick is just twice that of its width.

MODULAR LAPPING

Modular dimensions have resulted in standard overlapping techniques, and the expressions of ½ lap, ¾ lap, ¼ lap, and ⅓ lap have developed.

The most common method of laying brick with stretchers, most used in walls, is the ½-lap method. This is clearly shown in Fig. 8-5. One-half the brick length in each successive course overlaps ½ the brick length below it. On occasion, for a change of pace in appearance, ⅓ lap is used. One-third of the brick length overlaps ⅔ of the brick below it. However, with each reduction in overlap, the strength of the mortar bonding is reduced. The limit of reduction is no overlap at all, sometimes used for garden walls. To retain strength in such cases, a number of metal ties are used for bonding.

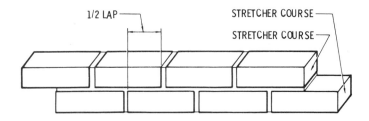

Fig. 8-5. Showing 1/2-lap bonding commonly used in stretcher courses.

The use of ¼ lap is usually with headers, or headers and stretchers. Fig. 8-6 shows two courses of headers with ¼ lap. This provides the maximum bonding for the shorter width of headers. The bonding is between half widths of headers but the ¼ lap is the width of a stretcher. Fig. 8-7 shows the ¼ lap used between courses of headers and stretchers. This is the method commonly used for a full course of headers. Where headers alternate with stretchers in the same course, the ¼ lap will look like Fig. 8-8. Note that headers have a full ½-lap bonding surface with facing stretchers above and below, but stretchers are reduced to ¼-lap and ¾-lap bonding. In rowlock construction, with alternate headers and stretchers, the result is ⅓ lap of all bonding surfaces (Fig. 8-9).

STRUCTURAL PATTERN BONDS

When brick is laid with good bonding in mind, a pattern is formed which, when repeated consistently, results in an appearance that is pleasing as well as being strong. This is especially true when header brick are involved in the structure.

Over the years standard patterns have been established, and each has its own particular name. There are six in wide general use. They are shown in Fig. 8-10.

Running Bond. All brick are laid in ½-lap stretcher bonding, without headers. A variation is the ⅓ running bond, as mentioned (Fig. 8-11). Because there are no headers for bonding to a

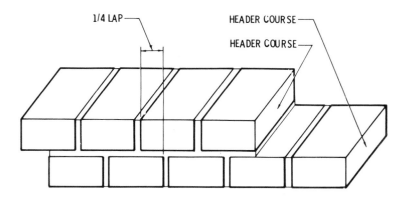

Fig. 8-6. Best bonding for courses of all headers is the 1/4 lap.

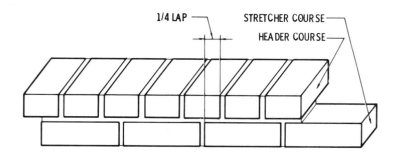

Fig. 8-7. A header course with 1/4 lap over a stretcher course.

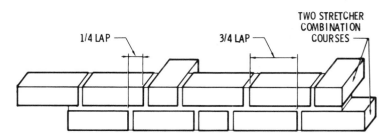

Fig. 8-8. Headers and stretchers in the same course result in a 1/4 lap and a 3/4 lap.

159

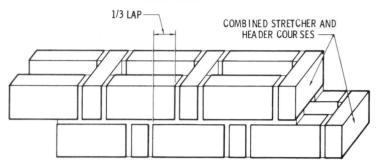

Fig. 8-9. In rowlock construction the result is 1/3-lap bonding.

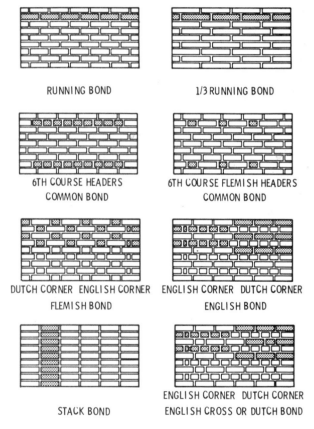

RUNNING BOND

1/3 RUNNING BOND

6TH COURSE HEADERS
COMMON BOND

6TH COURSE FLEMISH HEADERS
COMMON BOND

DUTCH CORNER ENGLISH CORNER
FLEMISH BOND

ENGLISH CORNER DUTCH CORNER
ENGLISH BOND

STACK BOND

ENGLISH CORNER DUTCH CORNER
ENGLISH CROSS OR DUTCH BOND

Fig. 8-10. The most popular bonding pattern in general use.

160

Fig. 8-11. Here 1/3-lap running bond is used in a single-tier brick-veneer form.

second thickness of brick behind, metal ties must be used between thicknesses. The metal ties allow for an air cavity between sections. The air space provides extra insulation against the transmission of heat through the brick wall. If strength is more important than insulation, the cavity may be filled with concrete. The running bond is also used for homes of brick-veneer construction, which are homes of 2″ × 4″ framing and a single layer of facing brick, essentially for appearance. A common method of home construction in some areas is an all-frame home with brick veneer in front and wood or stucco on the other three sides.

Common or American Bond. Perhaps the most frequently used bond pattern is the common or American bond. The pattern consists of several courses of stretchers only and headers every 5th, 6th, or 7th course, depending on the needs for structural strength. A continuous course of headers is used. A variation is the Flemish header course in which headers alternate with stretchers.

161

In the common-bond pattern, each row of headers must start their corners with a ¾ length of brick, to come out even with a symmetrical pattern, on the 4-in. module system. The half-size brick appearing at the ends of some stretcher courses is a full stretcher starting the adjacent wall around the corner.

Flemish Bond. This bonding pattern is obtained by using alternate header and stretcher brick with each header on one course centered with a stretcher in the course below and above. Where the strength of so many headers is not needed, many of the headers shown can be clipped brick or ½-bat size. However, there is no reduction in total brick used and labor can be saved by not clipping the brick.

English Bond. This pattern is formed with alternate courses of running stretchers and running headers. As with the Flemish bond, many of the header brick may be clipped if preferred.

English Cross or Dutch Bond. Similar to English bond, except the stretchers are spaced so each header faces the middle of a stretcher on one side and a joint between stretchers on the other. The joints form a series of overlapping X's, thus the name English "cross."

Stack Bond. Brick stacked vertically and horizontally with no overlap form a stacked bond. Bonding strength between brick and to the layer of brick behind is by means of metal straps or ties. This pattern is seldom used for a load-bearing wall, where the extra strength of overlapping is a must. Where stack bond is specified for load-bearing walls, a liberal use of steel reinforcing bars is important. This is also called block pattern.

CORNERS

Fig. 8-10 shows two types of corners for the English-, Flemish-, and Dutch-bond patterns. In the Dutch corner, the corners start with a ¾ bat. In the English corners, they start with a ¼ bat. The ¼ bat, however, must never be placed at the corner, but at least 4 in. from the corner. These corners allow for proper spacing to make the pattern come out as intended.

QUOINS

Brick that are cut for use on corners are called *quoins*, a word that is slowly passing out of popular use. Fig. 8-12 illustrates the various quoin shapes and how they are used. The diagonally clipped, or king quoin, will have the appearance of a ¼ bat visible in the brick face, but the structural strength of a larger size. Fig. 8-13 shows two methods of making the English corner. With a king closure or quoin, the brick will have the appearance of a ¼ -lap brick at the very corner, without the attendant weakness.

Fig. 8-14 shows the relation of the cut brick to the whole brick. Brick is easily cut by first scoring it with a chisel having one straight side and one beveled side (Fig. 8-15). After scoring, a whack with the flat of the mason's hammer head will break the brick off even with the scored lines. It takes experience to obtain the precision needed for making clean breaks.

CLOSURES

All brick work starts with the corners. The last brick placed to complete a course is called a *closure* or *closer*. Most often the closure is a full stretcher somewhere in the middle of the course. For many structural pattern bonds, closures require cutting brick as described above. Fig. 8-16 shows some of the frequently used closures, in addition to the full stretcher.

GARDEN WALL

A variation of the Flemish pattern bond was given the name of garden-wall pattern. If two or more stretchers are used between headers, the back and front of the brick wall will have the identical appearance. An early use of this design was for garden walls. Fig. 8-17 shows how brick is laid for two-, three-, and four-stretcher garden-wall construction.

Figs. 8-18 and 8-19 show two garden-wall patterns with intentional appearance design. Note the free use of ¼ -bat closures to

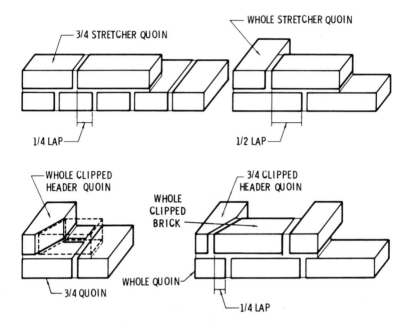

Fig. 8-12. Clipped brick used at corners are generally called quoins.

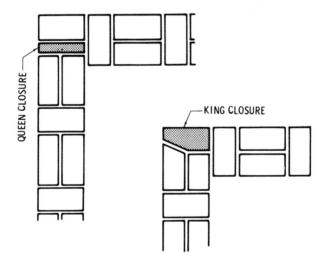

Fig. 8-13. Two methods of making an English corner.

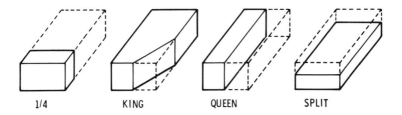

| 1/4 | KING | QUEEN | SPLIT |

Fig. 8-14. Cut brick used for closures and corners.

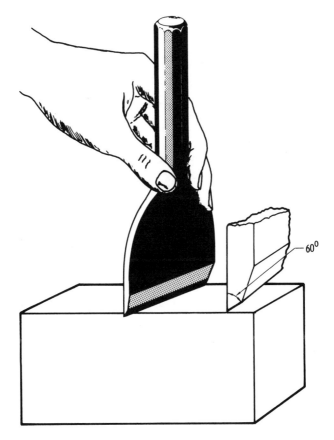

60°

Fig. 8-15. The type of beveled chisel used to score brick for cutting.

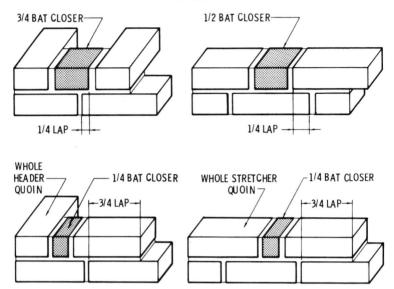

Fig. 8-16. Variations of cut brick used as closures.

make the diagonal designs come out even. Part of the design results from the use of brick with differing textures and shades of colors. See a brickyard for locally available textures and colors.

METAL TIES

There are certain conditions in which the use of header brick for bonding to the rear brick wall section is not appropriate. Metal ties are then used instead of header brick.

Fig. 8-20 illustrates the method of bonding a brick facing wall to a tile backing. Also shown are some of the shapes of metal ties available. While clay tile sizes follow the 4-in. module design concept, their larger size eliminates the use of headers for bonding. Metal ties do the job.

Where an air cavity between layers of brick is desired, metal ties provide the best means of bridging across the cavity for bonding. The center illustration of Fig. 8-20 shows this application with clay tile as the backing wall. Fig. 8-21 illustrates the use of metal ties in an all-brick wall with a cavity between.

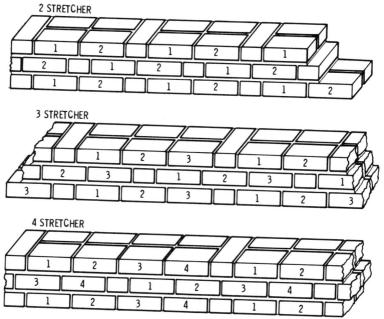

2 STRETCHER

3 STRETCHER

4 STRETCHER

Fig. 8-17. A variation of the Flemish pattern called garden wall. It has the same pattern on each side.

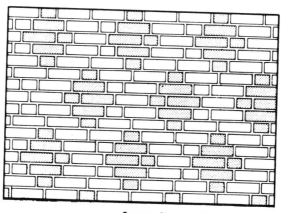

Courtesy Structural Clay Products Inst.

Fig. 8-18. Garden-wall design emphasizing diagonal lines.

167

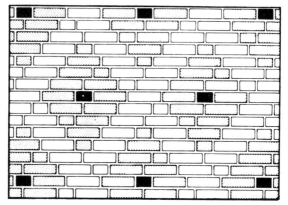

Courtesy Structural Clay Products Inst.

Fig. 8-19. A dovetailed garden-wall design.

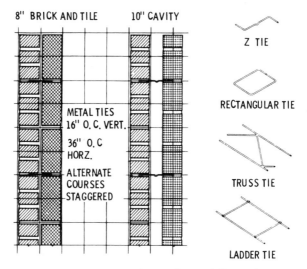

Courtesy Structural Clay Products Inst.

Fig. 8-20. Metal ties bonding together two walls. Various metal ties are also shown.

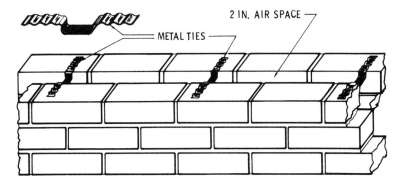

Fig. 8-21. Metal strip ties between two tiers of brick with an air cavity between.

While a flat metal tie is shown in Fig. 8-21, most ties are formed of round steel reinforcing rods. The diameter of the rods must be small enough to be fully embedded in the mortar. Since metal is subject to possible rust, a careful job of mortar work is necessary to reduce water seepage to an absolute minimum.

CHAPTER 9

Wall Types, Thickness, and Anchoring

Brick construction is used for many purposes but by far the greatest use is in wall construction. Brick features great strength, fire resistance, and some insulation properties.

Mainly because of its compressive strength, brick is still used to build multistorey structures with the brick carrying the loads of upper floors.

Perhaps the best example of all brick construction is the Monadnock Building in Chicago, which was built in 1893 and still stands. Six-foot thick walls at the bottom are tapered to 12 in. walls in the top floor. However, since the advent of the skyscraper, steel skeletons carry the loads and brick or other types of walls are used on the outside as well as for interior purposes. As a result, solid brick buildings have been specified only for buildings of three- or four-storey heights and less.

A number of years ago, engineers in Europe developed brickwall designs of only 6-in. thickness capable of carrying the loads of

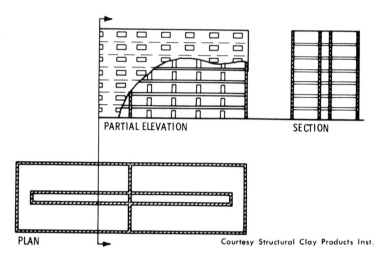

PARTIAL ELEVATION   SECTION

PLAN   Courtesy Structural Clay Products Inst.

Fig. 9-1. Design for longitudinal bearing walls offering high strength. A 6-in. wall can bear the load of a 16-storey building.

buildings up to 16 stories in height. This design is used in buildings in the United States. Fig. 9-1 shows the design used to provide good longitudinal strength. Fig. 9-2 shows the method for achieving good transverse strength. In actual practice, both methods are combined.

TYPES OF BRICK WALLS

There are three basic types of walls in common use. These are the *veneer* wall, which is one tier of brick and does not carry a load, used only as a facing on frame homes. The *solid* brick wall is from 8 to 24 in. thick, depending on the load they may carry. The *cavity* wall includes an air cavity between the first tier of brick and succeeding tiers of different thicknesses. Cavity walls may consist of brick in all tiers or a combination of brick and hollow clay tile.

VENEER WALLS

A typical home with a brick-veneer facing is of frame construction, using 2″ × 4″ wall studding. Ceiling joists and roof rafters are

172

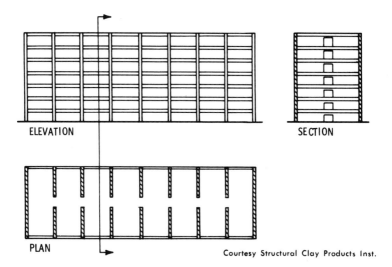

ELEVATION

SECTION

PLAN

Courtesy Structural Clay Products Inst.

Fig. 9-2. Design for high-strength transverse bearing walls.

supported on the sills over the 2″ × 4″ walls. The brick facing provides superior insulation and better appearance. The wide choice of brick textures and colors, and the beautiful patterns used in brick overlap, make it the most varied of home finishes. Many homes now use one or more interior walls of brick, especially the wall on which a fireplace is located.

There are two generally accepted methods of placing a brick-veneer wall against the frame construction of a home. These are shown in Fig. 9-3. The most frequently used method is to space the brick tier away from the sheathing on the frame studs. About a 1-in. air space is the usual practice. The brick wall is secured to the frame by corrugated metal tabs. There should be a metal tie for each 2 sq. ft. of wall area. The other method is the use of a paper-backed wire mesh against the studs, instead of sheathing. The brick is grouted right up against the wire mesh. A special grout or the regular mortar may be used.

Fig. 9-4 is a cutaway view of brick veneer with the metal ties plainly shown. Fig. 9-5 is a side view. The ties should be corrosion-resistant and should be placed with a slight slope downward to the brick, to allow any moisture that may be collected to run down toward the front and away from the sheathing.

173

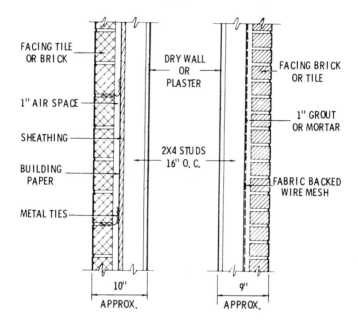

FACING TILE OR BRICK

DRY WALL OR PLASTER

FACING BRICK OR TILE

1" AIR SPACE

SHEATHING

2X4 STUDS 16" O. C.

1" GROUT OR MORTAR

BUILDING PAPER

METAL TIES

FABRIC BACKED WIRE MESH

10" APPROX.

9" APPROX.

Courtesy Structural Clay Products Inst.

Fig. 9-3. Two methods for placing brick veneer on a frame home.

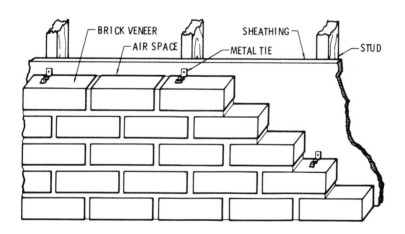

BRICK VENEER

AIR SPACE

SHEATHING

STUD

METAL TIE

Courtesy Structural Clay Products Inst.

Fig. 9-4. Metal tabs secure brick to sheathing.

174

Extremely important to the use of a brick-veneer facing on a frame is the sturdiness of the frame structure, with the least possible give to the pressures of wind or snow load. There is little elasticity to brick and any pressure from a change in position of the frame can cause serious cracks in the mortar. If there is any suspicion that there may be frame shift or vibration, sheer points should be included in the brick as it is laid up. Fig. 9-6 shows two methods of providing vibration joints (if considered necessary). One method is to omit the mortar from one course of brick, so there is no bond between the two courses. To maintain uniform patterns, a mortar line is laid on one course and allowed to set before the next course is laid. Extra ties must be used if this method is employed.

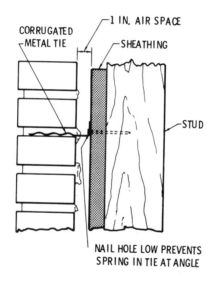

Fig. 9-5. Side view of the metal-tab holders.

Fig. 9-7 shows two methods for handling the details in laying up a brick-veneer wall. One method is for homes with raised floors—those with crawl space or basement. The other is for homes constructed on a concrete slab. Alternate foundations are also shown. A feature of brick-veneer construction is the excellent barrier formed against the passage of moisture from the outside. Moisture

175

which may get through the brick, will flow down the air space instead of passing through the frame construction. This calls for an outlet for the moisture in the form of weep holes at the bottom of the brick. Fig. 9-8 illustrates a typical weep hole. It consists of leaving the mortar out of the vertical joints of the bottom course about every 2 ft.

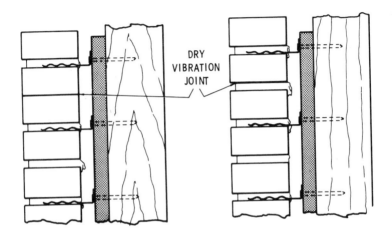

DRY
VIBRATION
JOINT

Fig. 9-6. Two methods of providing vibration joints.

SOLID BRICK WALLS

Solid brick walls are in common use in one- and two-storey homes and two- and three-storey apartment buildings. Floor and roof loads are borne principally by the outer walls, and to some extent by the inner room walls, usually made of 2″ × 4″ framing. Typical wall thicknesses are 8 and 12 in. and typical bonding patterns are those shown in Fig. 9-9. As mentioned, brick is still extensively used for multistorey structures.

The thickness of walls for a multistorey building will depend on the load they must bear. This will vary not only by the number of stories in the building but by the intended use. For example, industrial buildings may include floors for heavy machinery. The

176

design of walls for heavy loads is quite involved and is determined by engineers and architects. The engineering involved is beyond the scope of this book, but it is important that the apprentice bricklayer knows about the patterns used in laying thick walls since he may be called upon to construct them. What follows, for purposes of demonstration only, are minimum wall specifications. The actual specs would be much more detailed and would be based on more complicated calculations.

General: The minimum thickness of all masonry bearing or non-bearing walls shall be sufficient to resist or withstand all vertical or horizontal loads and the fire resistance requirement of any local code.

Thickness of Bearing Walls: The minimum thickness of masonry bearing walls shall be at least 12 in. in thickness for the uppermost 35 ft. of their height and shall be increased 4 in. in thickness for each successive 35 ft. or fraction thereof measured downward from the top of the wall.

Exceptions:

1. Stiffened Walls: Where solid masonry bearing walls are stiffened at distances not greater than 12 ft. apart by masonry cross walls or by reinforced concrete floors, they may be of 12-in. thickness for the uppermost 70 ft. measured downward from the top of the wall, and shall be increased 4 in. in thickness for each successful 70 ft. or fraction thereof.

2. Top-Storey Walls: The top-storey bearing wall of a building not exceeding 35 ft. in height may be of 8-in. thickness, provided it is not over 12 ft. in height and the roof construction imparts no lateral thrust to the walls.

3. One-Storey Walls: The walls of a one-storey building may be not less than 6 in. in thickness, provided the masonry units meet the minimum compressive strength requirement of 2500 psi for the gross area and that the masonry be laid in Type M, S or N mortar.

177

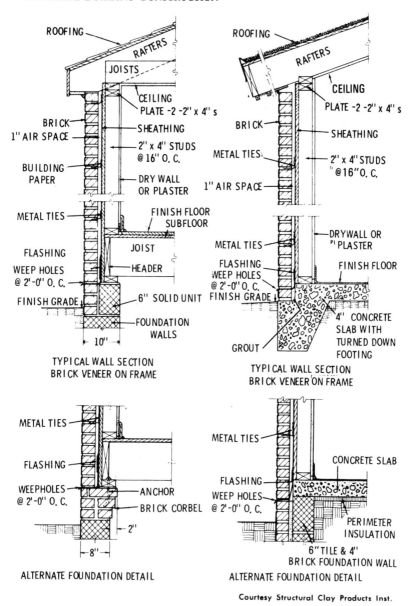

Fig. 9-7. Details for employing brick veneer.

178

4. Walls of Residence Buildings: In residence buildings not more than three stories in height, walls, other than coursed or rough or random rubble stone walls, may be of 8-in. thickness when not over 35 ft. in height. Such walls in one-storey residence buildings or private garages may conform to exception 3.

Fig. 9-8. Weep holes at the bottom of a brick-veneer wall.

5. Penthouses and Roof Structures: Masonry walls above roof level, 12 ft. or less in height, enclosing stairways, machinery rooms, shafts or penthouses, may be of 8-in. thickness and may be considered as neither increasing the height nor requiring any increase in the thickness of the wall below.

6. Walls of Plain Concrete: Plain concrete walls may be 2 in. less in thickness than required otherwise in this section, but not less than 8 in., except that they may be 6-in. in thickness when meeting the provisions of exception 3.

179

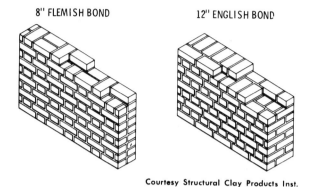

8" FLEMISH BOND   12" ENGLISH BOND

Courtesy Structural Clay Products Inst.

Fig. 9-9. Typical bonding patterns for 8- and 12-in. walls.

7. Cavity Walls: Cavity walls and hollow walls of masonry units shall not exceed 35 ft. in height, except that 10-in. cavity walls shall not exceed 25 ft. in height above the supports of such walls. The facing and backing of cavity walls shall each have a nominal thickness of at least 4 in. and the cavity shall be not less than 2 in. (actual) nor more than 3 in. in width.

8. Composite or Faced Walls: Neither the height of faced (composite) walls nor the distance between lateral supports shall exceed that prescribed for the masonry of either of the types forming the facing or the backing.

Thickness of Non-Bearing Walls:

1. Exterior Non-Bearing Walls: Non-bearing exterior masonry walls may be 4 in. less in thickness than required for bearing walls, but the thickness shall not be less than 8 in., except where 6-in. walls are specifically permitted.

2. Exterior Panel, Apron, or Spandrel Walls: Panel, apron, or spandrel walls that do not exceed 13 ft. in height above their support shall not be limited in thickness, provided they meet the fire resistive requirements of the code and are so anchored to the structural frame as to insure adequate lateral support and resistance to wind or other lateral forces.

Fig. 9-10 shows wall thicknesses based on number of brick lengths or widths. The thing to note is how the wall thicknesses fit the 4-in. module system. Fig. 9-11 shows the various combinations of brick that may be used to make these thicknesses.

Mason contractors for residential dwellings should be acquainted with the building codes in their city. Minimum wall requirements vary somewhat from city to city. The approximate requirements are about as follows:

Eight-Inch Walls—It is claimed that a thickness of 8 in. for brick walls of the usual home is ample yet there are numerous cities which do not allow walls under 12 in. thick. Some cities allow an 8-in. wall for both stories of a two-storey house and many thousands of dwellings have been constructed in these cities with 8-in. walls. Further, to our knowledge, no city that has adopted the 8-in. wall has changed back to the 12-in. walls.

The discriminating, however, who wish first class construction will insist on 12-in. walls. Some cities require 16-in. walls. The brick arrangement for 8-in. walls in the various bonds is shown in Fig. 9-12.

Twelve-Inch Walls—For ordinary dwellings, an objection to 12-in. walls is the extra space taken up as compared with 8-in.

THICKNESS MULTIPLE OF 4

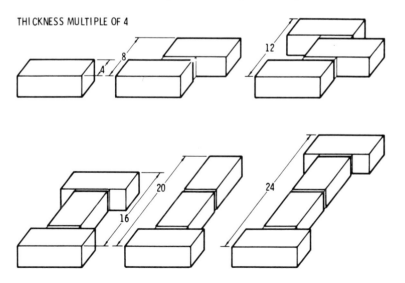

Fig. 9-10. Wall thickness based on 4-in. module system of brick sizes.

181

walls; the excess thickness reduces the area of the rooms in the house, which, in cities where land is very valuable must be taken into consideration. For a house 20′ × 30′, approximately 31 sq. ft. of area is lost on each storey, an area equal to a small bathroom or several good closets.

The extra thickness of the 12-in. walls, however, insulates a house better against cold or heat resulting in a warmer house in the

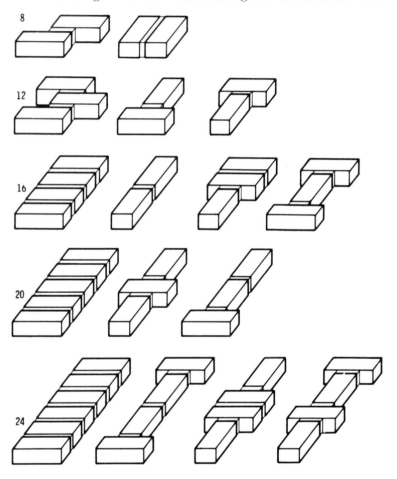

Fig. 9-11. The combinations in which brick may be laid for various wall thicknesses.

winter and a cooler house in the summer, though, of course, brick is not a sufficient insulator of itself. See Fig. 9-13 for brick arrangements in the various bonds.

Sixteen-Inch to Twenty-four-Inch Walls—For heavy duty, as in factory construction where the walls have to carry heavy loads of machinery and are subjected to more or less vibration, the walls may be 16 to 24 in. or more in thickness depending on conditions.

The arrangement of the brick is more complicated for these thick walls and the accompanying illustrations have been prepared with progressively extended courses like steps so that the brick arrangement in each course can be clearly seen. These details for 16- to 24-in. walls are shown in Figs. 9-14, 9-15, and 9-16.

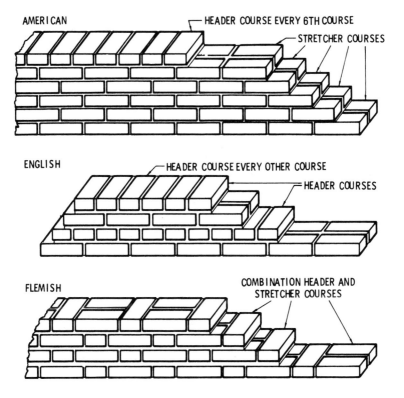

Fig. 9-12. Eight-inch walls in three popular bonding patterns.

The widest use of brick for interior walls has been in industrial plants. Brick has low heat conduction and high fire resistance, making it ideal for fire walls. Industrial fire walls of brick are partitions and usually are not load-bearing. A 4-in. thick wall is considered ample and construction is ½-lap running bond. They are not left free-standing but are bonded at the ends to adjacent perpendicular walls and to the ceiling above.

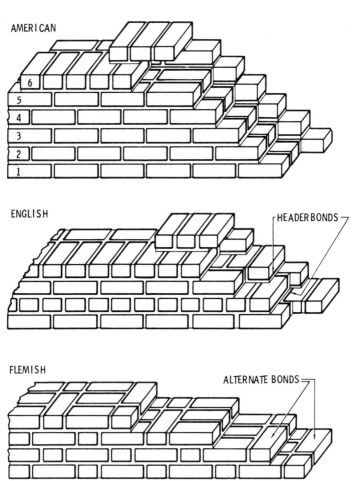

Fig. 9-13. Twelve-inch wall-bonding patterns.

AMERICAN

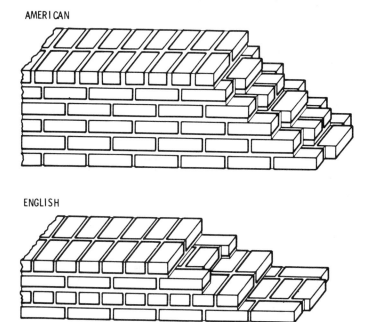

ENGLISH

FLEMISH

Fig. 9-14. A 16-in. wall-bonding pattern.

Partition walls of brick have many other applications, as well. They may be load bearing, sharing in the load of floors above. They may be needed for reducing sound transfer from one work area to another. Where interior walls may be subjected to bumping by heavy equipment, the strength of a thick solid brick wall cannot be equaled.

Hollow clay tile is frequently used for interior walls. They may even be combined with solid brick. Although larger in size than brick, they follow the 4-in. module size system for uniformity, and

185

AMERICAN

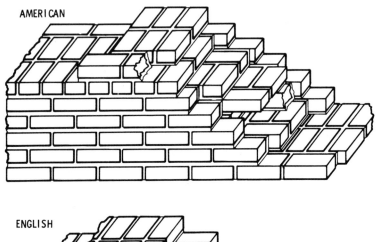

ENGLISH

FLEMISH

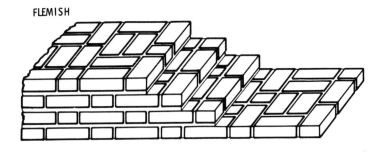

Fig. 9-15. A 20-in. wall-bonding pattern.

lend themselves to rapid construction. Brick or clay tile may be left unfinished or may be plastered.

Fig. 9-17 illustrates a number of typical interior brick and clay tile walls.

186

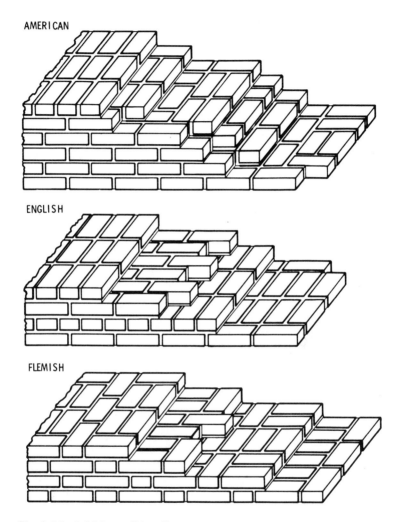

AMERICAN

ENGLISH

FLEMISH

Fig. 9-16. A 24-in. wall-bonding pattern.

CAVITY WALLS

Brick walls are not impervious to the seepage of water. The need to protect against the entry of water through a wall becomes important in areas of high wind, coupled with heavy rainfalls. The two maps in Fig. 9-18 show the possible wind velocities and annual

187

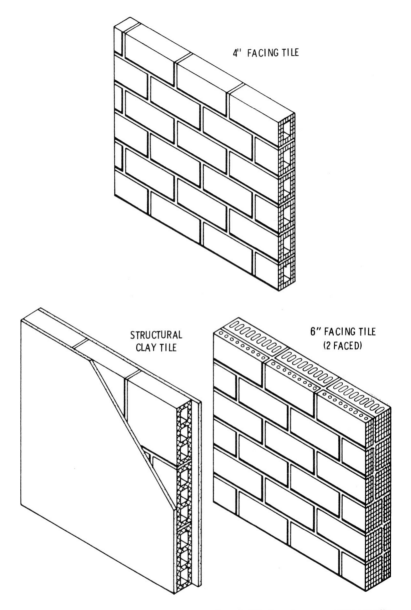

4" FACING TILE

STRUCTURAL CLAY TILE

6" FACING TILE (2 FACED)

Fig. 9-17. Interior or partition walls

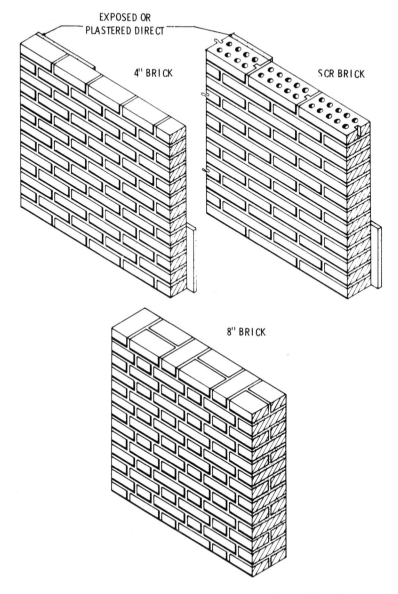

EXPOSED OR
PLASTERED DIRECT

4" BRICK

SCR BRICK

8" BRICK

Courtesy Structural Clay Products Inst.

of brick and hollow clay tile.

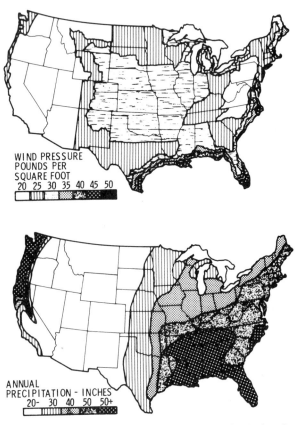

WIND PRESSURE
POUNDS PER
SQUARE FOOT
20 25 30 35 40 45 50

ANNUAL
PRECIPITATION - INCHES
20- 30 40 50 50+

Courtesy Structural Clay Products Inst.

Fig. 9-18. Wind and precipitation maps.

rainfall in various areas of the United States. The need to protect against water seepage through brick walls becomes greatest in those areas that combine both high winds and heavy rains, such as along the East and Gulf Coasts and nearly all of Florida. The best answer to the prevention of water seepage is the use of cavity wall construction, which is the "Cadillac" of brick wall construction.

Cavity or dual walls are those in which *two adjacent walls are separated by an air space.* A cavity wall, therefore, is made up of two walls or tiers of masonry, each nominally 4 in. thick with an air

190

space between the walls or tiers, and incorporates moisture resistance and insulation.

Cavity-wall construction is made up entirely of masonry materials, such as brick on the outside and brick on the inside; brick on the outside and tile or concrete block on the inside; tile on the outside and tile on the inside; or any of the other masonry materials generally used for load-bearing wall construction. In each case there are two walls separated with an air space. Because there are two walls, architects sometimes call cavity-wall construction dual- or barrier-wall masonry.

As illustrated in Fig. 9-19, the two walls are bonded or tied

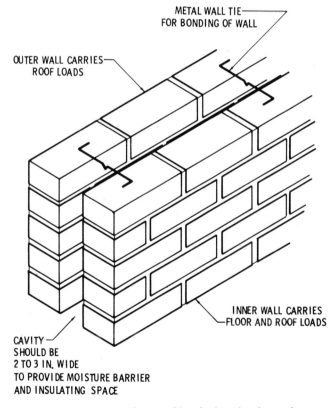

METAL WALL TIE
FOR BONDING OF WALL

OUTER WALL CARRIES
ROOF LOADS

INNER WALL CARRIES
FLOOR AND ROOF LOADS

CAVITY
SHOULD BE
2 TO 3 IN. WIDE
TO PROVIDE MOISTURE BARRIER
AND INSULATING SPACE

Fig. 9-19. Cavity-wall construction provides the best barrier against water seepage.

191

together with corrosion-proof, durable, and rigid metal ties. No brick headers are employed and the stretcher or running bond is generally used. Pattern bonds, such as Flemish bond, can also be used; they are, however, relatively expensive because the cutting of bricks results in higher labor cost.

Cavity-wall construction has been used extensively for both commercial and residential construction in continental Europe for over a century, and more recently in the United States where it has had an ever-increasing popularity.

Safety Considerations

Cavity walls are generally accepted throughout the United States by local and national codes. These codes as generally recognized are for general safety purposes. Cavity walls for residences and multistorey buildings are generally constructed with a 10-in. overall thickness. Institutional buildings like churches and schools of two-storey load-bearing construction are generally 14 in. in overall thickness. Regardless of the thickness of the cavity wall, it is limited to a height of 35 ft. and must be supported at right angles to the wall face at intervals not exceeding 14 times the nominal wall thickness.

The 10-in. cavity walls are usually limited to not more than 25 ft. in height and must be similarly supported laterally. This lateral support can be from the roof, floors, or partitions. Many buildings of multistorey design have been constructed of curtain walls having cavity-wall construction. By resting on spandrels or shelf angles they do not exceed the height limitations. Their interior treatment has varied from exposed masonry wall to plaster over lath and furring.

Keeping Water Out

The cavity wall was originally erected to provide a physical separation between the outside and the inner wall. It was determined many years ago that walls that permitted water penetration were generally of a character in which the mortar did not bond properly to the brick and as a consequence wind-driven rains would penetrate through the hairline cracks created by this lack of bond. Moisture appearing on the interior surface of the wall frequently caused a disintegration of wall finishes such as plaster or wood paneling.

Building with cavity walls prevents moisture from going through the wall. As shown in Fig. 9-20, any moisture that may gain entrance through the exterior tier of masonry will run down the inner side or cavity to the bottom of the cavity where it drains to the outside through weep holes provided for that purpose. Drops of water that may happen to reach the metal wall ties will drop off before they reach the inner wall if the wall tie has the drop loop. Thus the inner wall will generally remain dry. As in solid wall construction, however, the heads of windows, ventilating ducts, and other wall openings should be flashed. It is also very important to keep the cavity clean and clear of any solid material that may act

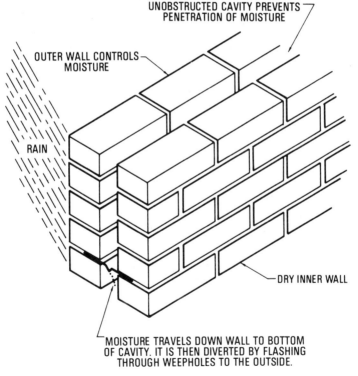

UNOBSTRUCTED CAVITY PREVENTS
PENETRATION OF MOISTURE

OUTER WALL CONTROLS
MOISTURE

RAIN

DRY INNER WALL

MOISTURE TRAVELS DOWN WALL TO BOTTOM
OF CAVITY. IT IS THEN DIVERTED BY FLASHING
THROUGH WEEPHOLES TO THE OUTSIDE.

Fig. 9-20. How cavity walls prevent the penetration of moisture from the outer wall through to the inner wall.

193

as a bridge for the passage of moisture from the outer to the inner wall, since a clogged cavity or one filled with mortar does not perform well or efficiently.

Weep Holes

Dual walls, as previously noted, are designed so that any moisture that may go through the outer wall will run down the inner side of the outer wall, as shown in Fig. 9-20. To drain off this moisture, weep holes are used, as shown in Fig. 9-21. As noted in the illustration, weep holes are located in the outer wall in the vertical joints of the bottom course, preferably two courses above grade, and are spaced from 3 to 5 brick apart.

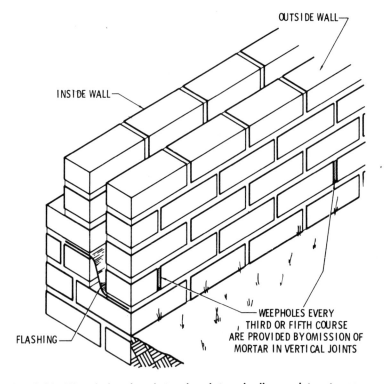

Fig. 9-21. Weep holes placed at various intervals allow moisture to escape from the wall cavity.

Weep holes are created by various means. The simplest is the omission of mortar from the vertical joint at predesigned intervals in the bottom course; or, preferably, not less than two courses above grade. Weep holes may also be made with ⅜ -in. oiled steel rods, pipe, or short lengths of sash cord or rubber hose, which are removed when the mortar sets up. Frequently, a piece of plywood the size of the joint will be put in the vertical joint and after the mortar sets up the plywood pieces are removed. At a later date, a 2½ " - 2½ " square piece of copper screening is wrapped around the piece of plywood and inserted in the weep hold, as precaution against the entrance of bugs into the cavity. The plywood is then removed leaving the screen in place. When the holes or sash cord are used, they should extend up into the cavity and project outwards from the base of the exterior wall for easy removal. This assures a clear and clean weep hole.

Weep holes are necessary for proper drainage to keep the bottom of the cavity dry. Dirt and gravel used in landscaping should not be allowed to pile up higher than the bottom of the cavity. Such material will block the weep holes and moisture may back up in the cavity so that it gets higher than the flashing and will then penetrate the inner tier of masonry.

Width of Cavity

The width of the cavity may vary considerably. The cavity in this type of wall construction is recommended to be not less than 2 in. or more than 3 in. wide, as shown in Fig. 9-22. The reason for this is that a cavity of less than 2 in. in width permits mortar to bridge the cavity by dripping on the wall ties as it falls down the wall.

A limitation of 3 in. in width on the cavity has been determined by tests made on the effective width in the cavity. Thus, it has been found that when the width of the cavity approaches 4 in., the width becomes great enough to cause air coming in from the weep holes to go up the inner side of the outer wall and come down the outer side of the inner wall, taking off warmth from the inner wall, thereby defeating a purpose of the construction. The foregoing action is created by an eddying current of air which is highly undesirable in cavity-wall construction. The cavities between walls may also be used for the installation of heating and air-conditioning ducts, as well as pipes, as shown in Fig. 9-23.

Figs. 9-24A & B show several methods of cavity-wall construction in addition to the cavity between two tiers (sometimes called wythes) of brick. In Fig. 9-24A is the brick veneer discussed

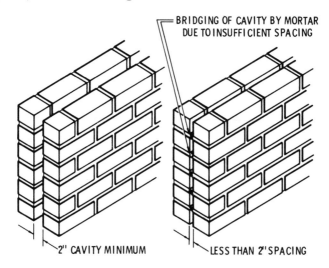

Fig. 9-22. The right and wrong width for cavity walls.

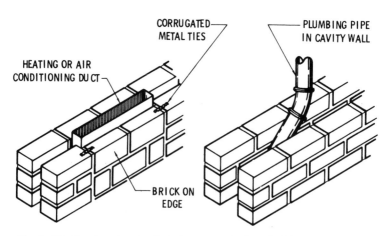

Fig. 9-23. Service pipes and air ducts may be installed in cavity.

before. While the cavity is small, it is sufficient for water protection. At lower right, SCR brick is used, a type developed by the Structural Clay Products Institute now called the Brick Institute of America. Its construction and width (6″ instead of the usual 4″) makes it especially suited to prevent or reduce water penetration. The cavity wall shown in Fig. 9-24B may include a tile back tier, as shown, or solid brick. The cavity may be left empty (an air cavity) or filled with water-resistant vermiculite or perlite for added insulation.

While bonding of the outer tier to the inner tier is usually done by metal straps, bonding may be accomplished by using header brick. The tiers are rowlock laid, with bricks on edge. The reduced tier thickness leaves a cavity across which a header brick can reach (lower left, Fig. 9-24B).

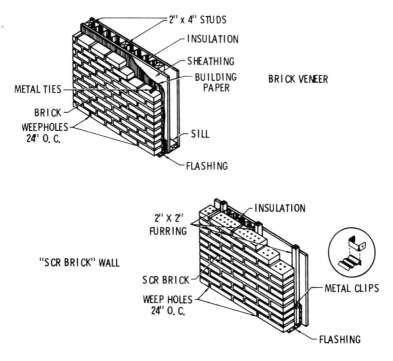

Courtesy Structural Clay Products Inst.

Fig. 9-24. (A) Brick-veneer and SCR brick cavity walls.

197

METAL TIES

Metal ties are used in cavity-wall construction to bond the walls and to furnish the necessary rigidity between the tiers. Fig. 9-25 shows the various types of wall ties that have been used in cavity-wall construction and their dimensions. Metal ties should be corrosion proof, rigid and durable, and not less than $3/16$ in. in diameter. The most frequently used is the Z-bar in which the wire is bent to provide a hook about 2 in. in length for embedment in the horizontal mortar joint of the inner and outer walls.

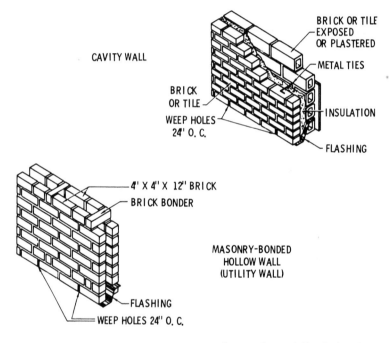

Courtesy Structural Clay Products Inst.

Fig. 9-24 (B) Standard insulated and masonry-bonded cavity walls.

Building codes usually call for a maximum spacing of ties as one tie for each 3 sq. ft. of wall area. In standard brick work the spacing accordingly will be about 24 in. on horizontal centers and each sixth course vertically. Mortar is spread on each wall section before the ties are placed, to provide a bond between the tie and the brick.

Regardless of the type of wall tie used, caution should be

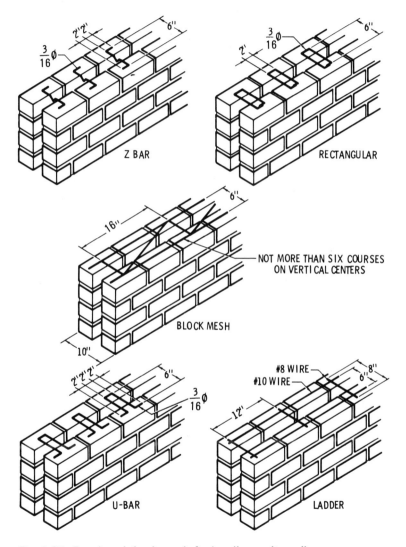

Fig. 9-25. Popular reinforcing rods for bonding cavity walls.

199

observed against placement with a pitch toward the inner wall. Wall ties must be placed within 12 in. of all wall openings and at the bottom of joists or slabs that rest on the wall.

EXPANSION JOINTS

No building is a perfectly rigid structure. There is always some movement, whether it is due to some settling, heavy traffic nearly, possible earth tremors, or other causes. One movement factor no building can avoid is that due to expansion and contraction resulting from temperature and moisture changes.

It takes a 100°F change in temperature to expand a 100-ft. long brick wall only ⅜ in. As little as the change seems to be, it does not take much of a temperature change to cause cracks in masonry walls. For this reason, large walls must not be bonded into a rigid unit but must include joints in the wall, especially vertically, to permit expansion and contraction.

An expansion joint is a complete separation between large sections of walls. To prevent the entrance of moisture, however, some filler must be used in the joint and the installation must be correctly made. Preformed copper sheeting with overlapping seams has been used for years. Also available are other types, such as premolded compressible and elastic fillers of rubber, neoprene, and other plastics. The old system of using fiberboard is not recommended, since fiberboard does not compress easily, and once compressed it does not return to its original size. Fig. 9-26 shows some of the newer acceptable types of expansion joints.

Fig. 9-27 shows the treatment of a number of wall structures. Note that it is important to use metal ties around the expansion joint. The most important thing for the brickmason to keep in mind is the importance of a good job in the installation of expansion joints. If anything in the joint prevents free movement, the effectiveness of the expansion joint is completely ruined. Mortar or pieces of broken brick must not be allowed to fall into the joints. The full length of the joint must be kept free of dirt or any type of rubble. If it is necessary to use ties across the joint, they must be of the flexible type, and the ties must not be bridged with mortar.

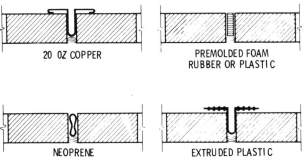

20 OZ COPPER

PREMOLDED FOAM
RUBBER OR PLASTIC

NEOPRENE

EXTRUDED PLASTIC

Courtesy Structural Clay Products Inst.

Fig. 9-26. Expansion-joint material in common use.

FOUNDATIONS

A necessity for laying up a wall with horizontal courses, and to provide long life with a minimum of settling, is to begin on a good foundation. While brick may be used as the foundation, modern practice is to pour a concrete footing.

The depth of the foundation must be at least below the frost line and the width should be twice the width of the wall. A trench dug with straight sides into which reinforcing steel rods have been placed is often a sufficient form. Wood forms may be placed if the earth is not solid enough to develop straight sides. It is important that the top surface of the foundation be as horizontal as possible. Adjustment can be made at the time of placing a mortar line and the first course of brick.

Detailed information on concrete forms appears in the chapter on that subject, dealing exclusively with all aspects of form construction. The following is a summary of information on foundation construction using such forms.

Foundation Materials—Foundations, due to the availability of material, are almost exclusively built of concrete. The materials necessary for making concrete are cement, sand, aggregate, water, and in some instances reinforcement. It has, however, become customary to refer to concrete as having only three ingredients, namely, cement, sand, and aggregate, the combination of

201

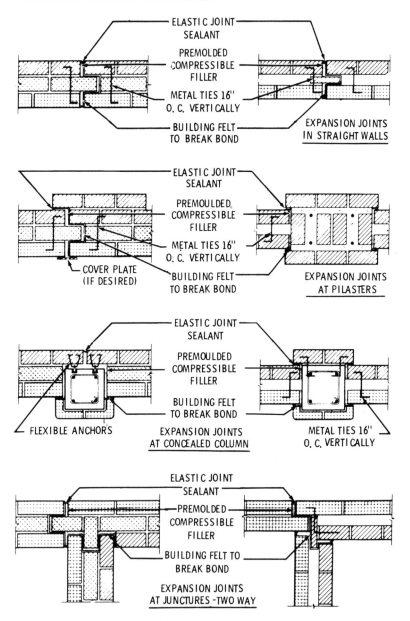

Fig. 9-27. Examples of placement of expansion

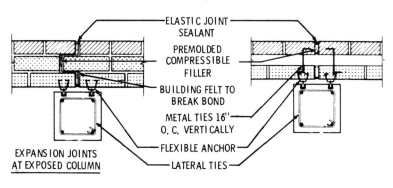

ELASTIC JOINT
SEALANT

PREMOLDED
COMPRESSIBLE
FILLER

BUILDING FELT TO
BREAK BOND

METAL TIES 16"
O. C. VERTICALLY

FLEXIBLE ANCHOR

LATERAL TIES

EXPANSION JOINTS
AT EXPOSED COLUMN

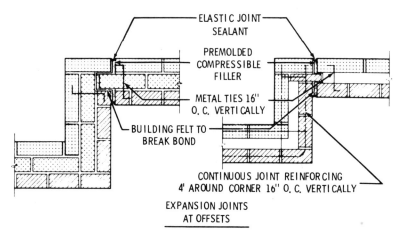

ELASTIC JOINT
SEALANT

PREMOLDED
COMPRESSIBLE
FILLER

METAL TIES 16"
O. C. VERTICALLY

BUILDING FELT TO
BREAK BOND

CONTINUOUS JOINT REINFORCING
4' AROUND CORNER 16" O. C. VERTICALLY

EXPANSION JOINTS
AT OFFSETS

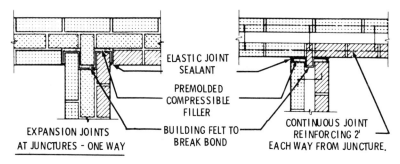

ELASTIC JOINT
SEALANT

PREMOLDED
COMPRESSIBLE
FILLER

BUILDING FELT TO
BREAK BOND

CONTINUOUS JOINT
REINFORCING 2'
EACH WAY FROM JUNCTURE.

EXPANSION JOINTS
AT JUNCTURES - ONE WAY

Courtesy Structural Clay Products Inst.

joints in various wall structures.

203

which is expressed as a mixture by volume in the order referred to. Thus, for example, a concrete mixture referred to as 1:2:4 actually means that the mixture contains 1 part of *cement*, 2 parts of *sand*, and 4 parts of *aggregate*, each proportioned by volume.

The general practice of omitting water from the ratio does not necessarily mean that the amount of water used in the mixture is less important but is omitted partly to simplify the formula and also because the amount of water used involves a consideration of both the degree of exposure and the strength requirements of the complete structure.

Cement Ratio—Table 9-1 gives recommended water-cement ratios on the basis of a definite minimum curing condition for concrete to meet different degrees of exposure in different classes of structures.

In determining the proportions of materials, it is desirable to arrive at those proportions which will give the most economical results consistent with proper placing. The relative proportions of fine and coarse aggregates and the total amount of aggregate that can be used with fixed amounts of cement and water will depend not only on the consistency of concrete required but also on the grading of each aggregate. A combination of aggregates made up largely of coarse particles presents less total surface to be coated with cement paste than aggregate of fine particles and is therefore more economical. For this reason it is desirable to use the lowest proportion of fine aggregate which will properly fill the "void" spaces in the coarse aggregate.

Aggregates graded so that they contain many sizes are more economical than aggregates in which one or two sizes predominate because the former contain fewer voids. The small particles fill the spaces between the larger particles, which otherwise must be filled with cement paste. A properly proportioned combination of well-graded fine and coarse aggregates contains all sizes between the smallest and the largest without an excessive amount of any one size. The best grading, however, is not necessarily one consisting of equal amounts of the various sizes because such a grading is seldom practicable. Satisfactory mixtures can usually be obtained with the commercial aggregates by proper combination of fine and coarse aggregates.

Table 9-1. Recommended Water-Cement Ratio for Different Applications and Conditions

Exposure	Water-cement ratio, U.S. gal. per sack[1]		
	Class of structure		
	Reinforced piles, thin walls, light structural members, exterior columns and beams in buildings	Reinforced reservoirs, water tanks, pressure pipes, sewers, canal linings, dams of thin sections	Heavy walls, piers, foundations, dams of heavy sections
Extreme: 1. In severe climates like in northern U. S., exposure to rain and snow and drying, freezing and thawing, as at the water line in hydraulic structures. 2. Exposure to sea and strong sulphate waters in both severe and moderate climates.	5½	5½	6
Severe: 3. In severe climates like in northern U. S., exposure to rain and snow and freezing and thawing, but not continuously in contact with water. 4. In moderate climates like southern U. S., exposure to alternate wetting and drying, as at water line in hydraulic structures.	6	6	6¾
Moderate: 5. In climates like southern U. S. exposure to ordinary weather, but not continuously in contact with water. 6. Concrete completely submerged, but protected from freezing.	6¾	6	7½
Protected: 7. Ordinary inclosed structural members; concrete below the ground and not subject to action of corrosive groundwaters or freezing and thawing.	7½	6	8¼

[1]Surface water or moisture carried by the aggregate must be included as part of the mixing water.

205

Increasing the proportion of coarse aggregate up to a certain point reduces the cement factor. Beyond this point the saving in cement is very slight, while the deficiency in mortar increases the labor cost of placing and finishing. Because coarser gradings are more economical, there has been a tendency to use mixtures that were undersanded and harsh. Harshness has been the principal cause for over-wet mixtures, resulting almost invariably in honeycombing in the finished work. While increasing the proportion of fine materials makes for smoother working mixes, excessive proportions of fine material present greater surface areas to be coated and more voids to be filled with cement paste. Under such conditions, the total amount of aggregate that can be used with fixed amounts of cement and water is greatly reduced.

The total amount of aggregate that can be used with given amounts of cement and water will depend on the consistency required by the conditions of the job. A stiffer mix permits more aggregates to be crowded into the cement paste and thus gives a larger volume of concrete. Stiffer mixes cost less for materials than the more fluid mixes but the cost of handling and placing increases when excessively dry mixes are used. On the other hand, mixes that are over-wet require high cement factors and cannot be placed without segregation of the materials. Such mixes are uneconomical in material and are seldom required for the conditions of placing. In many instances, where correct proportions of sand are used, it will be found practicable to use somewhat stiffer mixtures than have been used in the past without adding materially to the cost of handling or placing.

Because of the restrictions imposed by limiting the amount of water for each sack of cement, experienced foremen can generally be depended upon to obtain a proper balance between the various factors, with the result that the concrete will be neither harsh nor honeycombed on the one hand nor porous and over-wet on the other.

One of the important advantages to the contractor of the water-cement ratio method is that the materials may be proportioned to facilitate handling and placing, thereby reducing the cost of these items. With some latitude open in the matter of workability and proportions, he will be quick to select those mixes which give him the necessary workability at the lowest cost. At the same time, such a mixture will thoroughly fill the forms and reduce the cost of patching honeycomb spots to a minimum. Where the surfaces are to be given a special treatment, the process is invariably made easier.

The quantities of materials in a concrete mixture may be determined accurately by making use of the fact that the volume of concrete produced by any combination of materials, so long as the concrete is plastic, is equal to the sum of the absolute volume of the cement plus the absolute volume of the aggregate plus the volume of water. The absolute volume of a loose material is the actual total volume of solid matter in all the particles. This can be computed from the weight per unit volume and the apparent specific gravity as follows:

$$\text{Absolute Vol.} = \frac{\text{unit weight}}{\text{apparent specific gravity} \times \text{unit wt. of water}}$$
(62.5 lb. per.cu.ft.)

in which the unit weight is based on surface dry aggregate.

The method can best be illustrated by an example. Suppose the concrete batch consists of 1 sack of cement (94 lb.), 2.2 cu. ft. of dry fine aggregate weighing 110 lb. per cu. ft., and 3.6 cu. ft. of dry coarse aggregate weighing 100 lb. per cu. ft. which is to be mixed with a water-cement ratio of 7 gallons per sack. The apparent specific gravity of the cement is usually about 3.1 and of the more common aggregates about 2.65. The volume of concrete produced by the above mix is calculated as follows:

$$\text{Cement} = 1 \text{ cu. ft. at } \frac{94}{3.1 \times 62.5} = .49 \text{ cu. ft. abs. vol.}$$

$$\text{Fine Aggregate} = 2.2 \text{ cu. ft. at } \frac{110}{2.65 \times 62.5} = 1.46 \text{ cu. ft. abs. vol.}$$

$$\text{Coarse Aggregate} = 3.6 \text{ cu. ft. at } \frac{100}{2.65 \times 62.5} = 2.18 \text{ cu. ft. abs. vol.}$$

$$\text{Volume of Water} = \frac{7.0}{7.5} = .93 \text{ cu. ft. abs. vol.}$$

Total Volume of Concrete Produced = 5.06 cu. ft.

Thus 1 sack of cement produces 5.06 cu. ft., neglecting absorption or losses in manipulation. The cement required for 1 cu. yd. of concrete is, therefore,

$$\frac{27}{5.06} = 5.34 \text{ sacks}$$

The quantities of fine and coarse aggregate required can be found from a simple computation based on the number of cubic feet used with each sack of cement; thus, for fine aggregate:

$$\frac{5.34 \times 2.2}{27} = .43 \text{ cu. yd.}$$

for the course aggregate:

$$\frac{5.34 \times 3.6}{27} = .71 \text{ cu. yd.}$$

Placing of Concrete—No element in the whole cycle of concrete production requires more care than the final operation of placing concrete at the ultimate point of deposit. Before placing concrete, all debris and foreign matter should be removed from the places to be occupied by the concrete, and the forms, if used, should be thoroughly wetted or oiled. Temporary openings should be provided where necessary to facilitate cleaning and inspection immediately before depositing concrete. These should be placed so that excess water used in flushing the forms may be drained away.

Prevention of Segregation—With a well-designed mixture delivered with proper consistency and without segregation, placing of concrete is simplified; but even in this case care must be exercised to further prevent segregation and to see that the material flows properly into corners and angles of forms and around the reinforcement.

208

Constant supervision is essential to ensure such complete filling of the form and to prevent the rather common practice of depositing continuously at one point, allowing the material to flow to distant points.

Flowing over long distances will cause segregation, especially of the water and cement from the rest of the mass. An excessive amount of tamping or puddling in the forms will also cause the material to separate. When the concrete is properly proportioned, the entrained air will escape and the mass will be thoroughly consolidated with very little puddling. Light spading of the concrete next to the forms will prevent honeycombing and make surface finishing easier.

REINFORCED BRICK MASONRY

Reinforced brick masonry is sometimes called by its initials, RBM. Brick has tremendous compressive strength (as in the case of concrete) but the tensile or lateral strength depends on the bond between bricks. In areas where heavy winds or earth tremors are frequent, it may be necessary to increase the bond between bricks for greater lateral strength. This is done with steel reinforcing rods.

The actual number, size, and where they are placed becomes an engineering problem. It is important for the brick mason to know how to install them, to assure best bond of the rod to the brick structure. For example, it is important that a full layer of mortar be placed on the brick faces holding the rods. The rods must be imbedded in as much mortar as possible. Furthermore, there must be mortar on all sides of the rods—the rods must be suspended in the center of the mortar thickness. The mortar bed must be at least $\frac{1}{8}$ in. thicker than the rods, so there is a $\frac{1}{16}$-in. thickness of mortar on each side of the rods. Where large diameter rods are used, this may call for a thicker mortar joint. All joints must have the greater thickness to maintain a uniform pattern.

Fig. 9-28 is an example of steel rods laid the length of an 8-in. wall. A transverse rod bridging the two tiers of a cavity wall is shown in Fig. 9-29. One of the most important applications of reinforcing rods in brick construction is in the erection of columns and pillars made of brick. Fig. 9-30 is an example.

It is common practice to include horizontal reinforcing rods in the concrete poured as a footing for brick walls. Often this includes vertical-rod members placed every few feet along the footing. These rods are gauged to protrude into the open space of a cavity brick wall. Concrete is poured into the cavity for a couple of feet in height, to bond to the vertical rods. This results in a solid structural mass between the concrete footing and the brick wall. The remaining height may leave the air space in the cavity for bleeding water, as explained before. Weep holes must be included, of course. Fig. 9-31 shows a mason pouring concrete into the cavity, which has vertical-rod reinforcing.

STEEL REINFORCING BARS

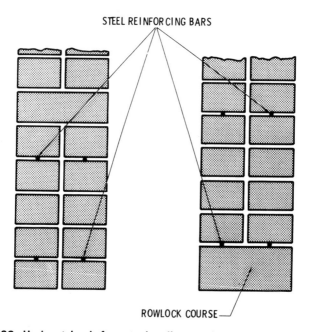

ROWLOCK COURSE

Fig. 9-28. Horizontal rods for extra bonding are shown in this end view.

USE OF ANCHORS

The metal ties for securing one tier of brick to another and the reinforcing rods mentioned above are called anchors. However, special heavy-duty forms of anchors are used:

1. To reinforce corners of brickwork.
2. To tie joists and roof plates to the brickwork.

The anchors are made in a multiplicity of shapes to meet the requirements of the service for which they are needed.

Anchoring Walls at Angles

An important feature in brickwork is that the walls should be anchored where they meet at corners; that is, the front and rear walls should be securely anchored as well as bonded to the side, partition, or partition walls. Fig. 9-32 shows some forms of the rods commonly used.

Courtesy Structural Clay Products Inst.

Fig. 9-29. Transverse rods used to bond tiers of brick in a cavity wall.

The provision for tying consists of an anchor placed at the center of a 4-in. recess or blocking. The T or pin anchor should be built into the center of the recess which should occur every thirteen courses. The anchor should project so as to give not less than 8 in. of holding on the wall to be tied. These anchors should never be

211

omitted when one wall is coursed up before the wall to be tied is built.

Fig. 9-33 shows an anchor in an 8-in. wall. It will secure this wall to the adjoining one when the next wall is laid up. In Fig. 9-34, anchors in a 12-in. wall are shown. An intersecting wall is tied to an outside wall with a nut and washer anchor in Fig. 9-35.

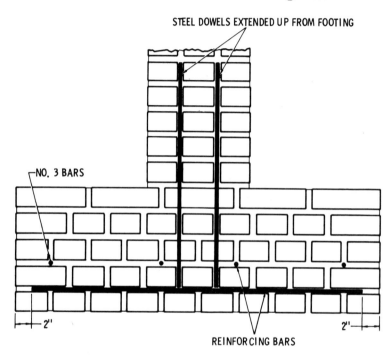

STEEL DOWELS EXTENDED UP FROM FOOTING

NO. 3 BARS

2"   2"

REINFORCING BARS

Fig. 9-30. A column of brick using vertical reinforcing rods.

Anchoring Floor Joists

In brickwork the courses can easily be adjusted so that the courses supporting joists will be at the exact height required. No "shims" or blocking under the joists are needed or should be allowed.

Joists and timbers should be set directly on the brick unless their bearing surface is so small that they transmit a load over the safe bearing capacity of the wall, which occurs very seldom but which would require bearing plates. These two conditions are shown in Fig. 9-36.

Fig. 9-31. Vertical rods extending from the footing tie the brick to the footing.

In the better construction of homes, floor joists are anchored to the walls. Some cities require this by ordinance. In the great majority of residence work outside of such cities, however, anchors are not used. Anchors are spaced approximately 6 ft. apart for both floor joists and roof plate. Great care should be exercised in placing these anchors as near the bottom of the joists as possible, in order to lessen the strain on the brick wall, in case a fire causes the joists to drop. Fig. 9-37 shows the right and wrong placement of joist anchors in solid walls, and Fig. 9-38 shows the correct placement in hollow walls.

In constructing the walls, the brickwork should be stopped at the point where the first floor joists are to rest on it. Care should be taken to have the top course perfectly level, so that the joists may be set without wedging or blocking. After the joists are placed, the

213

brickwork is continued up leaving a small "breathing" space around the joist to prevent dry rot. The same method of joisting is followed at the upper floors. On anchor joists the anchors are attached to the joists with spikes driven through the holes seen in the illustrations of anchors.

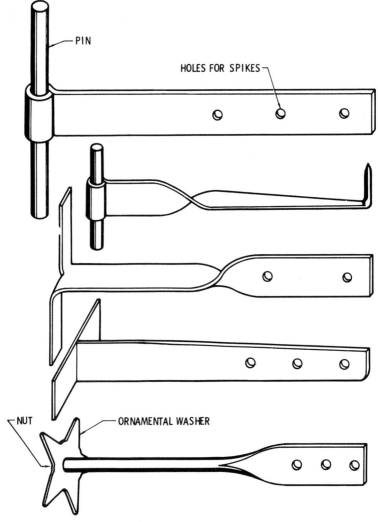

Fig. 9-32. Various heavy anchor rods used in brick work.

The ends of all joists are beveled whether they are anchor joists or intermediate joists so that in case of fire they will readily fall without injury to the wall.

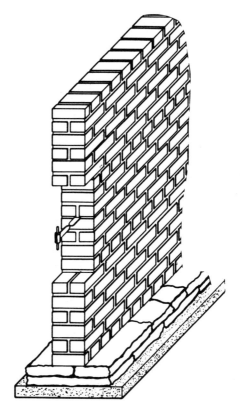

Fig. 9-33. A T-anchor in an 8-in. wall that will tie into a perpendicular wall.

Anchoring the Roof Plate

Before the top of the wall is reached, the anchors for bolting down the roof plate should be placed and the brickwork carried up around them as shown in Fig. 9-39. The bolt should be ½ in. in diameter and at least 12 in. long, with a tee or washer at the bottom and a nut and washer at the top, and should be set approximately every 6 ft. along the wall. After the carpenter has placed the roof

215

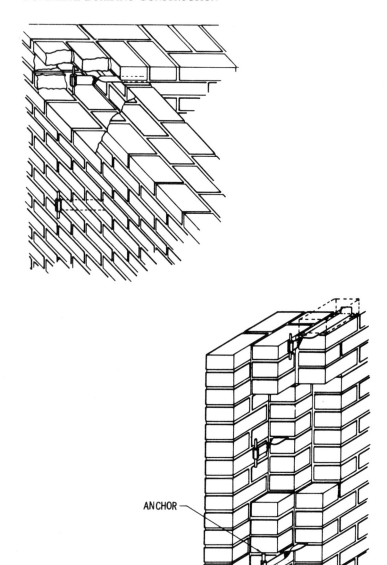

Fig. 9-34. T-anchors for bonding 12-in. walls.

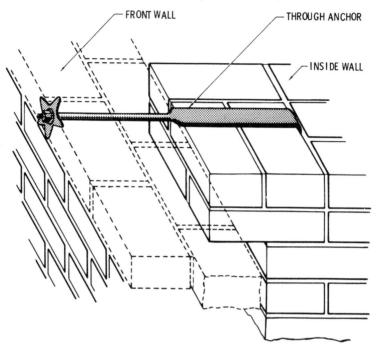

Fig. 9-35. A through anchor with nut and washer for intersecting walls.

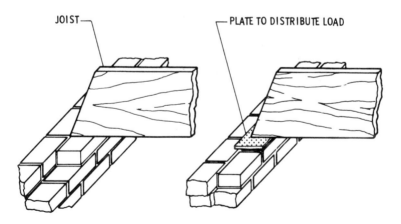

Fig. 9-36. A metal plate is placed under the floor joint in some instances to distribute weight.

217

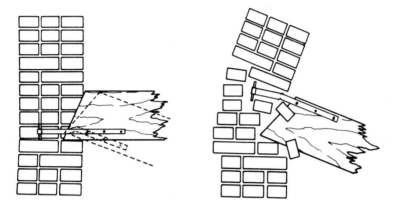

Fig. 9-37. Anchors installed to the bottom of beveled joist to prevent pulling out of brick in case of joist breakage.

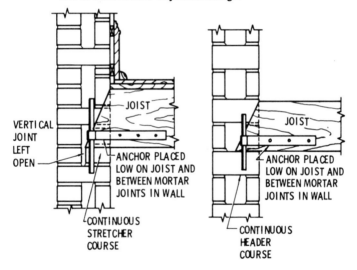

VERTICAL
JOINT
LEFT
OPEN

JOIST

ANCHOR PLACED
LOW ON JOIST AND
BETWEEN MORTAR
JOINTS IN WALL

CONTINUOUS
STRETCHER
COURSE

JOIST

ANCHOR PLACED
LOW ON JOIST AND
BETWEEN MORTAR
JOINTS IN WALL

CONTINUOUS
HEADER
COURSE

Fig. 9-38. Correct placement of anchors on joists in two types of hollow wall construction.

plate and before it is bolted down, the mason should place a bed of mortar under the plate.

When the wall is finally carried to the top and the roof rafters set, but before the roof boarding is in place, the mason should fill in between the roof rafters with one tier of brick as shown. This is

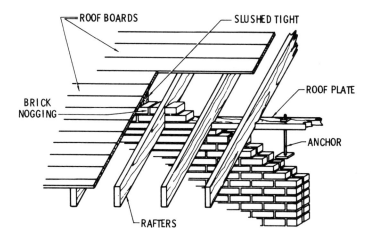

Fig. 9-39. Anchoring roof plate to brick wall.

called nogging or nudging. Its purpose is to effectively block the openings between the roof rafters and to prevent the wind from entering the walls and attic. This adds greatly to the comfort of the house in cold weather. In warm climates, nogging will not be necessary.

CHAPTER 10

Chimneys and Fireplaces

The term *chimney* generally includes both the chimney proper and (in house construction) the fireplace. There is no part of a house that is more likely to be a source of trouble than a chimney that is improperly constructed. Accordingly, it should be built so that it will be strong and designed and proportioned so that it gives adequate draft.

For strength, chimneys should be built of solid brick work and should have no openings except those required for the heating apparatus. If a chimney fire occurs, considerable heat may be engendered in the chimney, and the safety of the house will then depend on the integrity of the flue wall. A little intelligent care in the construction of fireplaces and chimneys will prove to be the best insurance. As a first precaution, all wood framing of floor and roof must be kept at least 2 in. away from the chimney and no woodwork of any kind should be projected into the brickwork surrounding the flues (Fig. 10-1).

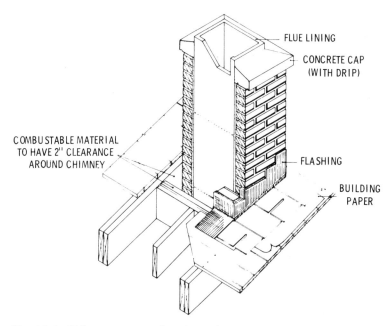

FLUE LINING

CONCRETE CAP
(WITH DRIP)

COMBUSTABLE MATERIAL
TO HAVE 2'' CLEARANCE
AROUND CHIMNEY

FLASHING

BUILDING
PAPER

Fig. 10-1. Chimney construction above the roof.

When it is understood that the only power available to produce a natural draft in a chimney is that due to the small difference in weight of the column of hot gases in the chimney and of a similar column of cold air outside, the necessity of properly constructing the chimney so that the flow of gases will encounter the least resistance should be clear.

The intensity of chimney draft is measured in inches of a water column sustained by the pressure produced and depends on:

1. The difference in temperature inside and outside the chimney.
2. The height of the chimney.

Theoretical draft in inches of water at sea level is as follows:
Let,

D = Theoretical draft,
H = Distance from top of chimney to grates,

T = Temperature of air outside the chimney,

T_1 = Temperature of gases in the chimney.

Then,

$$D = 7.00 \ H \ \left[\frac{1}{461 + T} - \frac{1}{461 + T_1} \right]$$

The results obtained represent the theoretical draft at sea level. For higher altitudes the calculations are subject to correction as follows:

For altitudes (in feet) of	Multiply by
1,000	0.966
2,000	0.932
3,000	0.900
5,000	0.840
10,000	0.694

A frequent cause of poor draft in house chimneys is that the peak of the roof extends higher than the chimney. In such case the wind sweeping across or against the roof will form eddy currents that drive down the chimney or check the natural rise of the gases as shown in Fig. 10-2. To avoid this, the chimney should be extended at least 2 ft. higher than the roof, as shown in Fig. 10-3.

In order to reduce to a minimum the resistance or friction due to the chimney walls the chimney should run as near straight as possible from bottom to top. This not only gives better draft but facilitates cleaning. If, however, offsets are necessary from one storey to another, they should be very gradual. The offset should never be displaced so much that the center of gravity of the upper portion falls outside the area of the lower portion. In other words, the center of gravity must fall within the width and thickness of the chimney below the offset.

FLUES

A chimney serving two or more floors should have a separate flue for every fireplace. The flues should always be lined with

223

BAD DRAUGHT

Fig. 10-2. How a roof peak higher than the top of the chimney can cause down drafts.

some fireproof material. In fact, the building laws of large cities provide for this. The least expensive way to build these is to make the walls 4 in. thick lined with burned clay flue lining. With walls of this thickness, never omit the lining and never replace the lining with plaster. The expansion and contraction of the chimney would cause the plaster to crack and an opening from the interior of the flue would be formed. See that all joints are completely filled with mortar.

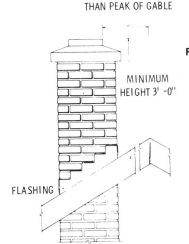

IF 10' OR LESS, CHIMNEY MUST BE 2'-0'' HIGHER THAN PEAK OF GABLE

MINIMUM HEIGHT 3'-0''

FLASHING

Fig. 10-3. Ample clearance is needed between peak of roof and top of chimney.

Courtesy Structural Clay Products Inst.

Flue Lining

Walls 8 in. or more in thickness may be used without a flue lining. However, walls under 8 in. must have a lining of fired clay. With the increased use of gas furnaces in homes, fired-clay linings are recommended for all chimneys because of chemical action of gas residue on common brick.

Clay lining for flues also follows the modular system of sizes. Table 10-1 lists currently available common sizes. The flue lining should extend the entire height of the chimney, projecting about 4 in. above the cap and a slope formed of cement to within 2 in. of the top of the lining, as shown in Fig. 10-3. This helps to give an upward direction to the wind currents at the top of the flue and tends to prevent rain and snow from being blown down inside the chimney.

The information given here is intended primarily for chimneys on residential homes. They will usually carry temperatures under 600°F. Larger chimneys, used for schools and other large buildings, have a temperature range between 600°F and 800°F. Indus-

225

Table 1. Standard Sizes of Modular Clay Flue Linings

Minimum Net inside Area (sq. in.)	Nominal[1] Dimensions (in.)	Outside[2] Dimensions (in.)	Minimum Wall Thickness (in.)	Approximate Maximum Outside Corner Radius (in.)
15	4x 8	3.5x 7.5	0.5	1
20	4x12	3.5x11.5	0.625	1
27	4x16	3.5x15.5	0.75	1
35	8x 8	7.5x 7.5	0.625	2
57	8x12	7.5x11.5	0.75	2
74	8x16	7.5x15.5	0.875	2
87	12x12	11.5x11.5	0.875	3
120	12x16	11.5x15.5	1.0	3
162	16x16	15.5x15.5	1.125	4
208	16x20	15.5x19.5	1.25	4
262	20x20	19.5x19.5	1.375	5
320	20x24	19.5x23.5	1.5	5
385	24x24	23.5x23.5	1.625	6

[1]Cross section of flue lining shall fit within rectangle of dimension corresponding to nominal size.

[2]Length in each case shall be 24 ± 0.5 in.

trial chimneys with temperatures above 800°F often are very high and require special engineering for their planning and execution. High-temperature brick chimneys must include steel reinforcing rods to prevent cracking due to expansion and contraction from the changes in temperature.

CHIMNEY CONSTRUCTION

Every possible thought must be given to providing good draft, leakproof mortaring, and protection from heat transfer to combustible material. Good draft means a chimney flue without obstructions. The flue must be straight from the source to the outlet. Metal pipes from the furnace into the flue must end flush with the inside of the chimney and not protrude into the flue, as shown in Fig. 10-4. The flue must be straight from the source to the

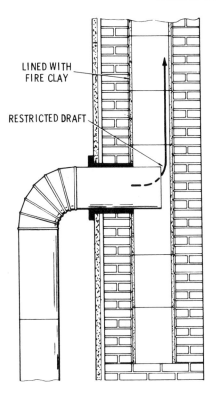

LINED WITH
FIRE CLAY

RESTRICTED DRAFT

Fig. 10-4. Furnace pipes must not project into the flue of a chimney.

outlet, without any bends, if at all possible. When two sources, such as a furnace and a fireplace feed the one chimney, they each must have separate flues.

To prevent leakage of smoke and gas fumes from the chimney into the house, and to improve the draft, a special job of careful mortaring must be observed. The layer of mortar on each course of brick must be even and completely cover the bricks. End buttering must be complete. However, it is best to mortar the flue lining lightly between the lining and the brick. Use just enough to hold the lining securely. The air space that is left acts as additional air insulation between the lining and the brick and reduces the transfer of heat.

No combustible material, such as the wood of roof rafters or floor joists, must abut the chimney itself. There should be at least a 2-in. space between the wood and the brick of the chimney, as shown in Fig. 10-1. Brick and flue lining are built up together. The lining clay is placed first and the bricks built up around it. Another section of lining is placed and brick built up, etc.

Chimneys carrying away the exhaust of oil- and coal-burning furnaces, where still used, need a cleanout trap. An airtight cast iron door is installed at a point below the entrance of the furnace smoke pipe.

Because of the heavier weight of the brick in a chimney, the base must be built to carry the load. A foundation for a residential chimney should be about 4 in. thick. If a fireplace is included, the foundation thickness should be increased to about 8 in.

After the chimney has been completed, it should be tested for leaks. Build a smudge fire in the bottom and wait for smoke to come out of the top. Cover the top and carefully inspect the rest of the chimney for leaks. If there are any, add mortar at the points of leakage.

Builders should become acquainted with local codes for the construction of chimneys. They should also check the national building code recommendations as set up by the American Insurance Association.

FIREPLACES

As a way of saving energy, a fireplace is definitely not the way to go—the fire uses house air to burn, and if that air is warmed by the furnace, you really *waste* fuel. But there is nothing like the flickering flames of a wood-burning fireplace to add cheer to the hearts of those sitting around it. Most new construction for single-family homes includes a fireplace when the house is built. A fireplace can be added to just about any home not so equipped.

There was a time when the fireplace, or an open fire, was the sole source for cooking and heating. In a few areas of the United States, where the winter climates are too mild for a heating system, the fireplace is still the only source of heat to take the chill off a cool evening.

The fireplace dates back to the earliest history of man. The first home fires, forerunners to the modern fireplace, were those kindled on the earth or on a conveniently placed slab stone around which the family gathered to prepare its food. In just what period in our history fires were first used will perhaps never be known. We have evidence that primitive man made use of caves in his first temporary dwelling and built fires at the mouth of these caves not only to prepare food but also to protect his family from enemies.

Later, when dwellings were constructed outside of caves, family life centered in one large room in the middle of which a wood was lighted. Here, the smoke was allowed to escape as best it could through a hole in the roof or crevices in the wall. This use of fire for heating and cooking was adopted even by the nomads, who built fires in the center of their tents and allowed the smoke to escape through a prepared opening at the top.

As more permanent and larger habitations were built and balconies or second floors were used for sleeping quarters, the hearthstone was moved to the corner of the room and an opening made in the wall to allow the smoke to escape. Later a stone hood that sloped back against the wall was added to aid in carrying the smoke out of the building.

Gradually the efficiency of the open fire was increased and eventually the fireplace was constructed in a recess in the center of one of the walls, with its own hood and enclosed flue, leading up to a chimney on top of the wall. As time passed, more consideration was given to the comforts of living. The fireplace was not only improved, but became the central decorative feature of the home.

The value of fireplaces was appreciated in England as early as the latter part of the fourteenth century, when they became an ornamental feature in the better homes. Count Rumford, an English scientist who published a series of essays on chimneys and fireplaces in 1796, is the one to whom we are most indebted for the improvement in fireplace design and for the rules governing the openings and flues. He spent a great deal of time studying the errors of fireplace construction and the principles governing the circulation of gases and combustion.

Probably in no other country have so many types and styles of fireplaces been constructed as in the United States. Although the ornamental mantel facings of fireplaces may be of other materials than brick, the chimney and its foundation are invariably of masonry construction. Fig. 10-5 shows a cross section of a fireplace and chimney stack suitable for the average home.

Fireplaces are generally built in the living room or family room. Whichever you choose, the location of the fireplace should allow the maximum of heat to be radiated into the room, with considera-

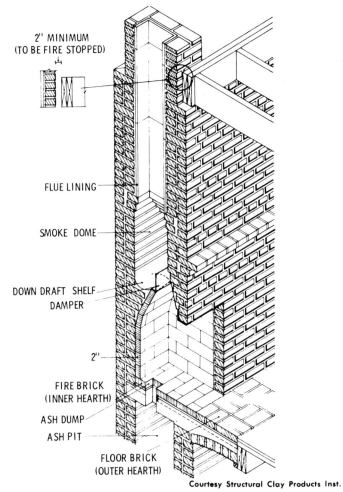

2" MINIMUM
(TO BE FIRE STOPPED)

FLUE LINING

SMOKE DOME

DOWN DRAFT SHELF
DAMPER

2"

FIRE BRICK
(INNER HEARTH)

ASH DUMP
ASH PIT

FLOOR BRICK
(OUTER HEARTH)

Courtesy Structural Clay Products Inst.

Fig. 10-5. Cross section of a typical fireplace and chimney of modern design.

tion given to making it the center of a conversation area. At one time its location was dictated by the location of the furnace, to make use of a common chimney. Today, with compact heating systems, the chimney is often a metal pipe from the furnace flue, straight through to the roof, not of the brick construction type. The brick fireplace chimney can then be placed to suit the best fireplace location in the home and room.

Fireplace styles vary considerably from a rather large one with a wide opening (Fig. 10-6) to a smaller corner fireplace. You can also get a variety of prefabricated units that the average do-it-yourselfer can install (Fig. 10-7).

Although large pieces of wood can be burned in the larger fireplaces, regardless of size, experience has indicated certain ratios of height, width, depth, etc., should be maintained for best flow of air under and around the burning wood. Recommended dimensions are shown in Table 10-2, which are related to the sketches in Fig. 10-8.

FIREPLACE CONSTRUCTION

Brick masonry is nearly always used for fireplace construction. Sometimes brick masonry is used around a metal fireplace form trademarked Heatilator. While any type of brick may be used for the outside of the fireplace, the fire pit must be lined with a high temperature fire clay or fire brick.

The pit is nearly always sloped on the back and the sides. This is to reflect forward as much of the heat as possible. The more surface exposure that is given to the hot gases given off by the fire, the more heat will be radiated into the room. Fig. 10-5 is a cutaway view of an all-brick fireplace for a home with a basement. The only nonbrick item is the adjustable damper. A basement makes possible a very large ash storage before cleanout is necessary. The ash dump opens into the basement cavity. A cleanout door at the bottom opens inward into the basement.

Fig. 10-9 is a side cutaway view of a typical fireplace for a home built on a concrete slab. It uses the metal form mentioned. The ash pit is a small metal box which can be lifted out, as shown in Fig. 10-10. In some slab home construction, the ash pit is a cavity

231

Courtesy Armstrong

Fig. 10-6. A large fireplace gives off heat and adds a homeyness to a room that nothing else can equal.

Table 2. Recommended Sizes of Fireplace Openings

Opening		Depth, d	Minimum back (horizontal) c	Vertical back wall, a	Inclined back wall, b	Outside dimensions of standard rectangular flue lining	Inside diameter of standard round flue lining
Width, w	Height, h						
Inches	Inches	Inches	Inches	Inches	Inches	Inches	Inches
24	24	16—18	14	14	16	8½ by 8½	10
28	24	16—18	14	14	16	8½ by 8½	10
24	28	16—18	14	14	20	8½ by 8½	10
30	28	16—18	16	14	20	8½ by 13	10
36	28	16—18	22	14	20	8½ by 13	12
42	28	16—18	28	14	20	8½ by 18	12
36	32	18—20	20	14	24	8½ by 18	12
42	32	18—20	26	14	24	13 by 13	12
48	32	18—20	32	14	24	13 by 13	15
42	36	18—20	26	14	28	13 by 13	15
48	36	18—20	32	14	28	13 by 18	15
54	36	18—20	38	14	28	13 by 18	15
60	36	18—20	44	14	28	13 by 18	15
42	40	20—22	24	17	29	13 by 13	15
48	40	20—22	30	17	29	13 by 18	15
54	40	20—22	36	17	29	13 by 18	15
60	40	20—22	42	17	29	18 by 18	18
66	40	20—22	48	17	29	18 by 18	18
72	40	22—28	51'	17	29	18 by 18	18

formed in the concrete foundation with an opening for cleanout at the rear of the house. A metal grate over the opening prevents large pieces of wood from dropping into the ash pit as shown in Fig. 10-11.

Importance of a Hearth

Every fireplace should include a brick area in front of it where hot wood embers may fall with safety. The plan view of Fig. 10-12 shows a brick hearth built 16 in. out from the fireplace itself. This should be about the minimum distance. Most often the hearth is raised several inches above the floor level. This raises the fireplace itself, all of which makes for easier tending of the fire.

Fig. 10-7. Today there are many types of prefab fireplaces.

In addition to the protection of the floor by means of a hearth, every wood burning fireplace should have a screen, to prevent flying sparks from being thrown beyond the hearth distance and onto a carpeted or plastic tile floor.

Ready-Built Fireplace Forms

There are a number of metal forms available, which make fireplace construction easier. Like the prefab units, they make a good starting point for the handy homeowner who can build his own fireplace addition to his home (Fig. 10-13). Many brands are available from fireplace shops and building supply dealers.

These units are built of heavy metal or boiler plate steel and designed to be set into place and concealed by the usual brick-work, or other construction, so that no practical change in the fireplace mantel design is required by their use. One claimed

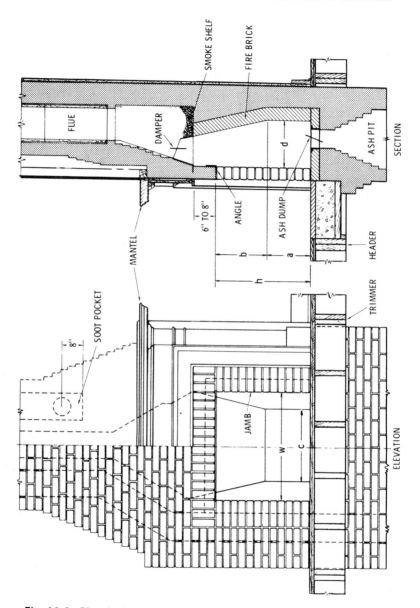

Fig. 10-8. Sketch of a basic fireplace. Letters refer to sizes recommended in Table 10-2.

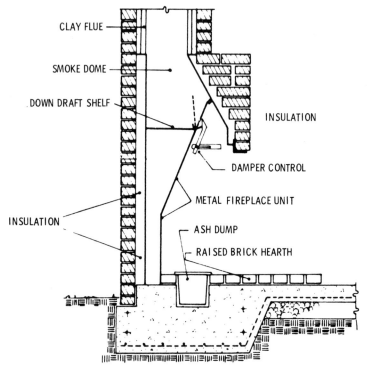

CLAY FLUE

SMOKE DOME

DOWN DRAFT SHELF

INSULATION

DAMPER CONTROL

METAL FIREPLACE UNIT

INSULATION

ASH DUMP

RAISED BRICK HEARTH

Courtesy Structural Clay Products Inst.

Fig. 10-9. Fireplace built on a concrete slab.

advantage for modified fireplace units is that the correctly designed and proportioned fire box manufactured with throat, damper, smoke shelf, and chamber provides a form for the masonry, thus reducing the risk of failure and assuring a smokeless fireplace.

There is, however, no excuse for using incorrect proportions; and the desirability of using a form, as provided by the modified unit, is not necessary merely to obtain good proportions. Each fireplace should be designed to suit individual requirements and if correct dimensions are adhered to, a satisfactory fireplace will be obtained.

Prior to selecting and erecting a fireplace, several suitable designs should be considered and a careful estimate of the cost

Fig. 10-10. Metal lift-out ash box used in many fireplaces built on a concrete slab.

Fig. 10-11. A cast-iron grate over the ash box to keep large pieces of burning wood from falling into the ash box.

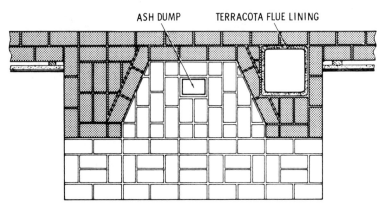

Fig. 10-12. A brick hearth in front of the fireplace catches hot embers that may fall out of the fire.

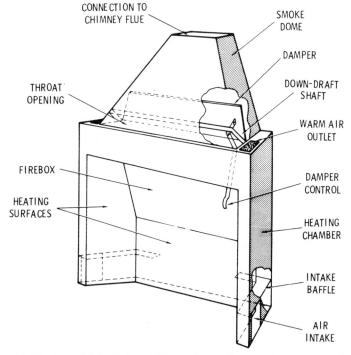

Fig. 10-13. A prefabricated metal form that makes fireplace construction easier.

238

should be made; and it should also be borne in mind that even though the unit of a modified fireplace is well designed, it will not operate properly if the chimney is inadequate. Therefore, for satisfactory operation, the chimney must be made in accordance with the rules for correct construction to give satisfactory operation with the modified unit as well as with the ordinary fireplace.

Manufacturers of modified units also claim that labor and materials saved tend to offset the purchase price of the unit and that the saving in fuel tends to offset the increase in first cost. A minimum life of 20 years is usually claimed for the type and thickness of metal commonly used in these units.

As illustrated in Figs. 10-14 and 10-15, and the sketches of Figs. 10-16 and 10-17, the brick work is built up around a metal fireplace form. The back view of Fig. 10-14 shows a layer of asbestos between the metal form and brick. The layer of fireproof asbestos wool battan or cementitious asbestos should be about 1 in. thick. Note the ash door, which gives access to the ash pit for the removal of ashes from the outside of the house.

Fig. 10-15 shows a partially built front view. By leaving a large air cavity on each side of the metal form and constructing the brick work with vents, some of the heat passing through the metal sides will be returned to the room. The rowlock stacked brick with no mortar, but an air space, permits cool air to enter below and warmed air to come out into the room from the upper outlet.

The front of the form includes a lintel for holding the course of brick just over the opening. A built-in damper is part of the form. Even with the use of a form, a good foundation is necessary for proper support as there is still quite a bit of brick weight. Chimney construction, following the illustrations and descriptions previously given, is still necessary.

Other Fireplace Styles

There are a number of other fireplace styles available. One such is the hooded type which permits the construction of the fireplace out into the room, rather than into the wall (Fig. 10-18).

Another style is two-sided, similar to that shown in the sketch of Fig. 10-19. It is used for building into a semi-divider type wall, such as between a living area and a dining area. Thus, the fire can be enjoyed from either room, or both at once, and what heat is

Fig. 10-14. Brick work around the back of a fireplace form.

Fig. 10-15. Front view of the brick work around a metal fireplace form.

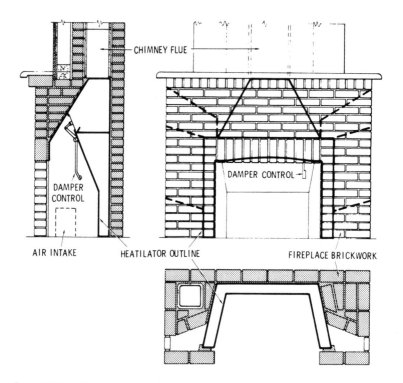

Fig. 10-16. Sketch of a typical fireplace built around a metal form.

given off is divided between the two areas.

Important to successful wood burning is good circulation of air under and around the sides. A heavy metal grate which lifts the burning logs above the floor of the fireplace is essential (Fig. 10-20).

SMOKY FIREPLACES

When a fireplace smokes, it should be examined to make certain that the essential requirements of construction as previously outlined, have been fulfilled. If the chimney has not been stopped up with fallen brick and the mortar joints are in good condition, a

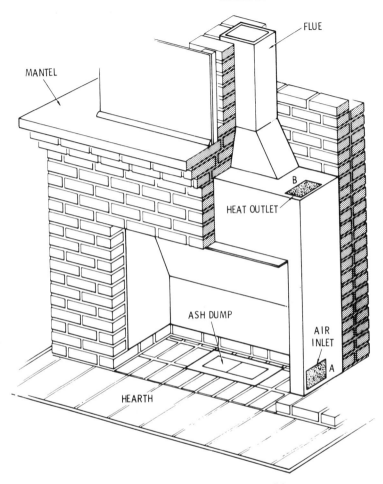

Fig. 10-17. Cutaway sketch of fireplace using a metal form.

survey should be made to ascertain that nearby trees or tall buildings do not cause eddy currents down the flue.

To determine whether the fireplace opening is in incorrect proportion to the flue area, hold a piece of sheet metal across the top of the fireplace opening and then gradually lower it, making the opening smaller until smoke does not come into the room. Mark the lower edge of the metal on the sides of the fireplace.

243

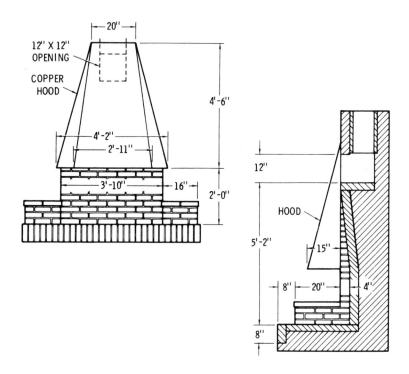

Fig. 10-18. A hood projecting out from the wall carries flue gases up through chimney.

The opening may then be reduced by building in a metal shield or hood across the top of the fireplace so that its lower edge is at the marks made during the test. Trouble with smoky fireplaces can also usually be remedied by increasing the height of the flue.

Uncemented flue-lining joints cause smoke to penetrate the flue joints and descend out of the fireplace. The best remedy is to tear out the chimney and join linings properly.

Where flue joints are uncemented and mortar in surrounding brick work disintegrated, there is often a leakage of air in the chimney causing poor draft. This prevents the stack from exerting the draft possibilities which its height would normally ensure.

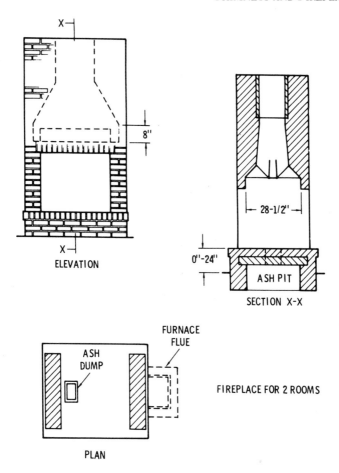

X—

8"

X—

ELEVATION

28-1/2"

0"-24"

ASH PIT

SECTION X-X

FURNACE
FLUE

ASH
DUMP

FIREPLACE FOR 2 ROOMS

PLAN

Fig. 10-19. A two-sided fireplace is ideal as a room divider.

Another cause of poor draft is wind being deflected down the chimney. The surroundings of a home may have a marked bearing on fireplace performance. Thus, for example, if the home is located at the foot of a bluff or hill or if there are high trees close at hand the result may be to deflect wind down the chimney in heavy gusts. A most common and efficient method of dealing with this type of difficulty is to provide a hood on the chimney top.

245

Fig. 10-20. A grate is used to hold logs above the base, allowing air to move under and through the burning wood.

Carrying the flue lining a few inches above the brick work with a bevel of cement around it can also be used as a means of promoting a clean exit of smoke from the chimney flue. This will effectively prevent wind eddies. The cement bevel also causes moisture to drain from the top and prevents frost troubles between lining and masonry.

CHAPTER 11

Woods

Wood is our most versatile, most useful building material, and a general knowledge of the physical characteristics of various woods used in building operations is important for carpenter and handyman alike.

Wood may be classified:

1. Botanically. All trees which can be sawed into lumber or timbers belong to the division called Spermatophytes. This includes softwoods as well as hardwoods.
2. With respect to its density, as
   a. Soft.
   b. Hard.
3. With respect to its leaves, as
   a. Needle or scale leaved, botanically Gymnosperms, or conifers, commonly called softwoods. Most of them, but not all, are evergreens.
   b. Broad-leaved, botanically Angiosperms, commonly called

hardwoods. Most are deciduous, shedding their leaves in the fall. Only one broad-leaved hardwood, the Chinese Ginkgo, belongs to the subdivision Gymno-sperms.

4. With respect to its shade or color, as
   a. White or very light.
   b. Yellow or yellowish.
   c. Red.
   d. Brown.
   e. Black, or nearly black.
5. In terms of grain, as
   a. Straight.
   b. Cross.
   c. Fine.
   d. Coarse.
   e. Interlocking.
6. With respect to the nature of the surface when dressed, as
   a. Plain—example, white pine.
   b. Grained—example, oak.
   c. Figured or marked—example, bird's-eye maple.

A section of a timber tree, as shown in Figs. 11-1 and 11-2, consists of:

1. Outer bark—living and growing only at the cambium layer. In most trees, the outside continually sloughs away.
2. In some trees, notably hickories and basswood, there are long tough fibers, called bast fibers, in the inner bark. In some trees such as the beech, they are notably absent.
3. Cambium layer—Sometimes this is only one cell thick. Only these cells are living and growing.
4. Medullary rays or wood fibers, which run radially.
5. Annual rings, or layers of wood.
6. Pith.

Around the pith, the wood substance is arranged in approximately concentric rings. The part nearest the pith is usually darker than the parts nearest the bark and is called the heartwood. The cells in the heartwood are dead. Nearer the bark is the sapwood, where the cells are living but not growing.

As winter approaches, all growth ceases, and thus each annual ring is separate and in most cases distinct. The leaves of the

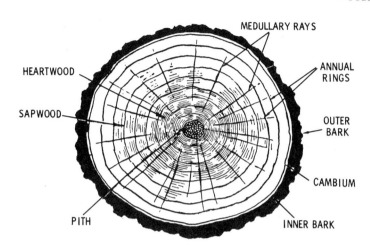

Fig. 11-1. Cross section of a 9-year-old oak showing pith, concentric rings comprising the woody part, cambium layer, and bark. The arrangement of the wood in concentric rings is due to the fact that one layer was formed each year. These rings, or layers, are called annual rings. That portion of each ring formed in spring and early summer is softer, lighter colored, and weaker than that formed during the summer and is called spring wood. The denser, stronger wood formed later is called summer wood. The cells in the heartwood of some species are filled with various oils, tannins, and other substances, which make these timbers rot-resistant. There is practically no difference in the strength of heartwood and sapwood if they weigh the same. In most species, only the sapwood can be readily impregnated with preservatives.

deciduous trees, or those which shed their leaves, and the leaves of some of the conifers, such as cypress and larch, fall, and the sap in the tree may freeze hard. The tree is dormant but not dead. With the warm days of the next spring, growth starts again strongly, and the cycle is repeated. The width of the annual rings varies greatly, from 30 to 40 or more per inch in some slow-growing species, to as few as 3 or 4 per inch in some of the quick-growing softwoods. The woods with the narrowest rings, because of the large percentage of summer wood, are generally strongest, although this is not always the case.

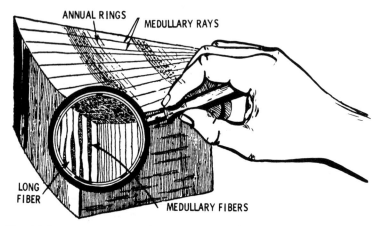

Fig. 11-2. A piece of wood magnified slightly to show its structure. The wood is made up of long, slender cells called fibers, which usually lie parallel to the pith. The length of these cells is often 100 times their diameter. Transversely, bands of other cells, elongated but much shorter, serve to carry sap and nutrients across the trunk radially. Also, in the hardwoods, long vessels or tubes, often several feet long, carry liquids up to the tree. There are no sap-carrying vessels in the softwoods, but spaces between the cells may be filled with resins.

CUTTING AT THE MILL

When logs are taken to the mill, they may be cut in a variety of ways. One way of cutting is quartersawing. Fig. 11-3 shows four variations of this method. Each quarter is laid on the bark and ripped into quarters, as shown in the figure. Quartersawing is rarely done this way, though, because only a few wide boards are yielded; there is too much waste. More often, when quartersawed stock is required, the log is started as shown in Fig. 11-4 and 11-5, sawed until a good figure (pattern) shows, then turned over and sawed. This way, there is little waste, and the boards are wide. In other words, most quartersawed lumber is resawed out of plain sawed stock.

The plain sawed stock, as shown, is simply flat sawn out of

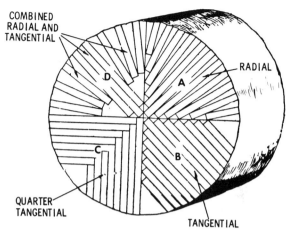

COMBINED
RADIAL AND
TANGENTIAL

RADIAL

D

A

C

B

QUARTER
TANGENTIAL

TANGENTIAL

Fig. 11-3. Methods of quartersawing.

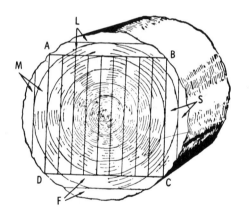

L

A

B

M

S

D

C

F

**Fig. 11-4. Plain or bastard sawing, called flat or slash sawing. The log is first
squared by removing boards *MS* and *LF*, giving the rectangular
section *ABCD*. This is necessary to obtain a flat surface on the
log.**

quartersawed. Quartersawed stock has its uses. Boards shrink
most in a direction parallel with the annual rings, and door stiles
and rails are often made of quartersawed material.

Lumber is sold by the board foot, meaning one 12-in. 1-in.-thick
square of wood. Any stock under 2 in. thick is known as lumber;

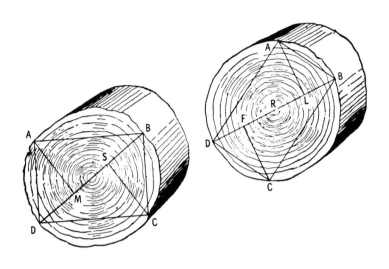

Fig. 11-5. Obtaining beams from a log.

over 4 in., it is timber. The terms have become interchangeable, however, and are used interchangeably in this book.

Lumber, of course, is sold in nominal and actual size, the actual size being what the lumber is after being milled. As the years have gone by, the actual size has gotten smaller. A 2″ × 4″, for example, used to be an actual size of 1 9/16″ × 3 9/16″. It is now 1½″ × 3½″, and other boards go up or down in size in half-inch increments.

DEFECTS

The defects found in manufactured lumber, as shown in Figs. 11-6 and 11-7, are grouped in several classes:

1. Those found in the natural log, as
   a. Shakes.
   b. Knots.
   c. Pitchpockets.
2. Those due to deterioration, as

a. Rot.

b. Dote.

3. Those due to imperfect manufacture, as

 a. Imperfect machining.

 b. Wane.

 c. Machine burn.

 d. Checks and splits from imperfect drying.

Heart shakes, as shown in Fig. 11-6, are radial cracks that are wider at the pith of the tree than at the outer end. This defect is most commonly found in those trees which are old, rather than in young vigorous saplings; it occurs frequently in hemlock.

A wind or cup shake is a crack following the line of the porous part of the annual rings and is curved by a separation of the annual rings. A wind shake may extend for a considerable distance up the trunk. Other explanations for wind shakes are expansion of the sap wood and wrenchings received due to high winds (hence the name). Brown ash is especially susceptible to wind shakes.

A star shake resembles the wind shake but differs in that the crack extends across the center of the trunk without any appearance of decay at that point; it is larger at the outside of the tree.

Dry rot, to which timber is so subject, is due to fungi; the name is misleading as it only occurs in the presence of moisture and the absence of free air circulation.

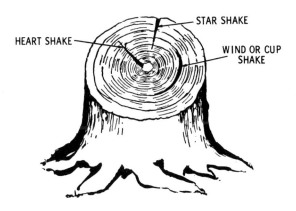

Fig. 11-6. Various defects that can be found in lumber.

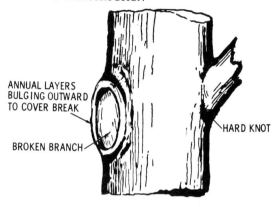

ANNUAL LAYERS
BULGING OUTWARD
TO COVER BREAK

HARD KNOT

BROKEN BRANCH

Fig. 11-7. Hard knot and broken branch show nature's way of covering the break.

SELECTION OF LUMBER

A variety of factors must be considered when picking lumber for a particular project. For example, is it seasoned or not, that is, has it been dried naturally—the lumber is stacked up with air spaces between, as shown in Fig. 11-8—or artificially, as it is when dried in kilns? The idea is to produce lumber with a minimum amount of moisture that will warp least on the job. If your project requires nonwarping material, ask for kiln-dried lumber. If not, so-called green lumber will suffice. Green lumber is often used for framing (outside) where the slight warpage that occurs after it is nailed in place is not a problem. Green lumber is less costly than kiln-dried, of course.

Another factor to consider is the grade of the lumber. The best lumber you can buy is Clear, which means the material is free of defects. Following this is Select, which has three subdivisions—Nos. 1, 2, and 3—with No. 1 the best of the Select with only a few blemishes on one side of the board and few, if any, on the other. Last is Common. This is good wood, but it will have blemishes and knots that can interfere with a project if you want to finish it with a clear material.

Rough lumber also has a grading system, and it reflects the material, which comes green or kiln-dried.

254

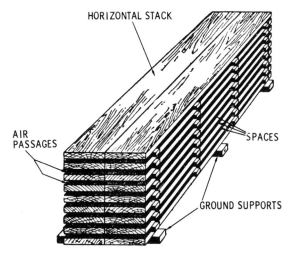

HORIZONTAL STACK

AIR
PASSAGES

SPACES

GROUND SUPPORTS

Fig. 11-8. Horizontal stack of lumber for air drying.

Lumber has two grading systems, numerical and verbal. Numerically there are Nos. 1, 2, and 3 with No. 1 the best and No. 3 the least desirable. Roughly corresponding to these numbers are Construction, Standard, and Utility.

Wood may also be characterized as hardwood or softwood. These designations do not refer to the physical hardness of the wood. Although hardwoods are normally harder than softwoods. The designation refers to the kinds of trees the wood comes from: cone-bearing trees are softwoods, leaf-bearing trees are hardwoods. By far the most valuable softwood is pine, a readily available material in all sections of the country. (Of course the type of pine will depend on the particular area.) Hardwoods, which come in random lengths from 8 to 16 ft. long and 4 to 16 in. wide are all Clear—these would be mahogany, birch, oak, or maple. Hardwoods are usually much more expensive than softwoods.

In addition to the above, there are overall characteristics of the particular wood to consider. What follows is a round-up of individual characteristics that will make the material more or less suitable for your project.

Brown Ash—Not a framing timber, but an attractive trim wood. Brown heart, lighter sapwood. The trees often wind shake so

255

badly that the heart is entirely loose. Attractive veneers are sliced from stumps and forks.

White Cedar, Northern—Light-brown heart, sapwood thin and nearly white. Light, weak, soft, decay resistant, holds paint well.

Western Red Cedar—Also called canoe cedar or shinglewood. Light, soft, straight-grained, small shrinkage, holds paint well. Heart is light brown, extremely rot resistant. Sap quite narrow, nearly white. Used for shingles, siding, boat building.

Eastern Red Cedar, or Juniper—Pungent aromatic odor said to repel moths. Red or brown heartwood, extremely rot-resistant white sapwood. Used for lining clothes closets and chests and for fence posts.

Cypress—Probably our most durable wood for contact with the soil. Wood moderately light, close-grained, heartwood red to nearly yellow, sapwood nearly white. Does not hold paint well, but otherwise desirable for siding and outside trim. Attractive for inside trim.

Gum, Red—Moderately heavy, interlocking grain; warps badly in seasoning; heart is reddish brown, sapwood nearly white. The sapwood may be graded out and sold as white gum, the heartwood as red gum, or together as unselected gum. Cuts into attractive veneers.

Hickory—A combination of hardness, weight, toughness, and strength found in no other native wood. A specialty wood, almost impossible to nail when dry. Not rot-resistant.

Hemlock, Eastern—Heartwood is pale brown to reddish, sapwood not distinguishable from heart. May be badly wind shaken. Brittle, moderately weak, not at all durable. Used for cheap, rough framing veneers.

Hemlock, Western—Heartwood and sapwood almost white with purplish tinge. Moderately strong, not durable, mostly used for pulpwood.

Locust, Black—Heavy, hard, strong; the heart-wood is exceptionally durable. Not a framing timber. Used mostly for posts and poles.

Maple, Hard—Heavy, strong, hard, and close grained; color light brown to yellowish. Used mostly for wear-resistant floors, and furniture (Fig. 11-9). Circularly growing fibers cause the

Fig. 11-9. Oak and maple (here) are two fine woods for furniture making.

attractive "birds-eye" grain in some trees. One species, the oregon maple, occasionally contains the attractive "quilted" grain.

Maple, Soft—Softer and lighter than hard maple; lighter colored. Box elder is sometimes marketed with soft maple. Used for much the same purposes as hard maple, but not nearly so desirable.

257

Oak, White—Several species are marketed together, but the woods are practically identical. Hard, heavy, tough, strong, and somewhat rot-resistant. Brownish heart, lighter sapwood. Desirable for trim and flooring, and one of our best hardwood framing timbers.

Oak, Red—Several species are marketed together. They cannot be distinguished one from the other, but can be distinguished from the white oaks. Good framing timber, but not rot-resistant.

Pine, White, Western—also called idaho white pine. Creamy or light-brown heartwood, sapwood thick and white. Used mostly for millwork and siding. Moderately light, moderately strong, easy to work, holds paint well.

Pine, Red or Norway—Resembles the lighter weight specimens of southern yellow pine. Moderately strong and stiff, moderately soft, heartwood pale red to reddish brown. Used for millwork, siding, framing, and ladder rails.

Pine, Long-Leaf Southern Yellow—Not a species but a grade. All southern yellow pine that has six or more annual rings per inch is marketed as long-leaf, and it may contain lumber from any of the several species of southern pine. Heavy, hard, and strong, but not especially durable in contact with the soil. The sapwood takes creosote well. One of our most useful timbers for light framing.

Pine, Short-Leaf Southern Yellow—Contains timber from any of several related species of southern pine having less than six annual rings per inch. Quite satisfactory for light framing, and the sapwood is attractive as an interior finish.

Douglas Fir—Our most plentiful commercial timber. Varies greatly in weight, color, and strength. Strong, moderately heavy, splintery, splits easily. Used in all kinds of construction; much is rotary cut for plywood.

Poplar, Yellow or Tulip—Our easiest-working native wood. Old growth has a yellow to brown heart. Sapwood and young trees are tough and white. Not a framing lumber, but used for siding.

Redwood—One of our most durable and rot-resistant timbers. Light, soft, moderately high strength, heartwood reddish brown, sapwood white. Does not paint exceptionally well, as it oftentimes "bleeds" through. Used mostly for siding and outside trim, decks, furniture (Figs. 11-10 and 11-11).

Fig. 11-10. For building outdoor furniture and items such as planters, redwood is great.

Spruce, Sitka—Light, soft, medium strong, heart is light reddish brown, sapwood is nearly white, shading into the heartwood. Usually cut into boards, planing-mill stock, and boat lumber.

Spruce, Eastern—Stiff, strong, hard, and tough. Moderately light weight, light color, little difference between heart and sapwood. Commercial eastern spruce includes wood from three related species. Used for pulpwood, framing lumber, millwork, etc.

Spruce, Engelmann—Color, white with red tint. Straight grained, light weight, low strength. Used for dimension lumber and boards, and for pulpwood. Extremely low rot-resistance.

Tamarack, or Larch—Small to medium sized trees; not much is

Fig. 11-11. Redwood is a favorite, but expensive, material for deck
building.

sawed into framing lumber, but much is cut into boards. Yellowish
brown heart, sapwood white. Much is cut into posts and poles.

Walnut, Black—Our most attractive cabinet wood. Heavy,
hard, and strong, heartwood is a beautiful brown, sap nearly
white. Mostly used for fine furniture, but some is used for fine
interior trim. Somewhat rot-resistant. Used also for gunstocks.

260

Walnut, White or Butternut—Sapwood light to brown, heart light chestnut brown with an attractive sheen. The cut is small, mostly going into cabinet work and interior trim. Moderately light, rather weak, not rot-resistant.

Decay of Lumber

Decay of lumber is the result of one cause, and one cause only: the work of certain low-order plants called fungi. All these organisms require water and air to live, grow, and multiply; consequently, wood that is kept dry, or that is dried quickly after wetting, will not decay. Similarly, wood kept submerged in water will soften but will not decay, for the air supply is shut off, and timber set deep in the ground, such as piling, which is shut off from the air, is practically permanent.

There is no such thing as "dry rot"; however, the term is rather loosely used sometimes when speaking about any rot or any dry and decayed wood. Although rotten wood may be dry when observed, it was wet while decay was progressing. This kind of decay is often found inside living, growing trees, but it occurs only in the presence of water.

OTHER MATERIALS

Plywood

In addition to boards and lumber, carpenters and handymen have come to rely on other materials. Among the most important is plywood (Fig. 11-12).

The most familiar plywood used in the United States is made from douglas fir. Short logs are chucked into a lathe and a thin, continuous layer of wood is peeled off. This thin layer is straightened, cut to convenient sizes, covered with glue, laid up with the grain in successive plies crisscrossing, and subjected to heat and pressure. This is the plywood of commerce, one of our most useful building materials.

All plywood has an odd number of plies, allowing the face plies to have parallel grain while the lay-up is "balanced" on each side of a center ply. This process equalizes stresses set up when the board

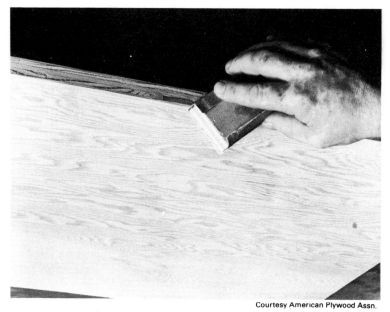

Courtesy American Plywood Assn.

Fig. 11-12. Plywood is indispensable to the builder. It comes in forms from utilitarian to elegant.

dries or when it is subsequently wetted and dried.

Plywood is also graded, and the grading system has changed over the years. Most important, however, there is exterior or interior plywood, which refers to the type of glue used to bond plies, and both come in various grades. The face of the plywood has a letter grade—A, B, C, or D. Grade A means the face has no defects—it is perfect. Grade B means there are some defects; perhaps an area has a small patch. Grade C allows checks (splits) and small knotholes. Grade D allows large knotholes.

In theory, you should be able to get a wide variety of grades and types in the lumberyard, but in reality you commonly find, at this writing, AC interior, AD interior, AC exterior, and CD, which is sheathing.

In selecting plywood, the rule is simple. Just pick what is right for the job at hand. If one side is going to be hidden, for example, you do not need a high grade.

Particleboard

In the manufacture of particleboard, the structure of the wood is not broken down, but simply reduced to flakes or particles, which are bound together with a synthetic resin, often reaformaldehyde or phenol-formaldehyde. The boards are then cured under heat and moderate pressure.

Particleboard is much less costly than plywood, but it is rough on saw blades and difficult to nail at the edges. Indeed, it cannot be nailed at the edges. For saving money, though, it cannot be beaten. Like plywood, particleboard is available in 4' × 8' sheets of various thicknesses.

Hardboard

Hardboard is made from wood pulp, and it usually contains no binder other than the slightly thermoplastic lignin in the wood. The board is formed under heat and heavy pressure. One side of the board is smooth; the other may or may not have a textured surface. It may be tempered by impregnating the board with oil and then baking it.

Hardboard is good for light jobs and for lining floors prior to installing flooring.

CHAPTER 12

Framing

KNOWLEDGE OF LUMBER

The basic construction material in carpentry is lumber. There are many kinds of lumber varying greatly in structural characteristics. Here, we deal with the lumber common to construction carpentry, its application, the standard sizes in which it is available, and the methods of computing lumber quantities in terms of *board feet*.

Standard Sizes of Bulk Lumber

Lumber is usually sawed into standard lengths, widths, and thickness. This permits uniformity in planning structures and in ordering material. Table 12-1 lists the common widths and thickness of wood in rough, and in dressed dimensions in the United States. Standards have been established for dimension differences between nominal size and the standard size (which is actually the reduced size when dressed). It is important that these dimension

TABLE 1. Your Guide to New Sizes of Lumber

WHAT YOU ORDER	WHAT YOU GET		WHAT YOU USED TO GET
	* Dry or Seasoned	** Green or Unseasoned	Seasoned or Unseasoned
1 x 4	¾ x 3½	$^{25}\!/_{32}$ x $3^{9}\!/_{16}$	$^{25}\!/_{32}$ x 3⅝
1 x 6	¾ x 5½	$^{25}\!/_{32}$ x 5⅝	$^{25}\!/_{32}$ x 5½
1 x 8	¾ x 7¼	$^{25}\!/_{32}$ x 7½	$^{25}\!/_{32}$ x 7½
1 x 10	¾ x 9¼	$^{25}\!/_{32}$ x 9½	$^{25}\!/_{32}$ x 9½
1 x 12	¾ x 11¼	$^{25}\!/_{32}$ x 11½	$^{25}\!/_{32}$ x 11½
2 x 4	1½ x 3½	$1^{9}\!/_{16}$ x $3^{9}\!/_{16}$	1⅝ x 3⅝
2 x 6	1½ x 5½	$1^{9}\!/_{16}$ x 5⅝	1⅝ x 5½
2 x 8	1½ x 7¼	$1^{9}\!/_{16}$ x 7½	1⅝ x 7½
2 x 10	1½ x 9¼	$1^{9}\!/_{16}$ x 9½	1⅝ x 9½
2 x 12	1½ x 11¼	$1^{9}\!/_{16}$ x 11½	1⅝ x 11½
4 x 4	3½ x 3½	$3^{9}\!/_{16}$ x $3^{9}\!/_{16}$	3⅝ x 3⅝
4 x 6	3½ x 5½	$3^{9}\!/_{16}$ x 5⅝	3⅝ x 5½
4 x 8	3½ x 7¼	$3^{9}\!/_{16}$ x 7½	3⅝ x 7½
4 x 10	3½ x 9¼	$3^{9}\!/_{16}$ x 9½	3⅝ x 9½
4 x 12	3½ x 11¼	$3^{9}\!/_{16}$ x 11½	3⅝ x 11½

*19% Moisture Content or under.
**Over 19% Moisture Content.

differences be taken into consideration when planning a structure. A good example of the dimension difference may be illustrated by the common 2 × 4. As may be seen in the table, the familiar quoted size (2 × 4) refers to a rough or nominal dimension, but the actual standard size to which the lumber is dressed is 1½″ × 3½″.

Grades of Lumber

Lumber as it comes from the sawmill is divided into three main classes; yard lumber, structural material, and factory or shop lumber. In keeping with the purpose of this book, only yard lumber will be considered. Yard lumber is manufactured and classified, on a quality basis, into sizes, shapes, and qualities required for ordinary construction and general building purposes. It is then further subdivided into classifications of select lumber and common lumber.

Select Lumber—Select lumber is of good appearance and finished or dressed. It is identified by the following grade names:

Grade A. Grade A is suitable for natural finishes, of high quality, and is practically clear.

Grade B. Grade B is suitable for natural finishes, of high quality, and is generally clear.

Grade C. Grade C is adapted to high-quality paint finish.

Grade D. Grade D is suitable for paint finishes and is between the higher finishing grades and the common grades.

Common Lumber—Common lumber is suitable for general construction and utility purposes and is identified by the following grade name:

No. 1 common. No. 1 common is suitable for use without waste. It is sound and tight knotted, and may be considered watertight material.

No. 2 common. No. 2 common is less restricted in quality than No. 1, but of the same general quality. It is used for framing, sheathing, and other structural forms where the stress or strain is not excessive.

No. 3 common. No. 3 common permits some waste with prevailing grade characteristics larger than in No. 2. It is used for footings, guardrails, and rough subflooring.

No. 4 common. No. 4 common permits waste, and is of low quality, admitting the coarsest features such as decay and holes. It is used for sheathing, subfloors, and roof boards in the cheaper types of construction. The most important industrial outlet for this grade is for boxes and shipping crates.

FRAMING LUMBER

The frame of a building consists of the wooden form constructed to support the finished members of the structure. It includes such items as posts, girders (beams), joists, subfloor, sole plate, studs, and rafters. Softwoods are usually used for lightwood framing and all other aspects of construction carpentry considered in this book. One of the classifications of softwood lumber cut to standard sizes is called yard lumber and is manufactured for general building purposes. It is cut into the standard sizes required for light framing, including 2×4, 2×6, 2×8, 2×10, 2×12, and all other sizes required for framework, with the exception of those sizes classed as structural lumber.

Although No. 1 and No. 3 common are sometimes used for framing, No. 2 common is most often used, and is therefore most often stocked and available in retail lumber yards in the common sizes for various framing members. However, the size of lumber required for any specific structure will vary with the design of the building, such as light-frame or heavy-frame, and the design of the particular members, such as beams or girders.

The exterior walls of a frame building usually consist of three layers—sheathing, building paper, and siding. Sheathing lumber is usually 1 × 6s or 1 × 8s, No. 2 or No. 3 softwood, finished plain, tongue-and-groove, or shiplap. The siding lumber may be grade C, which is most often used, or grade D. Siding is usually available in bundles consisting of a given number of square feet per bundle, and it comes in various lengths up to a maximum of 20 ft. Sheathing grade (CD) plywood is also extensively used and styrofoam faced with foil is used; the latter has fair insulating properties.

COMPUTING BOARD FEET

The arithmetic method of computing the number of board feet in one or more pieces of lumber is by the use of the following formula:

$$\frac{\text{Pieces} \times \text{Thickness (inches)} \times \text{Width (inches)} \times \text{Length (feet)}}{12}$$

Example 1—Find the number of board feet in a piece of lumber 2 inches thick, 10 inches wide, and 6 feet long.

$$\frac{1 \times 2 \times 10 \times 6}{12} = 10 \text{ board feet.}$$

Example—Find the number of board feet in 10 pieces of lumber 2 inches thick, 10 inches wide, and 6 feet long.

$$\frac{10 \times 2 \times 10 \times 6}{12} = 100 \text{ board feet.}$$

Example—Find the number of board feet in a piece of lumber 2 inches thick, 10 inches wide, and 18 inches long.

$$\frac{2 \times 10 \times 18}{144} = 2\frac{1}{2} \text{ board feet.}$$

(NOTE: If all three dimensions are expressed in inches, the same formula applies except the divisor is changed to 144.)

Board feet can also be calculated by use of a table normally found of the back of the tongue of a steel framing square. The inch graduations on the outer edge of the square are used in combination with the values in the table to get a direct indication of the number of board feet in a particular board. Complete instructions come with the square.

METHODS OF FRAMING

Good material and workmanship will be of very little value unless the underlying framework of a building is strong and rigid. The resistance of a house to such forces as tornadoes and earthquakes, and control of cracks due to settlement, all depends on a good framework.

Although it is true that no two buildings are put together in exactly the same manner, disagreement exists among architects and carpenters as to which method of framing will prove most satisfactory for a given condition. *Light framed construction* may be classified into three distinct types known as:

1. Balloon frame.
2. Plank and beam.
3. Western frame (also identified as platform frame).

Balloon Frame Construction

The principal characteristic of *balloon framing* is the use of studs extending in one piece from the foundation to the roof, as shown in Fig. 12-1. The joists are nailed to the studs and also supported by a ledger board set into the studs. Diagonal sheathing may be used instead of wall board to eliminate corner bracing.

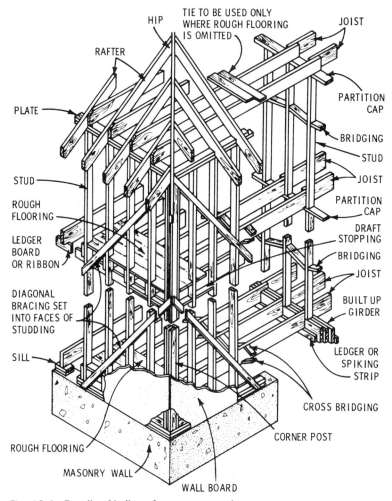

TIE TO BE USED ONLY
HIP WHERE ROUGH FLOORING
IS OMITTED

JOIST

RAFTER

PLATE

STUD

ROUGH
FLOORING

LEDGER
BOARD
OR RIBBON

DIAGONAL
BRACING SET
INTO FACES OF
STUDDING

SILL

PARTITION
CAP

BRIDGING

STUD

JOIST

PARTITION
CAP

DRAFT
STOPPING

BRIDGING

JOIST

BUILT UP
GIRDER

LEDGER OR
SPIKING
STRIP

CROSS BRIDGING

CORNER POST

ROUGH FLOORING

MASONRY WALL

WALL BOARD

Fig. 12-1. Details of balloon frame construction.

Plank and Beam Construction

The *plank-and-beam construction* is said to be the oldest
method of framing in the country, having been imported from
England in colonial times. Although in a somewhat modified
form, it is still being used in certain states, notably in the east.
Originally, this type of framing was characterized by heavy

270

timber posts at the corners, as shown in Fig. 12-2, and often with intermediate posts between, which extended continuously from a heavy foundation sill to an equally heavy plate at the roof line.

Western Frame Construction

This type of framing is characterized by platforms independently framed, the second or third floor being supported by the studs from the first floor, as shown in Fig. 12-3. The chief advantage in this type of framing (in all-lumber construction) lies in the fact that if there is any settlement due to shrinkage, it will be uniform throughout and will not be noticeable.

FOUNDATION SILLS

The foundation sill consists of a plank or timber resting on the foundation wall. It forms the support or bearing surface for the outside of the building and, as a rule, the first floor joists rest upon it. Shown in Fig. 12-4 is the balloon-type construction of first and second floor joist and sills. In Fig. 12-5 is shown the joist and sills used in plank-and-beam framing, and in Fig. 12-6 is shown the western-type construction.

Size of Sills

The size of sills for small buildings of light frame construction consists of $2'' \times 6''$ lumber, which has been found to be large enough under most conditions. For two-storey buildings, and especially in locations subject to earthquakes or tornadoes, a double sill is desirable, as it affords a larger nailing surface for diagonal sheathing brought down over the sill, and ties the wall framing more firmly to its sills. In cases where the building is supported by posts or piers, it is necessary to increase the sill size, since the sill supported by posts acts as a girder. In balloon framing, for example, it is customary to build up the sills with two or more planks 2 or 3 in. thick, which are nailed together.

In most types of construction, since it is not necessary that the sill be of great strength, the foundation will provide uniform solid bearing throughout its entire length. The main requirements are: resistance to crushing across the grain; ability to withstand decay and attacks of insects; and to furnish adequate nailing area for studs, joists, and sheathing.

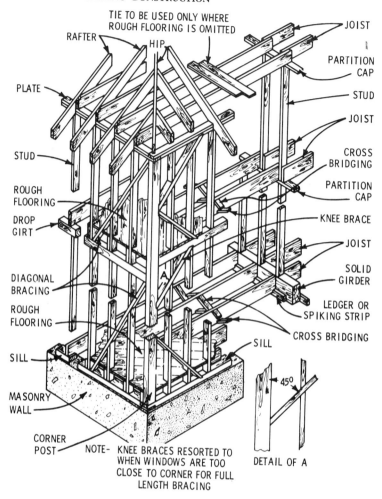

Fig. 12-2. Details of plank-and-beam construction.

Length of Sill

The length of the sill is determined by the size of the building, and hence the foundation should be laid out accordingly. Dimension lines for the outside of the building are generally figured from the outside face of the subsiding or sheathing, which is about the same as the outside finish of unsheathed buildings.

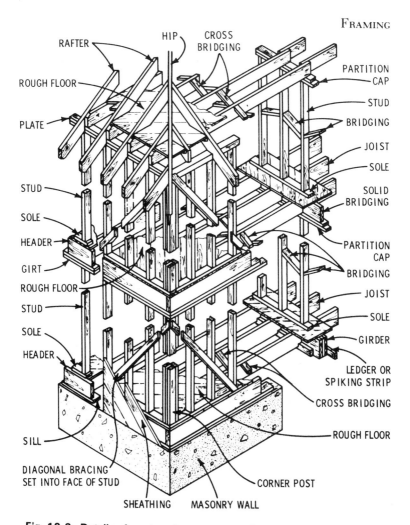

Fig. 12-3. Details of western frame construction.

Anchorage of Sill

It is important, especially in locations of strong winds, that buildings be thoroughly anchored to the foundation. Anchoring is accomplished by setting at suitable intervals (6 to 8 ft.) ½-in. bolts that extend at least 18 in. into the foundation. They should project above the sill to receive a good size washer and nut. With hollow

273

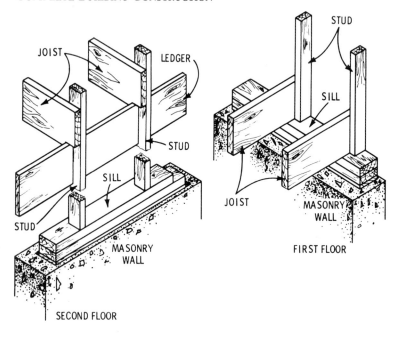

Fig. 12-4. Details of balloon framing of sill plates and joists.

tile, concrete blocks, and material of cellular structure, particular care should be taken in filling the cells in which the bolts are placed solidly with concrete.

Splicing of Sill

As previously stated, a 2″ × 6″ sill is large enough for small buildings under normal conditions, if properly bedded on the foundations. In order to properly accomplish the splicing of a sill, it is necessary that special precaution be taken. A poorly fitted joint weakens rather than strengthens the sill frame. Where the sill is built up of two planks, the joints in the two courses should be staggered.

Placing of Sill

It is absolutely essential that the foundation be level when placing the sill. It is good practice to spread a bed of mortar on the top foundation blocks and lay the sill upon it at once, tapping it gently to secure an even bearing throughout its length. The nuts of the anchoring bolts can then be put in place over the washers and tightened lightly. The nut is securely tightened only after the mortar has had time to harden. In this manner a good bearing of the sill is provided, which prevents air leakage between the sill and the foundation wall.

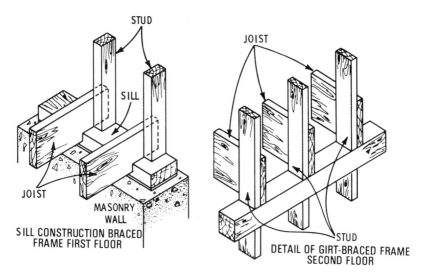

Fig. 12-5. Details of plank-and-beam framing of sill plates and joists.

GIRDERS

A girder in small house construction consists of a large beam at the first-storey line, which takes the place of an interior foundation wall and supports the inner ends of the floor joists. In a building where the space between the outside walls is more than 14 to 15 ft., it is generally necessary to provide additional support near the center to avoid the necessity of excessively heavy floor joists. When a determination is made as to the number of girders and their location, consideration should be given to the required length of the joists, to the room arrangement, as well as to the location of the bearing partitions.

275

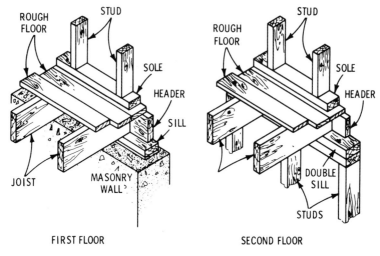

FIRST FLOOR   SECOND FLOOR

BOX-SILL CONSTRUCTION WESTERN FRAME

Fig. 12-6. Details of western framing of sill plates and joists.

LENGTH OF JOISTS

In ordinary cases one girder will generally be sufficient, but if the joist span exceeds 15 ft., a $2'' \times 10''$ joist is usually required, making another girder necessary.

Bridging Between Joists

Cross bridging consists of diagonal pieces (usually $1'' \times 3''$ or $1'' \times 4''$) formed in an X pattern and arranged in rows running at right angles to the joists, as shown in Fig. 12-7. Their function is to provide a means of stiffening the floors and to distribute the floor load equally. Each piece should be nailed with two or three nails at the top, with the lower ends left loose until the subflooring is in place.

Solid bridging is sometimes used in place of cross bridging. Solid bridging serves the same function as cross bridging, but is made of solid lumber the size of the joist, Fig. 12-8.

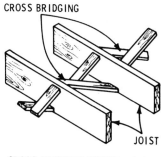

CROSS BRIDGING

JOIST

CROSS-BRIDGING BETWEEN JOISTS

Fig. 12-7. Cross-bridging between floor joists.

INTERIOR PARTITIONS

An interior partition differs from an outside partition, in that it seldom rests on a solid wall. Its supports therefore require careful consideration, making sure they are large enough to carry the required weight. The various interior partitions may be bearing or nonbearing, and may run at either right angles or parallel to the joists upon which they rest.

Partitions Parallel to Joists

Here the entire weight of the partition will be concentrated upon one or two joists, which perhaps are already carrying their full share of the floor load. In most cases, additional strength should be provided. One method is to provide double joists under such partitions—to put an extra joist beside the regular ones. Computation shows that the average partition weighs nearly three times as much as a single joist should be expected to carry. The usual (and approved method) is to double the joists under nonbearing partitions. An alternative method is to place a joist on each side of the partition.

Partitions at Right Angles to Joists

For nonbearing partitions, it is not necessary to increase the size

277

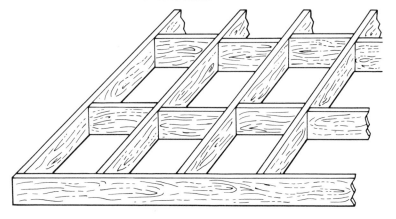

Fig. 12-8. Solid bridging between joists.

or number of the joists. The partitions themselves may be braced, but even without bracing, they have some degree of rigidity.

PORCHES

All porches not resting directly upon the ground are always supported by various posts or piers which are termed *porch supports*. Unless porches are entirely enclosed, and of two or more stories in height, their weight will not require a massive foundation.

In pier foundations used to carry concentrated loads, the following sizes will be sufficient:

For very small porches, such as stoops 4 to 6 ft. square, concrete footings 12 in. square and 6 in. thick should be used.

For the usual type of front or side porch, the supports of which are not over 10-ft. apart, concrete footings 18 in. square and 8 in. thick should be used.

For large porches (especially if they are enclosed), the piers of which exceed a 10-ft. spacing, footings 21 or 24 in. square and 10 or 12 in. thick should be used.

Large enclosed porches, especially if more than one storey, should have footings and foundations similar to those of the main building.

Porch Joists

Porch joists differ from floor joists, because of their need for greater weather resisting qualities due to their exposed location. In order that rain water may run off freely, the porch should slope approximately $\frac{1}{8}$-in. per foot away from the wall of the building. Since there is no subfloor, this requires the flooring to run in the direction of the slope (at right angles to the wall). Otherwise, water will stand on the floor wherever there are slight irregularities.

This construction requires an arrangement comprising a series of girders running from the wall to the piers. The joists run at right angles to these girders and rest on top, or are cut in between them. To protect the tops of the joists from rot, place a strip of 45-lb. roofing on the top edge under the flooring.

FRAMING AROUND OPENINGS

It is necessary that some parts of the studs be cut out around windows or doors in outside walls or partitions. It is imperative to insert some form of a header to support the lower ends of the top studs that have been cut off. There is a member that is termed a rough sill at the bottom of the window openings. This sill serves as a nailer, but does not support any weight.

HEADERS

Headers are of two classes, namely:

1. Nonbearing headers which occur in the walls which are parallel with the joists of the floor above, and carry only the weight of the framing immediately above.
2. Load-bearing headers, in walls which carry the end of the floor joists either on plates or rib bands immediately above the openings, and must therefore support the weight of the floor or the floors above.

279

Size of Headers

The determining factor in header sizes is whether they are load bearing or not. In general, it is considered good practice to use a double 2 × 4 header placed on edge unless the opening in a nonbearing partition is more than 3 ft. wide. In cases where the trim inside and outside is too wide to prevent satisfactory nailing over the openings, it may become necessary to double the header to provide a nailing base for the trim.

CORNER STUDS

These are studs which occur at the intersection of two walls at right angles to each other. A satisfactory arrangement of corner studs is shown in Fig. 12-9.

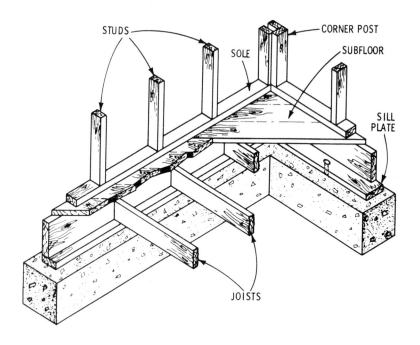

Fig. 12-9. Detail view of a corner stud.

ROOFS

Generally, it may be said that rafters serve the same purpose for a roof as joists do for the floors. They provide a support for sheathing and roof material. Among the various kinds of rafters used, regular rafters extending without interruption from the eave to the ridge are the most common.

Spacing of Rafters

Spacing of rafters is determined by the stiffness of the sheathing between rafters, by the weight of the roof, and by the rafter span. In most cases, the rafters are spaced 16 or 24 in. on center.

Size of Rafters

The size of the rafters will depend upon the following factors:

1. The span.
2. The weight of the roof material.
3. The snow and wind loads.

Span of Rafters

In order to avoid any misunderstanding as to what is meant by the span of a rafter, the following definition is given:

The *rafter span* is the horizontal distance between the supports and not the overall length from end to end of the rafter. The span may be between the wall plate and the ridge, or from outside wall to outside wall, depending on the type of roof.

Length of Rafters

Length of rafters must be sufficient to allow for the necessary cut at the ridge and to allow for the protection of the eaves as determined by the drawings used. This length should not be confused with the span as used for determining strength.

Collar Beams

Collar beams may be defined as ties between rafters on opposites sides of a roof. If the attic is to be utilized for rooms, collar beams may be lathed as ceiling rafters, providing they are spaced properly. In general, collar beams should not be relied upon as ties. The closer the ties are to the top, the greater the leverage action. There is a tendency for the collar beam nails to pull out, and also for the rafter to bend if the collar beams are too low. The function of the collar beam is to stiffen the roof. These beams are often, although not always, placed at every rafter. Placing them at every second or third rafter is usually sufficient.

Size of Collar Beams

If the function of a collar beam were merely to resist a thrust, it would be unnecessary to use material thicker than 1-in. lumber. But, as stiffening the roof is their real purpose, the beams must have sufficient body to resist buckling, or else must be braced to prevent bending.

HIP RAFTERS

The hip roof is built in such a shape that its geometrical form is that of a pyramid, sloping down on all four sides. A hip is formed where two adjacent sides meet. The hip rafter is the one which runs from the corner of the building upward along the same plane of the common rafter. Where hip rafters are short and the upper ends come together at the corner of the roof (lending each other support), they may safely be of the same size as that of the regular rafters.

For longer spans, however, and particularly when the upper end of the hip rafter is supported vertically from below, an increase of size is necessary. The hip rafter will necessarily be slightly wider than the jacks, in order to give sufficient nailing surface. A properly sheathed hip roof is nearly self-supporting. It is the strongest roof of any type of framing in common use.

DORMERS

The term *dormer* is given to any window protruding from a roof. The general purpose of a dormer may be to provide light or to add to the architectural effect.

In general construction, there are three types of dormers, as follows:

1. Dormers with flat sloping roofs, but with less slope than the roof in which they are located, as shown in Fig. 12-10.
2. Dormers with roofs of the gable type at right angles to the roof (Fig. 12-11).

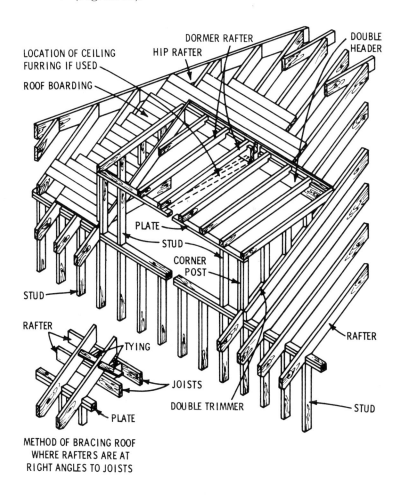

Fig. 12-10. Detail view of a flat-roof dormer.

3. A combination of the above types, which gives the hip-type dormer.

When framing the roof for a dormer window, an opening is provided in which the dormer is later built in. As the spans are usually short, light material may be used.

STAIRWAYS

The well built stairway is something more than a convenient means of getting from one floor to another. It must be placed in the right location in the house. The stairs themselves must be so designed that traveling up or down can be accomplished with the least amount of discomfort.

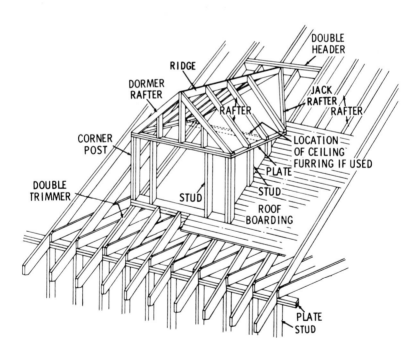

Fig. 12-11. Detail view of gable-roof dormer.

The various terms used in stairway building (Fig. 12-12) are as follows:

1. The *rise* of a stairway is the height from the top of the lower floor to the top of the upper floor.
2. The *run* of the stairs is the length of the floor space occupied by the construction.
3. The *pitch* is the angle of inclination at which the stairs run.
4. The *tread* is that part of the horizontal surface on which the foot is placed.
5. The *riser* is the vertical board under the front edge of the tread.
6. The *stringer* is the framework on either side, which is cut to support the treads and risers.

A commonly followed rule in stair construction is that the *tread* should not measure less than 9 in. deep, and the *riser* should not be more than 8 in. high. The width measurement of the tread and height of the riser combined should not exceed 17 in. Measure-

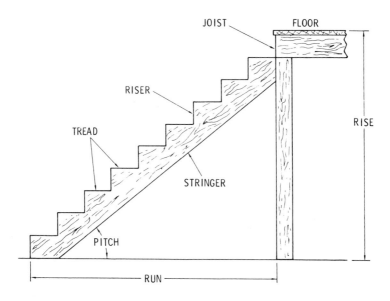

Fig. 12-12. Parts of a stairway.

ments are for the cuts of the stringers, not the actual width of the boards used for risers and treads. Treads usually have a projection, called a *nosing*, beyond the edge of the riser.

FIRE AND DRAFT STOPS

It is known that many fires originate on lower floors. It is therefore important that fire stops be provided in order to prevent a fire from spreading through the building by way of air passages between the studs. Similarly, fire stops should be provided at each floor level to prevent flames from spreading through the walls and partitions from one floor to the next. Solid blocking should be provided between joists and studs to prevent fire from passing across the building.

In the platform frame and plank-and-beam framing, the construction itself provides stops at all levels. In this type of construction, therefore, fire stops are needed only in the floor space over the bearing partitions. Masonry is sometimes utilized for fire stopping, but is usually adaptable in only a few places. Generally, obstructions in the air passages may be made of 2-in. lumber, which will effectively prevent the rapid spread of fire. Precautions should be made to insure the proper fitting of fire stops throughout the building.

CHIMNEY AND FIREPLACE CONSTRUCTION

Although the carpenter is ordinarily not concerned with the building of the chimney, it is necessary, however, that he be acquainted with the methods of framing around the chimney.

The following minimum requirements are recommended:

1. No wooden beams, joists, or rafters shall be placed within 2 in. of the outside face of the chimney. No woodwork shall be placed within 4 inches of the back wall of any fireplace.
2. No studs, furring, lathing, or plugging should be placed against any chimney or in the joints thereof. Wooden construction shall either be set away from the chimney or the

plastering shall be directly on the masonry or on metal lathing or on incombustible furring material.

3. The walls of fireplaces shall never be less than 8 in. thick if of brick or 12 in. if built of stone.

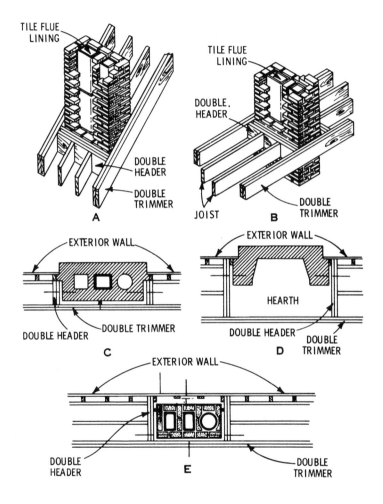

Fig. 12-13. Framing around chimneys and fireplaces: (A) roof framing around chimney; (B) floor framing around chimney; (C) framing around chimney above fireplace; (D) floor framing around fireplace; (E) framing around concealed chimney above fireplace.

Formerly, it was advised to pack all spaces between chimneys and wood framing with incombustible insulation. It is now known that this practice is not as fire-resistant as the empty air spaces, since the air may carry away dangerous heat while the insulation may become so hot that it becomes a fire-hazard itself. Fig. 12-13 shows typical framing around chimneys and fireplaces.

CHAPTER 13

Girders and Sills

Chapter 12 gave an overall idea of the several kinds of framing. The details of each vary greatly. There are many ways of constructing each part, and in this connection, buyers or those contemplating having a house built should be acquainted not only with the right construction methods but also with the methods that are not good.

It is poor economy to specify inexpensive and inferior construction as houses so built are often not satisfactory. In this and following chapters the various parts of the frame, such as *girders*, *sills*, *corner posts*, and *studding* are considered in detail, showing the numerous ways in which each part is treated.

GIRDERS

By definition, a *girder* is a principal beam extended from wall to wall of a building affording support for the joists or floor beams where the distance is too great for a single span. Girders may be either solid or built up, as shown in Figs. 13-1 and 13-2.

Construction of Girders

Girders can be built up of wood if select stock is used. Be sure it is straight and sound. If the girders are to be built up of 2 × 8 or 2 × 10 stock, place the pieces on the sawhorses and nail them together. Use the piece of stock that has the least amount of warp for the center piece and nail other pieces on the sides of the center stock. Use a common nail that will go through the first piece and nearly through the center piece. Square off the ends of the girder after the pieces have been nailed together. If the stock is not long enough to build up the girder the entire length, the pieces must be built up by staggering the joints. If the girder supporting post is to be built up, it is to be done in the same manner as described for the girder.

Table 13-1 gives the size of built-up wood girders for various loads and spans, based on douglas fir 4-square framing lumber.

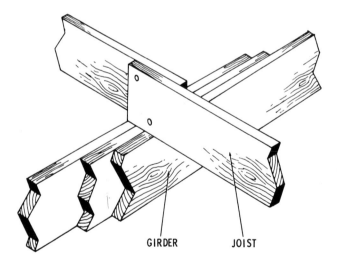

GIRDER   JOIST

Fig. 13-1. Girder and joist construction.

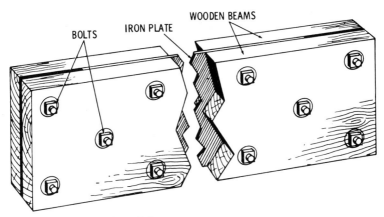

Fig. 13-2. A flitch plate girder.

Placing Basement Girders

Basement girders must be lifted into place on top of the piers and walls built for them, and set perfectly level and straight from end to end. Some carpenters prefer to give the girders a slight crown of approximately 1 in. in the entire length, which is a wise plan, because the piers generally settle more than the outside walls. When there are posts instead of brick piers used to support the girder, the best method is to temporarily support the girder by

Table 1. Nominal Size of Girder Required

Load per linear foot of girder	Length of Span				
	6-foot	7-foot	8-foot	9-foot	10-foot
750	6x8 in	6x8 in	6x8 in	6x10 in	6x10 in
900	6x8	6x8	6x10	6x10	8x10
1050	6x8	6x10	8x10	8x10	8x12
1200	6x10	8x10	8x10	8x10	8x12
1350	6x10	8x10	8x10	8x12	10x12
1500	8x10	8x10	8x12	10x12	10x12
1650	8x10	8x12	10x12	10x12	10x14
1800	8x10	8x12	10x12	10x12	10x14
1950	8x12	10x12	10x12	10x14	12x14
2100	8x12	10x12	10x14	12x14	12x14
2250	10x12	10x12	10x14	12x14	12x14
2400	10x12	10x14	10x14	12x14	x
2550	10x12	10x14	12x14	12x14	x
2700	10x12	10x14	12x14	x	x
2850	10x14	12x14	12x14	x	x
3000	10x14	12x14	x	x	x
3150	10x14	12x14	x	x	x
3300	12x14	12x14	x	x	x

uprights made of 2 × 4s resting on blocks on the ground below. When the superstructure is raised, these can be knocked out after the permanent posts are placed. The practice of temporarily shoring the girders, and not placing the permanent supports until after the superstructure is finished, is favored by many builders, and it is well for carpenters to know just how it should be done. Permanent supports are usually made by using 3- or 4-in. steel pipe set in a concrete foundation, or footing, that is at least 12 in. deep. They are called lally columns.

SILLS

A *sill* is that part of the side walls of a house that rests horizontally upon, and is securely fastened to, the foundation.

Sills may be divided into two general classes:

1. Solid.
2. Built-up.

The built-up sill has become more or less a necessity because of the scarcity and high cost of timber, especially in the larger sizes. The work involved in sill construction is very important for the carpenter. The foundation wall is the support upon which the entire structure rests. The sill is the foundation on which all the framing structure rests, and is the real point of departure for actual carpentry and joinery activities. The sills are the first part of the frame to be set in place. They either rest directly on the foundation piers or on the ground, and may extend all around the building.

The type of sill used depends upon the general type of construction; they are called:

1. Box sills.
2. T-sills.
3. Braced framing sills.
4. Built-up sills.

Box Sills—Box sills are often used with the very common style of platform framing, either with or without the sill plate. In this type of sill the part that lies on the foundation wall or ground is called the sill plate. The sill is laid edgewise on the outside edge of the sill plate.

T-Sills—There are two types of T-sill construction, one commonly used in the south or dry warm climate, and one used in the north and east where it is colder. Their construction is similar except in the east the T-sills and joists are nailed directly to the studs, as well as to the sills. The headers are nailed in between the floor joists.

Braced Framing Sills—The floor joists are notched out and nailed directly to the sills and studs.

Built-up Sills—Where built-up sills are used, the joists are staggered. If piers are used in the foundation, heavier sills are used. These sills are of single heavy timber or are built up of two or more pieces of timber. Where heavy timber or built-up sills are used, the joints should occur over the piers. The size of the sill depends on the load to be carried and on the spacing of the piers. Where earth floors are used, the studs are nailed directly to the sill plates.

Setting the Sills

After the girder is in position, the sills are placed on top of the foundation walls, are fitted together at the joints, and leveled throughout. The last operation can either be done by a sight level, or by laying them in a full bed of mortar and leveling them with the anchor bolts, as shown in Fig. 13-3. Or it can be left loose and then

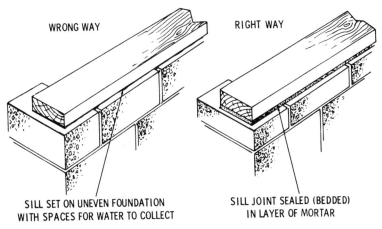

WRONG WAY   RIGHT WAY

SILL SET ON UNEVEN FOUNDATION
WITH SPACES FOR WATER TO COLLECT

SILL JOINT SEALED (BEDDED)
IN LAYER OF MORTAR

Fig. 13-3. Wrong and right way to set sills. The top solid block serves as a termite barrier.

all the bolts can be tightened as needed to bring the sill level.

Sills that are to rest on a wall of masonry should be kept up at least 18 in. above the ground, as decaying sills are a frightful source of trouble and expense in wooden buildings. Sheathing, paper, and siding should therefore be very carefully installed to exclude all wind and wet weather.

CHAPTER 14

Floor Framing

After the girders and sills have been placed, the next operation consists in sawing to size the floor beams or joists of the first floor, and placing them in position on the sills and girders. If there is a great variation in the size of timbers, it is necessary to cut the joists ½ in. narrower than the timber, so that their upper edges will be in alignment. This sizing should be made from the top edge of the joist as shown in Fig. 14-1. When the joists have been cut to the correct dimension they should be placed upon the sill and girders, and spaced 16 in. between centers, beginning at one side or end of a room. This is done to avoid wasting the material.

CONNECTING THE JOISTS TO SILLS AND GIRDERS

Joists are connected to sills and girders by several methods. In modern construction, the method that requires the least time and labor and yet gives the maximum efficiency is used. In joining

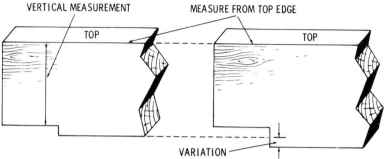

VERTICAL MEASUREMENT   MEASURE FROM TOP EDGE

TOP

TOP

VARIATION

Fig. 14-1. Showing the variation in joist width.

joists to sills, always be sure that the connection is able to hold the load that the joists will carry.

The placing of girders is an important factor in making the connection. The joists must be level; therefore, if the girder is not the same height as the sill, the joists must be notched. In placing joists, always have the crown up since this counteracts the weight on the joists; in most cases there will be no sag below a straight line. When a joist is to rest on plates or girders, the joist is cut long enough to extend the full width of the plate or girder.

BRIDGING

To prevent joists springing sideways under load, which would reduce their carrying capacity, they are tied together diagonally by 1×3 or 2×3 strips. This reinforcement is called *bridging*. The 1×3 ties are used for small houses and the 2×3 stock on larger work (Fig. 14-2).

Rows of bridging should not be more than 8 ft. apart. Bridging pieces may be cut all in one operation with a miter box, or the pieces may be cut to fit. Bridging is put in before the subfloor is laid, and each piece is fastened with two nails at the top end. The subfloor should be laid before the bottom end is nailed.

A more rigid (less vibrating) floor is made by nailing in solid 2-in. blocking the same depth as the joists. It should be cut perfectly square and a little full, say $\frac{1}{16}$ inch longer than the inside distance between the joists. First, set one in every other space, then go back and put in the intervening ones. This prevents spreading

296

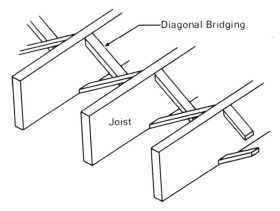

Fig. 14-2. Diagonal bridging.

the joists and allows the second ones to be driven in with the strain the same in both directions. This solid blocking is much more effective than cross bridging. The blocks should be toenailed and not staggered and nailed through the joists.

HEADERS AND TRIMMERS

The foregoing operations would complete the first-floor framework in rooms having no framed openings, such as those for stair ways, chimneys, or elevators. What follows are tips for framing openings.

The definition of a *header* is a short transverse joist which supports the ends of one or more joists (called tail beams) where they are cut off at an opening. A *trimmer* is a carrying joist which supports an end of a header.

Shown in Fig. 14-3 are typical floor openings used for chimneys and stairways. For these openings, the headers and trimmers are set in place first, then the floor joists are installed. Hanger irons of the type shown in Fig. 14-4 are nearly obsolete, but many other metal hangers are available, such as the one shown in Fig. 14-5.

Headers run at right angles to the direction of the joists and are doubled. Trimmers run parallel to the joists and are actually doubled joists. The joists are framed to the headers where the headers form the opening frame at right angles to the joists. These

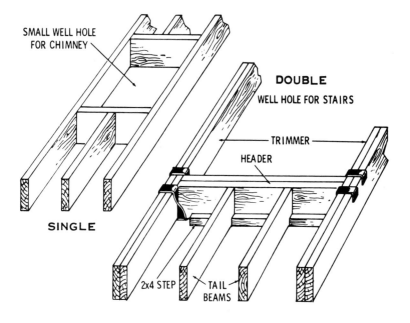

Fig. 14-3. Single and double header and trimmer construction for floor openings.

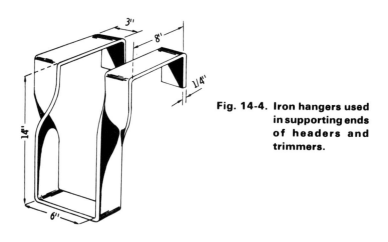

Fig. 14-4. Iron hangers used in supporting ends of headers and trimmers.

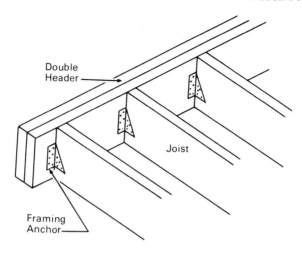

Fig. 14-5. One of the many metal hangers in common use.

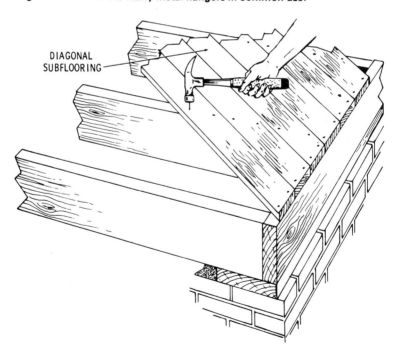

Fig. 14-6. The subflooring is laid diagonally for greater strength.

Courtesy American Plywood Assn.

Fig. 14-7 Plywood is the most popular subflooring. It is secured at right angles to joists.

shorter joists framed to the headers are called *tail beams*. The number of headers and trimmers required at any opening depends upon the shape of the opening.

SUBFLOORING

With the sills and floor joists completed, it is necessary to install the subflooring. The subflooring is permanently laid before erecting any wall framework, since the wall plate rests on it. In most cases, when boards are used, the subfloor is laid diagonally, as shown in Fig. 14-6, to give strength and to prevent squeaks in the floor. This floor is called the rough floor, or subfloor, and may be visioned as a large platform covering the entire width and length of the building. Two layers or coverings of flooring material (subflooring and finished flooring) are placed on the joists. The subfloor may be $1'' \times 4''$ square edge stock, $1'' \times 6''$ or $1'' \times 8''$ shiplap, or $4' \times 8'$ plywood panels (Fig. 14-7).

FINISH FLOORING

A finished floor is in most cases ¾-in. tongue-and-groove hardwood, such as oak. Prefinished flooring can be obtained. In addition, there are non-wood materials, which will be covered in detail later.

CHAPTER 15

Outer Wall Framing

Where the construction permits, it is advisable to permanently lay the subfloor before erecting the wall studding. Besides saving time in laying and removing a temporary floor, it also furnishes a surface on which to work and a sheltered place in the basement for storage of tools and materials.

BUILT-UP CORNER POSTS

There are many ways in which corner posts may be built up, using studding or larger sized pieces. Some carpenters form corner posts with two 2 × 4 studs spiked together to make a piece having a 4 × 4 section. Except for very small structures, such a flimsily built-up piece should not be used as a corner post. Fig. 15-1 shows various arrangements of built-up posts commonly used.

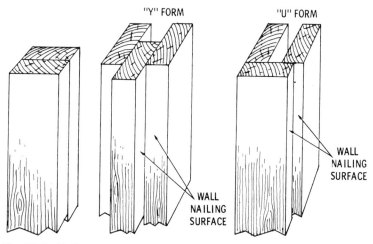

"Y" FORM   "U" FORM

WALL
NAILING
SURFACE

WALL
NAILING
SURFACE

Fig. 15-1. Various ways to build up a corner post.

BALLOON BRACING

In balloon framing, the bracing may be temporary or permanent. Temporary bracing means strips nailed on (as in Fig. 15-2) to hold the frame during construction which are removed when the permanent bracing is in place. There are two kinds of permanent bracing. One type is called *cut-in* bracing, and is shown in Fig. 15-3. A house braced in this manner withstood one of the worst hurricanes ever to hit the eastern seaboard, and engineers, after an inspection, gave the bracing the full credit for the survival. The other type of bracing is called *plank* bracing and is shown in Fig. 15-4. This type is put on from the outside, the studs being cut and notched so that the bracing is flush with the outside edges of the studs. This method of bracing is commonly used and is very effective.

PREPARING THE CORNER
POSTS AND STUDDING

In laying out the posts and studs, a pattern should be used to ensure that all will be of the same length, with the gains, and any other notches or mortises to be cut, at the same elevation. The pattern can be made from a ⅞ -in. board, or a 2 × 4 stud, and must be cut to the exact length and squared at both ends. The pattern should be made from a selected piece of wood having straight edges.

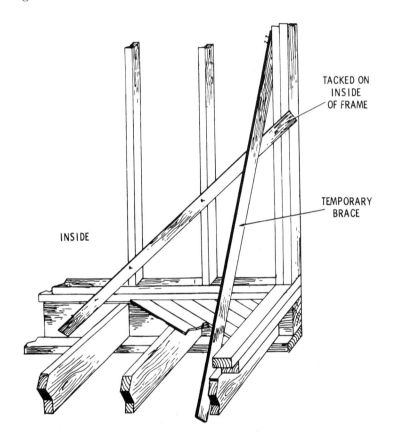

Fig. 15-2. Temporary frame bracing.

If a stud is selected for the pattern, it may be used later in the building and thus counted in with the total number of pieces to be framed. The wall studs are placed on their edges on the horses in quantities of 6 to 10 at a time and marked from the pattern, as shown in Fig. 15-5.

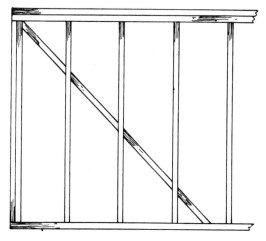

Fig. 15-3. Permanent 2 × 4 cut between stud bracing. This type of bracing is called cut-in braces.

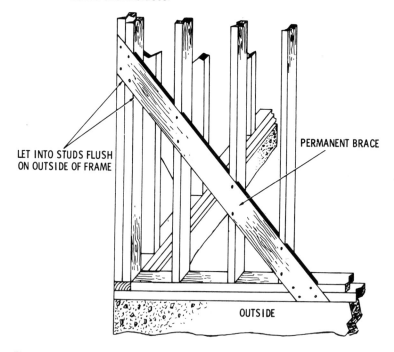

LET INTO STUDS FLUSH ON OUTSIDE OF FRAME

PERMANENT BRACE

OUTSIDE

Fig. 15-4. Permanent plank-type bracing.

ERECTING THE FRAME

If the builder is short-handed, or working by himself, a frame can be *crippled* together, one member at a time. An experienced carpenter can erect the studs and toenail them in place with no assistance whatever. Then the corners are plumbed and the top plates nailed on from a stepladder. Where enough men are available, most contractors prefer to nail the sole plates and top plates as the entire wall is lying down on the rough floor. Some even cut all the door and window openings, after which the entire gang raises the assembly and nails it in position. Needless to say, the platform frame with the subfloor in place is best adapted to this kind of construction. It is probably the speediest of any possible method, but it cannot be done by only one or two men. Some contractors even put on the outside sheathing before raising the wall and then use a lift truck or a highlift excavating machine to raise it into position.

There are many ways to frame the openings for windows and doors. Some builders erect all of the wall studding first, and then cut out the studs for the openings. It is much easier to saw studs lying level on horses than when nailed upright in the frame. The openings should be laid out and framed complete. Studding at all openings must be double to furnish more area for proper nailing of the trim, and must be plumb, as shown in Fig. 15-6.

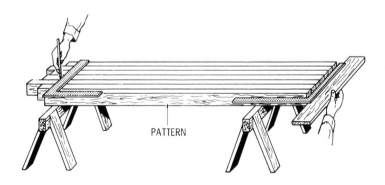

PATTERN

Fig. 15-5. Marking wall studs from pattern.

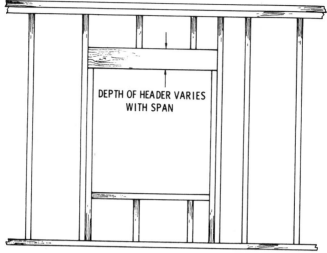

DEPTH OF HEADER VARIES WITH SPAN

Fig. 15-6. An approved framing for window opening.

OPENING SIZES FOR WINDOWS AND DOORS

Standard double-hung windows are listed by unit dimensions, giving the width first. A 2′ 0″ × 3′ 0″ double-hung window has an overall glass size of 28½ in. The rough opening is 2′ 10″, and the sash opening is 2′ 8″.

It is common to frame the openings with dimensions 2 in. over the unit dimension. This will allow for the plumbing and leveling of the window.

For door openings, allow a full 1 inch over the door size the thickness of the jambs, with about 1 inch additional for blocking and plumbing. This will make the opening width 3 in. over the door size. Standard oak thresholds are ⅝ in. thick, making the overall height of the door-frame butt ⅝ in. above the finish floor line. If you wish to allow for the lugs on the frame (which is a good idea), allow 5 in. over the door size for the height of the opening above the finish floor line. Door casings are set to show a slight rabbet, or set back, to allow for clearance of the hinges, while windows are usually cased flush with the inside edges of the jambs. If the heights of the window and door head casings are to be the same (as they often are), the window frames must be set the

308

height of this rabbet *higher* than the door frames. Continuous head trim is almost a thing of the past but where it is to be used, the carpenter can easily get into an almost uncorrectable situation if this difference is neglected.

PARTITIONS

Interior walls which divide the inside space of the buildings into rooms, halls, etc., are known as partitions. These are made up of studding covered with plaster board and plaster, metal lath and plaster, or dry wall. Fig. 15-7 shows the studding of an ordinary partition, each joist being 16 in. on center. Where partitions are placed between and parallel to floor joist, bridges must be placed between the joists to provide a means of fastening the partition plate. The construction at the top is shown in Fig. 15-8.

Openings over 30 in. wide in partitions or outside walls must have heavy headers (Fig. 15-9). Partition wall studs are arranged in a row with their ends bearing on a long horizontal member called a

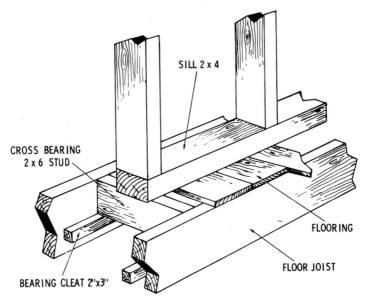

SILL 2 x 4

CROSS BEARING
2 x 6 STUD

FLOORING

FLOOR JOIST

BEARING CLEAT 2'x3"

Fig. 15-7. A method of constructing a partition between two floor joists.

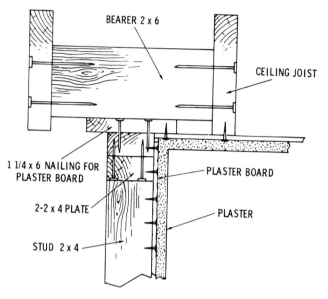

BEARER 2 x 6

CEILING JOIST

1 1/4 x 6 NAILING FOR PLASTER BOARD

PLASTER BOARD

2-2 x 4 PLATE

PLASTER

STUD 2 x 4

Fig. 15-8. A method of constructing a partition between two ceiling joists.

bottom plate, and their tops capped with another plate, called a top plate. Double top plates are used in bearing walls and partitions. The bearing strength of stud walls is determined by the strength of the studs.

Walls and partition coverings are divided into two general types—wet-wall material (generally plaster) and drywall material (also called plasterboard, sheetrock, and gypsum board). Drywall material comes in 4′ × 8′ and 4′ × 12′ sheets and in various thicknesses from ⅜ in. up; ⅜, ½, and ⅝ in. are the most common sizes. It is normally applied in single thickness. When covering both walls and ceiling, it is best to start at the ceiling.

You can build two T-braces to hold the panels in place, as shown in Fig. 15-10. Raise the first panel in position and hold it in place with the T-brackets. Each panel should be nailed at least every 4 in. starting at the center of the board and working outward. Due to the weight of drywall boards, a special nail can be purchased that will not pull out and let the ceiling sag. Dimple the nail below the surface of the panel, being careful not to break the surface of the board by the blow of the hammer.

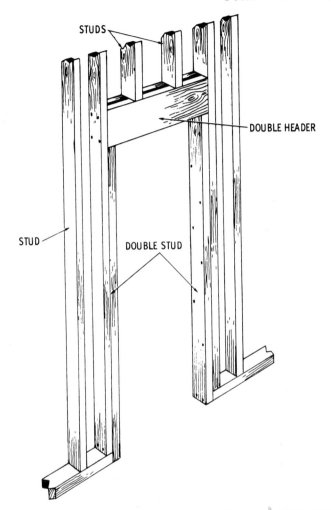

STUDS

DOUBLE HEADER

STUD

DOUBLE STUD

Fig. 15-9. Construction of headers when openings over 30 in. between studs appear in partitions or outside walls.

SHEETROCKING WALLS

In general, you should install drywall from the top down. This reduces the amount of taping to be done; you will have short joints instead of long ones.

311

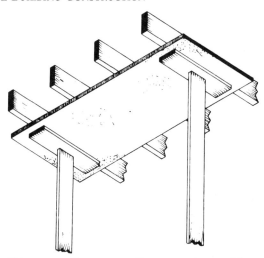

Fig. 15-10. T-braces used to temporarily support sheets of drywall material on the ceiling.

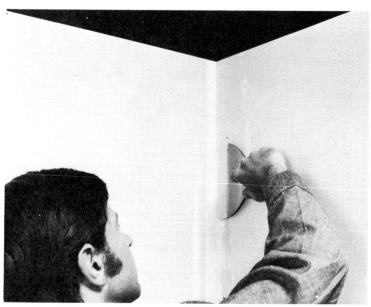

Courtesy National Gypsum Products

Fig. 15-11. Joint tape is applied over bed of compound. Successive coats of compound are applied, smoothed, feathered out gradually.

312

We favor a horizontal application of drywall because the joints are at waist height rather than higher up. Mark stud locations on the ceiling before covering walls; you will know where they are for nailing.

First, pick up a panel, hike it into position, then tack it in place to hold it there. Now nail it home. Put in the panel below it in the same way. (You will, of course, have to cut the panels to fit.)

TAPING

To tape, use a broad-edged knife. Apply the material as indicated in Fig. 15-11. First, apply a coat about 16 in. wide. Into this set the tape, smoothing it in place with the knife. When dry, apply a 12-in.-wide coat, let it dry, then finish with an 18-in. coat. Also, use the joint cement over the nailheads. Together, all coats should be no more than $\frac{1}{16}$ in. thick.

A good taping job should not require sanding. But if it does, sand it after priming. The paint will reveal any fuzziness that must be removed. (For a highly detailed discussion of installing drywall, see Chapter 22.)

Roof Framing

In order to be accurate and reasonably proficient in the many phases of carpentry, particularly in roof framing and stair building, a good understanding of how to use the steel square is necessary. This tool is invaluable to the carpenter in roof framing; the reader is urged to purchase a quality square and thoroughly study the instructions for its use. A knowledge of how to use the square is assumed here.

TYPES OF ROOFS

There are many forms of roofs and a great variety of shapes. The carpenter and the student, as well as the architect, should be familiar with the names and features of each of the various types.

Shed or Lean-to Roof

This is the simplest form of roof (shown in Fig. 16-1), and is usually employed for small sheds and out buildings. It has a single slope and is not a thing of beauty.

Saw-Tooth Roof

This is a development of the shed or lean-to roof, being virtually a series of lean-to roofs covering one building, as in Fig. 16-2. It is used on factories, principally because of the extra light which may be obtained through windows on the vertical sides.

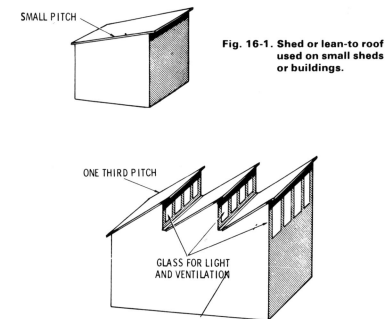

SMALL PITCH

Fig. 16-1. Shed or lean-to roof used on small sheds or buildings.

ONE THIRD PITCH

GLASS FOR LIGHT AND VENTILATION

Fig. 16-2. A saw-tooth roof used on factories for light and ventilation.

Gable or Pitch Roof

This is a very common, simple, and efficient form of roof, and is used extensively on all kinds of buildings. It is of triangular section,

316

having two slopes meeting at the center or *ridge* and forming a *gable,* as in Fig. 16-3. It is popular because of the ease of construction, economy, and efficiency.

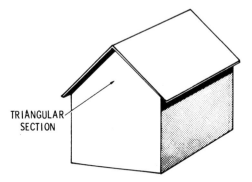

Fig. 16-3. Gable or pitch roof that can be used on all buildings.

Gambrel Roof

This is a modification of the gable roof, each side having two slopes, as shown in Fig. 16-4.

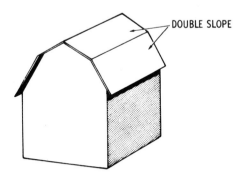

Fig. 16-4. Gambrel roof used on barns.

Hip Roof

A hip roof is formed by four straight sides, all sloping toward the center of the building, and terminating in a ridge instead of a deck, as in Fig. 16-5.

317

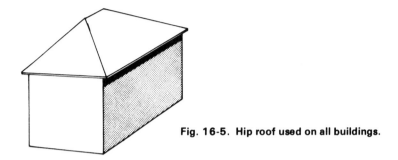

Fig. 16-5. Hip roof used on all buildings.

Pyramid Roof

A modification of the hip roof in which the four straight sides sloping toward the center terminate in a point instead of a ridge as in Fig. 16-6. The pitch of the roof on the sides and ends is different. This construction is not often used.

Hip-and-Valley Roof

This is a combination of a hip roof and an intersecting gable roof covering a T- or L-shaped building, as in Fig. 16-7, and so called because both hip and valley rafters are required in its construction. There are many modifications of this roof. Usually the intersection

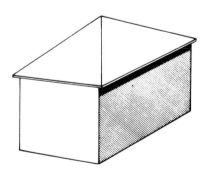

Fig. 16-6. Pyramid roof, which is not often used.

318

is at right angles, but it need not be; either ridge may rise above the other and the pitches may be equal or different, thus giving rise to an endless variety, as indicated in Fig. 16-7.

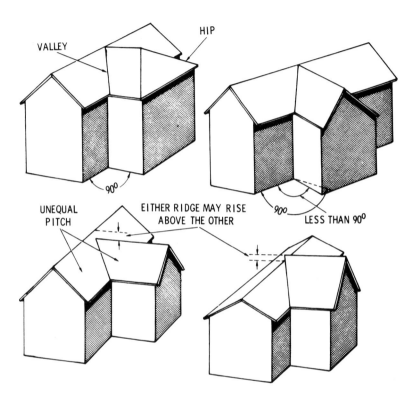

Fig. 16-7. Various hip and valley roofs.

Double-Gable Roof

This is a modification of a gable or a hip-and-valley roof in which the extension has two gables formed at its end, making an M-shape section, as in Fig. 16-8.

Ogee Roof

A pyramidal form of roof having steep sides sloping to the

319

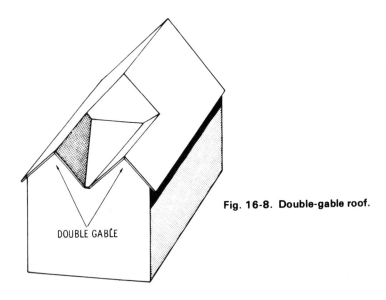

DOUBLE GABLE

Fig. 16-8. Double-gable roof.

center, each side being ogee-shaped, lying in a compound hollow and round curve, as in Fig. 16-9.

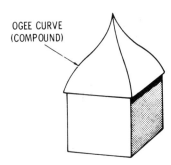

OGEE CURVE
(COMPOUND)

Fig. 16-9. Ogee roof.

Mansard Roof

The straight sides of this roof slope very steeply from each side of the building toward the center, and the roof has a nearly flat deck on top, as in Fig. 16-10. It was introduced by the architect whose name it bears.

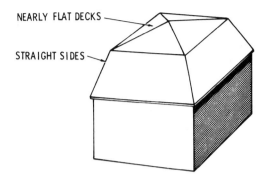

Fig. 16-10. Mansard roof.

French or Concave Mansard Roof

This is a modification of the Mansard roof, its sides being concave instead of straight, as in Fig. 16-11.

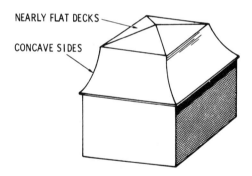

Fig. 16-11. French or concave Mansard roof.

Conical Roof or Spire

A steep roof of circular section which tapers uniformly from a circular base to a central point. It is frequently used on towers, as in Fig. 16-12.

Dome

A hemispherical form of roof (Fig. 16-13) used chiefly on observatories.

321

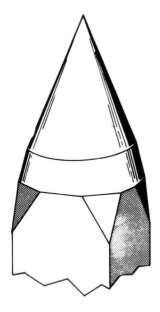

Fig. 16-12. Conical or spire roof.

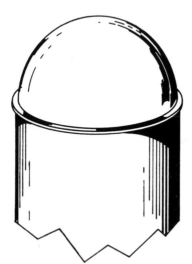

Fig. 16-13. Dome roof.

ROOF CONSTRUCTION

The frame of most roofs is made up of timbers called rafters. These are inclined upward in pairs, their lower ends resting on the top plate, and their upper ends being tied together with a ridge board. On large buildings, such frame work is usually reinforced by interior supports to avoid using abnormally large timbers.

The primary object of a roof in any climate is to keep out the rain and the cold. The roof must be sloped to shed water. Where heavy snows cover the roof for long periods of time, it must be constructed more rigidly to bear the extra weight. Roofs must also be strong enough to withstand high winds.

The most commonly used types of roof construction include.

1. Gable.
2. Lean-to or shed.
3. Hip.
4. Gable and valley.

Terms used in connection with roofs are, *span, total rise, total run, unit of run, rise in inches, pitch, cut of roof, line length,* and *plumb and level lines.*

Span—The *span* of any roof is the shortest distance between the two opposite rafter seats. Stated in another way, it is the measurement between the outside plates, measured at right angles to the direction of the ridge of the building.

Total Rise—The *total rise* is the vertical distance from the plate to the top of the ridge.

Total Run—The term *total run* always refers to the level distance over which any rafter passes. For the ordinary rafter, this would be one-half the span distance.

Unit of Run—The unit of measurement, 1 ft. or 12 in. is the same for the roof as for any other part of the building. By the use of this common unit of measurement, the framing square is employed in laying out large roofs.

Rise in Inches—The *rise in inches* is the number of inches that a roof rises for every foot of run.

Pitch—*Pitch* is the term used to describe the amount of slope of a roof.

Cut of Roof—The *cut of a roof* is the *rise in inches* and *the unit of run* (12 in.).

Line Length—The term *line length* as applied to roof framing is the hypotenuse of a triangle whose base is the total run and whose altitude is the total rise.

Plumb and Level Lines—These terms have reference to the direction of a line on a rafter and not to any particular rafter cut. Any line that is vertical when the rafter is in its proper position is called a *plumb line*. Any line that is level when the rafter is in its proper position is called a *level line*.

RAFTERS

Rafters are the supports for the roof covering and serve in the same capacity as joists do for the floor or studs do for the walls. According to the expanse of the building, rafters vary in size from ordinary 2×4s to 2×10s. For ordinary dwellings, 2×6 rafters are used, spaced from 16 to 24 in. on centers.

The various kinds of rafters used in roof construction are:

1. Common.
2. Hip.
3. Valley.
4. Jack (hip, valley, or cripple).
5. Octagon.

The carpenter should thoroughly know these various types of rafters, and be able to distinguish each kind as they are briefly described.

Common Rafters

A rafter extending at right angles from plate to ridge, as shown in Fig. 16-14.

Hip Rafter

A rafter extending diagonally from a corner of the plate to the ridge, as shown in Fig. 16-15.

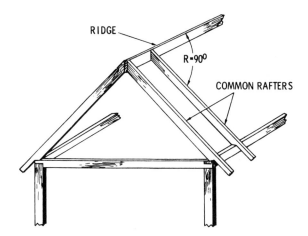

Fig. 16-14. Common rafters.

Valley Rafter

A rafter extending diagonally from the plate to the ridge at the intersection of a gable extension and the main roof (Fig. 16-16).

Jack Rafter

Any rafter which does not extend from the plate to the ridge.

Hip Jack Rafter

A rafter extending from the plate to a hip rafter, and at an angle of 90° to the plate, as shown in Fig. 16-15.

Valley Jack Rafter

A rafter extending from a valley rafter to the ridge and at an angle of 90° to the ridge, as shown in Fig. 16-16.

Cripple Jack Rafter

A rafter extending from a valley rafter to hip rafter and at an angle of 90° to the ridge, as shown in Fig. 16-17.

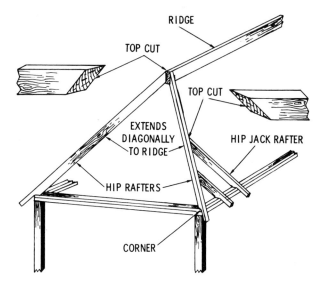

Fig. 16-15. Hip roof rafters.

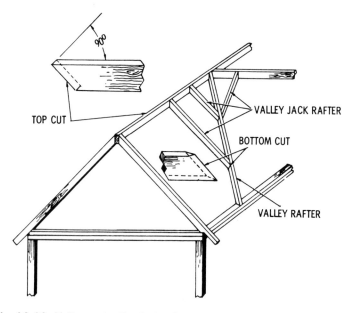

Fig. 16-16. Valley and valley jack rafters.

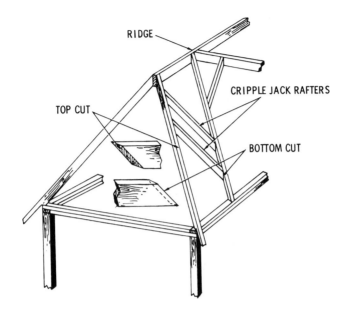

Fig. 16-17. Cripple jack rafters.

Octagon Rafter

Any rafter extending from an octagon-shaped plate to a central apex, or ridge pole.

A rafter usually consists of a main part or rafter proper, and a short length called the *tail*, which extends beyond the plate. The rafter and its tail may be all in one piece, or the tail may be a separate piece nailed on to the rafter.

LENGTH OF RAFTER

The length of a rafter may be found in several ways:

1. By calculation.
2. With steel framing square.
   a. Multi-position method.
   b. By scaling.
   c. By aid of the framing table.

Example—What is the length of a common rafter having a run of 6 feet and rise of 4 inches per foot?

1. *By calculation* (See Fig. 16-18)

The total rise = 6 × 4 = 24 in. = 2 ft.

Since the edge of the rafter forms the hypotenuse of a right triangle whose other two sides are the run and rise, then the length of the rafter = $\sqrt{run^2 + rise^2}$ = $\sqrt{6^2 + 2^2}$ = $\sqrt{40}$ = 6.33 ft., as illustrated in Fig. 16-18.

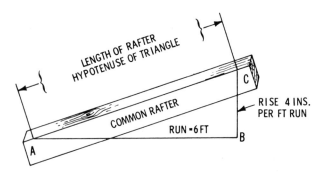

Fig. 16-18. Method of finding the length of a rafter by calculation.

Practical carpenters would not consider it economical to find rafter lengths in this way, because it takes too much time and there is chance of errors. It is to avoid both of these objections that the *framing square* has been developed.

2. *With steel framing square*

The steel framing square considerably reduces the mental effort and chances of error in finding rafter lengths. An approved method of finding rafter lengths with the square is by the aid of the rafter table included on the square for that purpose. However, some carpenters may possess a square which does not have rafter tables. In such case, the rafter length can be found either by the *multi-position* method shown in Fig. 16-19, or by *scaling* as in Fig. 16-20. In either of these methods, the measurements should be made with care because, in the *multi-position* method, a separate measurement must be made for each foot run with a chance for error in each measurement. The following refers to Fig. 16-19.

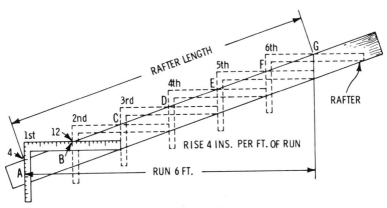

Fig. 16-19. Multi-position method of finding rafter length.

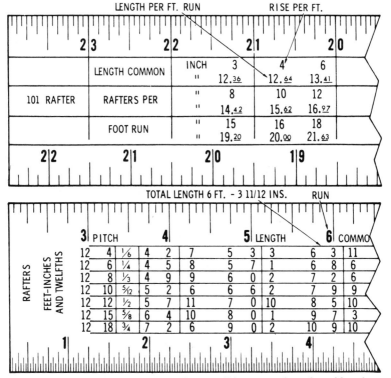

Fig. 16-20. Rafter table readings of two well-known makes of steel framing squares.

Problem: Lay off the length of a common rafter having a run of 6 ft. and a rise of 4-in. per ft. Locate a point *A* on the edge of the rafter, leaving enough stock for a lookout, if any is used. Place the steel framing square so that division 4 coincides with *A*, and 12 registers with the edge of *B*. Evidently, if the run were 1-ft., distance *AB* thus obtained would be the length of the rafter *per foot run*. Apply the square six times for the 6-ft. run, obtaining points *C*, *D*, *E*, *F*, and *G*. The distance *AG*, then, is the length of the rafter for a given run.

Fig. 16-20 shows readings of rafter tables of two well-known makes of squares for the length of the rafter in the preceding example, one giving the length per foot run, and the other the total length for the given run.

Problem 1: *Given the rise per ft. in inches.* Use two squares, or a square and a straightedge scale, as shown in Fig. 16-21. Place the straightedge on the square so as to be able to read the length of the diagonal between the rise of 4-in. on the tongue and the 1-ft. (12-in.) run on the body as shown. The reading is a little over 12 inches. To find the fraction, place dividers on 12 and a point *A*, as in Fig. 16-22. Transfer to the hundredths scale and read .65, as in Fig. 16-23, making the length of the rafter 12.65 in. *per ft. run*, which for a 6-ft. run =

$$\frac{12.65 \times 6}{12} = 6.33 \text{ ft.}$$

Problem 2: *Total rise and run given in feet.* Let each inch on the tongue and body of the square = 1 ft. The straightedge should be divided into inches and 12ths of an inch so that on a scale, 1 in. = 1 ft. Each division will therefore equal 1 in. Read the diagonal length between the numbers representing the run and rise (12 and 4), taking the whole number of inches as

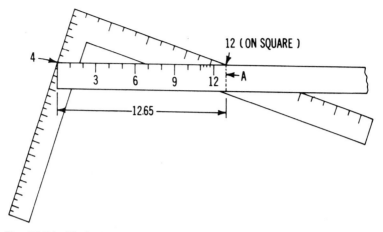

12 (ON SQUARE)

Fig. 16-21. Method of finding rafter length by scaling.

Fig. 16-22 Reading the straightedge in combination
with the carpenter's
square.

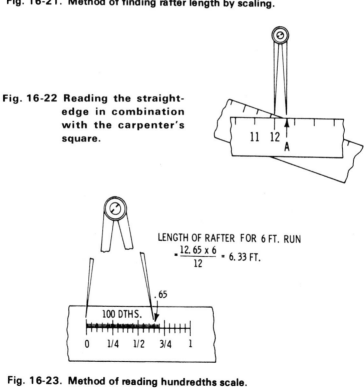

LENGTH OF RAFTER FOR 6 FT. RUN

$$= \frac{12.65 \times 6}{12} = 6.33 \text{ FT.}$$

Fig. 16-23. Method of reading hundredths scale.

feet, and the fractions as inches. Transfer the fraction with dividers and apply the 100th scale, as was done in Problem 1, Figs. 16-22 and 16-23.

In estimating the total length of stock for a rafter having a tail, the run of the tail or length of the lookout must of course be considered.

RAFTER CUTS

All rafters must be cut to the proper angle or bevel at the points where they are fastened and, in the case of overhanging rafters, also at the outer end. The various cuts are known as:

1. Top or plumb.
2. Bottom, seat, or heel.
3. Tail or lookout.
4. Side, or cheek.

Common Rafter Cuts

All of the cuts for common rafters are made at right angles to the sides of the rafter; that is, not beveled as in the case of jacks. Fig. 16-24 shows a common rafter from which the nature of two of these various cuts are seen.

In laying out cuts for common rafters, one side of the square is always placed on the edge of the stock at 12, as shown in Fig. 16-24. This distance 12 corresponds to 1 ft. of the run; the other side of the square is set with the edge of the stock to the rise in inches *per foot run*. This is virtually a repetition of Fig. 16-19, but it is very important to understand why one side of the square is set to 12 for common rafters—not simply to know that 12 must be used. On rafters having a full tail, as in Fig. 16-25B, some carpenters do not cut the rafter tails, but wait until the rafters are set in place so that they may be lined and cut while in position. Certain kinds of work permit the ends to be cut at the same time the remainder of the rafter is framed.

The method of handling the square in laying out the bottom and lookout cuts is shown in Fig. 16-26. In laying out the top or plumb cut, if there is a ridge board, one-half of the thickness of the ridge

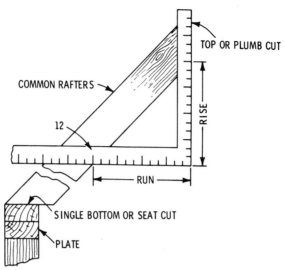

Fig. 16-24. Placement of steel square for proper layout of plumb cut; seat cut has already been laid out and made, using the opposite leg of the square.

must be deducted from the rafter length. If a lookout or a tail cut is to be vertical, place the square at the end of the stock with the rise and run setting as shown in Fig. 16-26, and scribe the cut line *LF*. Lay off *FS* equal to the length of the lookout, and move the square up to *S* (with the same setting) and scribe line *MS*. On this line, lay off *MR*, the length of the vertical side of the bottom cut. Now apply the same setting to the bottom edge of the rafter, so that the edge of the square cuts *R*, and scribe *RN*, which is the horizontal side line of the bottom cut. In making the bottom cut, the wood is cut out to the lines *MR* and *RN*. The lookout and bottom cuts are shown made in Fig. 16-25B, *RN* being the side which rests on the plate, the *RM* the side which touches the outer side of the plate.

Hip-and-Valley Rafter Cuts

The hip rafter *lies in the plane of the common rafters* and forms *the hypotenuse of a triangle*, of which one leg is the adjacent common rafter and the other leg is the portion of the plate intercepted between the feet of the hip and common rafters, as in Fig. 16-27.

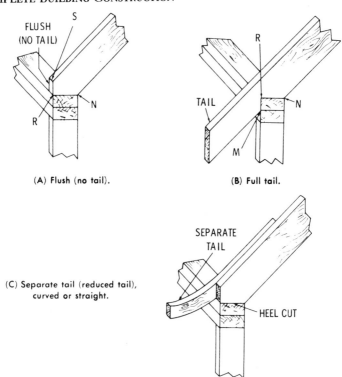

FLUSH
(NO TAIL)

S

R

N

R

TAIL

N

M

(A) Flush (no tail).

(B) Full tail.

SEPARATE
TAIL

(C) Separate tail (reduced tail),
curved or straight.

HEEL CUT

Fig. 16-25. Various common rafter nails.

Problem: In Fig. 16-27, take the run of the common rafter as 12, which may be considered as 1 ft. (12 in.) of the run or the total run of 12 ft. ($\frac{1}{2}$ the span). Now for 12 ft., intercept on the plate the hip run inclined 45° to the common run, as in the triangle ABC. Thus, $AC^2 = \sqrt{AB^2 + BC^2} = \sqrt{12^2 + 12^2} = 16.97$, or approximately 17. Therefore, the run of the hip rafter is to the run of the common rafter as 17 is to 12. Accordingly, in laying out the cuts, use figure 17 on one side of the square and the given rise in *inches per foot* on the other side. This also holds true for top and bottom cuts of the valley rafter when the plate intercept AB = the run BC of the common rafter.

The line of measurement for the length of a hip and valley rafter is along the middle of the back or top edge, as on common and jack rafters. The rise is the same as that of a common rafter, and the run of a hip rafter is the horizontal distance from the plumb line of its rise to the outside of the plate at the foot of the hip rafter, as shown in Fig. 16-28.

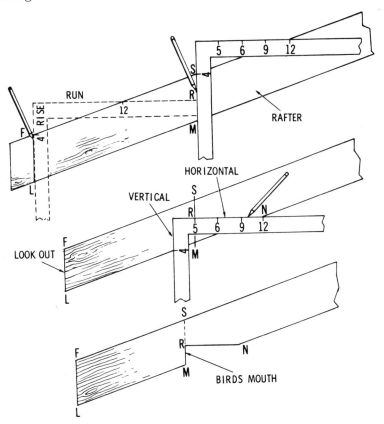

Fig. 16-26. Method of using the square in laying out the lower or end cut of the rafter.

In applying the source for cuts of hip or valley rafters, *use the distance 17 on the body of the square the same way as 12 was used for common rafters.* When the plate distance between hip and common rafters is equal to half the span or to the run of the

335

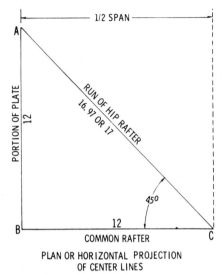

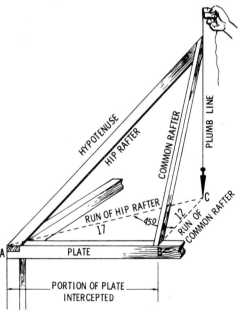

Fig. 16-27. View of hip and common rafters in respect to each other.

common rafter, the line of run of the hip will lie at 45° to the line of run of the common rafter, as indicated in Fig. 16-27.

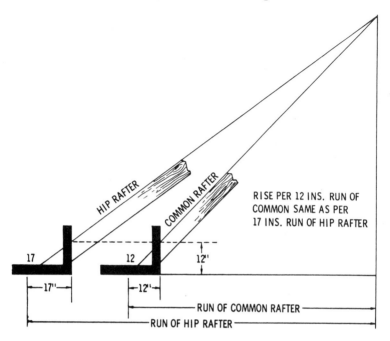

RISE PER 12 INS. RUN OF
COMMON SAME AS PER
17 INS. RUN OF HIP RAFTER

Fig. 16-28. Hip and common rafters shown in the same plane, illustrating the use of 12 for the common rafter and 17 for the hip rafter.

The length of a hip rafter, as given in the framing table on the square, is the distance from the ridge board to the outer edge of the plate. In practice, deduct from this length one-half the thickness of the ridge board, and add for any projection beyond the plate for the eave. Fig. 16-29A shows the correction for the table length of a hip rafter to allow for a ridge board, and Fig. 16-29B shows the correction at the plate end. Fig. 16-30 shows the correction at the plate end of a valley rafter.

The table length, as read from the square, must be reduced an amount equal to *MS*. This is equal to the hypotenuse (*ab*) of the little triangle abc, which in value =

$$\sqrt{ac^2 + bc^2} = \sqrt{ac^2 + (\text{half thickness of ridge})^2}$$

337

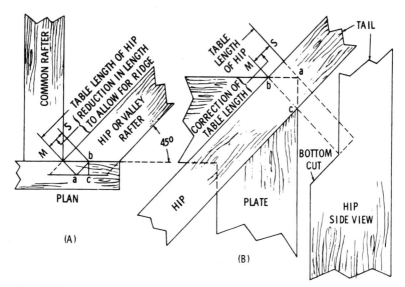

Fig. 16-29. Correction in table length of hip to allow for half-thickness of ridge board.

In ordinary practice, take *MS as equal to* half the thickness of the ridge. The plan and side view of the hip rafter shows the table length and the correction *MS*, which must be deducted from the table length so that the sides of the rafter at the end of the bottom cut will intersect the outside edges of the plate. The table length of the hip rafter, as read on the framing square, will cover the span from the ridge to the outside cover *a* of the plate, but the side edges of the hip intersect the plates at *b* and *c*. The distance that *a* projects beyond a line connecting *bc* or *MS*, must be deducted; that is, measured backward toward the ridge end of the hip. In making the bottom cut of a valley rafter, it should be noted that a valley rafter differs from a hip rafter in that the correction distance for the table length must be *added* instead of subtracted, as for a hip rafter. A distance *MS* was subtracted from the table length of the hip rafter in Fig. 16-29B, and an equal distance (*LF*) was added for the valley rafter in Fig. 16-30.

After the plumb cut is made, the end must be mitered outward for a hip, as in Fig. 16-31, and inward for a valley, as in Fig. 16-32, to receive the *fascia*. A *fascia* is the narrow vertical member

338

fastened to the outside ends of the rafter tails. Other miter cuts are shown with full tails in Fig. 16-33, which also illustrates the majority of cuts applied to hip and valley rafters.

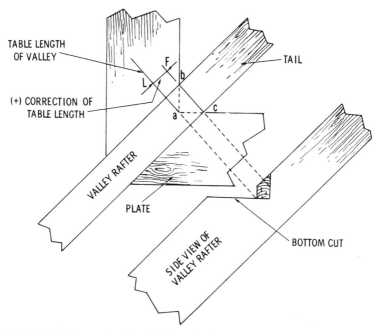

Fig. 16-30. Correction in table length of valley rafter to allow for half-thickness of ridge. Correction is added, not subtracted.

Side Cuts of Hip and Valley Rafters

These rafters have a side or cheek cut at the ridge end. In the absence of a framing square, a simple method of laying out the side cut for a 45° hip or valley rafter is as follows:

Measure back on the edge of the rafter from point A of the top cut, as shown at left in Fig. 16-34. Distance AC is equal to the thickness of the rafter. Square across from C to B on the opposite edge, and scribe line AB, which gives the side cut. At right in Fig. 16-34, FA is the top cut, and AB is the side cut. The plumb and side cuts should be made at the same time by sawing along lines FA and AB, in order to save extra labor.

This rule for laying out hip side cuts does not hold for any angle other than 45°.

BACKING OF HIP RAFTERS

By definition, the term *backing* is the bevel upon the top side of a hip rafter which allows the roofing boards to fit the top of the rafter without leaving a triangular hole between it and the back of the roof covering. The height of the hip rafter, measured on the outside surface vertically upward from the outside corner of the plate, will be the *same as that of the common rafter measured from the same line*, whether the hip is backed or not. This is not true for an unbacked valley rafter when the measurement is made at the center of the timber.

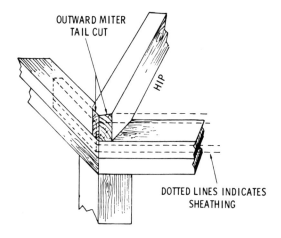

OUTWARD MITER
TAIL CUT

HIP

DOTTED LINES INDICATES
SHEATHING

Fig. 16-31. Flush hip rafter miter cut.

The graphical method of finding the backing of hip rafters is shown in Fig. 16-35. Let *AB* be the span of the building, and *OD* and *OC* the plan of two unequal hips. Lay off the given rise as shown. Then *DE* and *CF* are the lengths of the two unequal hips. Take any point, such as *G* on *DE*, and erect a perpendicular cutting *DF* at *H*. Resolve *GH* to *J*, that is, make *HJ = GH*, draw *NO* perpendicular to *OD* and through *H*. Join *J* to *N* and *O*, giving a

340

bevel angle *NJO*, which is the backing for rafter *DE*. Similarly, the bevel angle *NJO* is found for the backing of rafter *CF*.

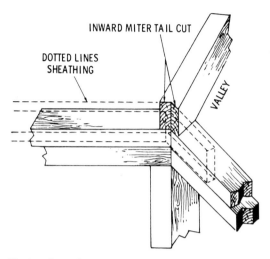

Fig. 16-32. Flush valley miter cut.

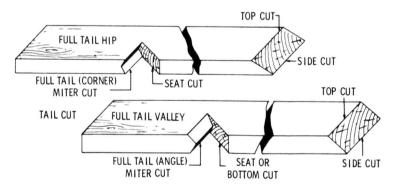

Fig. 16-33. Flush tail hip and valley rafters showing all cuts.

JACK RAFTERS

As outlined in the classification, there are several kinds of jack rafters as distinguished by their relation with other rafters of the roof. These various jack rafters are known as:

1. Hip jacks.
2. Valley jacks.
3. Cripple jacks.

The distinction between these three kinds of jack rafters, as shown in Fig. 16-36, is as follows: *Rafters which are framed between a hip rafter and the plate are* hip jacks; *those framed between the ridge and a valley rafter are* valley jacks; *those framed between hip and valley rafters are* cripple jacks.

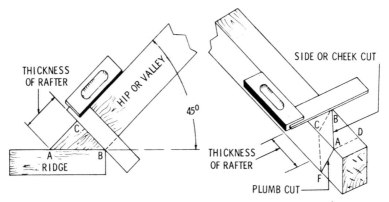

Fig. 16-34. A method of obtaining a side cut of 45° hip or valley rafter without the aid of a framing square.

The term cripple is applied because the ends or *feet* of the rafters are cut off—the rafter does not extend the full length from ridge to plate. From this point of view, a valley jack is sometimes erroneously called cripple; it is virtually a semi-cripple rafter, but confusion is avoided by using the term cripple for rafters framed between the hip and valley rafters, as above defined.

Jack rafters are virtually *discontinuous common rafters.* They are cut off by the intersection of a hip or valley, or both, before reaching the full length from plate to ridge. Their lengths are found in the same way as for common rafters—the number 12 being used on one side of the square and the rise in inches per foot run on the other side. This gives the length of jack rafter per foot run, and is true for all jacks—hip, valley, and cripple.

In actual practice, carpenters usually measure the length of hip or valley jacks from the long point to the plate or ridge, instead of

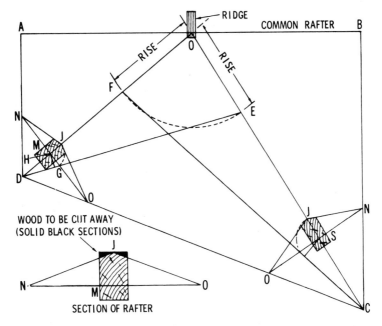

Fig. 16-35. Graphical method of finding length of rafters and backing of hip rafters.

along the center of the top, no reduction being made for one-half the diagonal thickness of the hip or valley rafter. Cripples are measured from long point to long point, no reduction being made for the thickness of the hip or valley rafter.

As no two jacks are of the same length, various methods of procedure are employed in framing, as:

1. Beginning with shortest jack.
2. Beginning with longest jack.
3. Using framing table.

Shortest Jack Method

Begin by finding the length of the shortest jack. Take its spacing from the corner, measured on the plates, which in the case of a 45° hip is equal to the jacks run. The length of this first jack will be the *common difference* which must be added to each jack to get the length of the next longer jack.

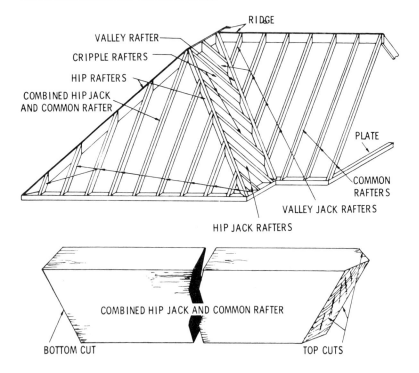

Fig. 16-36. A perspective view of hip and valley roof showing the various jack rafters, and enlarged detail of combined hip jack and common rafters showing cuts.

Longest Jack Method

Where the longest jack is a full-length rafter (that is, a common rafter), first find the length of the longest jack, then count the spaces between jacks and divide the length of the longest jack by number of spaces. The quotient will be the common difference. Then frame the longest jack and make each jack shorter than the preceding jack by this common difference.

Framing Table Method

On various steel squares, there are tables giving the length of the shortest jack rafters corresponding to the various spacings, such as 16, 20, and 24 in. between centers for the different pitches. This length is also the common difference and thus serves for obtaining the length of all the jacks.

Example—Find the length of the shortest jack or the common difference in the length of the jack rafters, where the rise of the roof is 10 in. per foot and the jack rafters are spaced 16 in. between centers; also, when spaced 20 in. between centers. Fig. 16-37 shows the reading of the jack table on one square for 16-in. centers, and Fig. 16-38 shows the reading on another square for 20-in. centers.

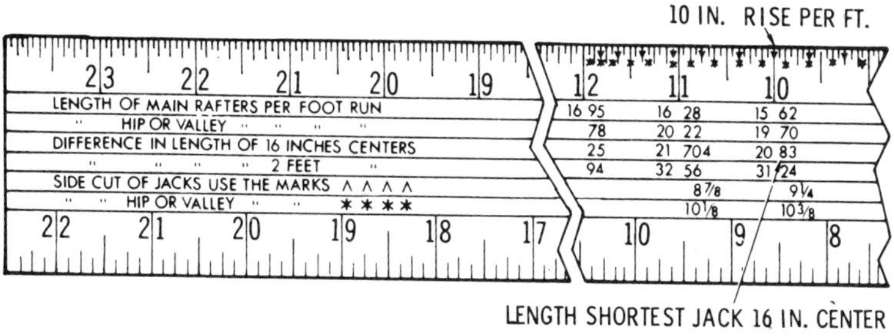

Fig. 16-37. Square showing table for shortest jack rafter at 16 in. on center.

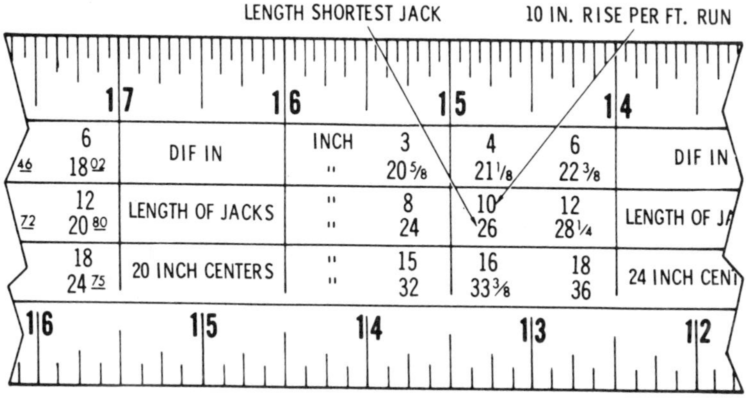

Fig. 16-38. Square showing table for shortest jack rafter at 20 in. on center.

Jack-Rafter Cuts

Jack rafters have top and bottom cuts which are laid out the same as for common rafters, and also side cuts which are laid out

345

the same as for a hip rafter. To lay off the top or plumb cut with a square, take 12 on the body and the rise in inches (of common rafter) per foot run on the tongue, and mark along the tongue as in Fig. 16-39. The following example illustrates the use of the framing square in finding the side cut.

Example—Find the side cut of a jack rafter framed to a 45° hip or valley for a rise of 8 inches per foot run. Fig. 16-40 shows the reading on the jack side-cut table of the framing square, and Fig.

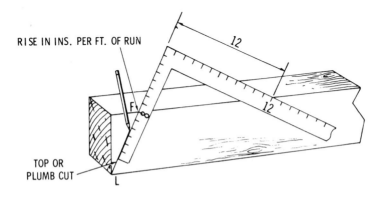

RISE IN INS. PER FT. OF RUN

12

12

F

TOP OR
PLUMB CUT

L

Fig. 16-39. Method of finding plumb and side cuts of jack framed to 45° hip or valley.

JACK SIDE CUT   8 IN. RISE PER FT. RUN

11			10			9			8	
4 25¼	6 26⅞	FIG'S GIVING	INCH ''	3 7¾	8	4 99¼	6 9 10	FIG'S GIVIN		
10 31¼	12 34	SIDE CUT	'' ''	8 10 12		10 10 13	12 12 17	SIDE CUT OF HIP ON		
16 40	12 43¼	OF JACKS	'' ''	15 10 16		16 9 15	18 10 18	VALLEY RAFTER		

1	0	9	8	7	6

Fig. 16-40. A framing square showing readings for side cut of jack corresponding to 8-in. rise per ft. run.

16-41 shows the method of placing the square on the timber to obtain the side cut. It should be noted that different makers of squares use different setting numbers, but the ratios are always the same.

METHOD OF TANGENTS

The tangent value is made use of in determining the side cuts of jack, hip, or valley rafters. By taking a circle with a radius of 12 in., the value of the tangent can be obtained in terms of the constant of the common rafter run.

Considering rafters with zero pitch, as shown in Fig. 16-42, if the common rafter is 12 ft. long, the tangent MS of a 45° hip is the same length. Placing the square on the hip, setting to 12 on the tongue and 12 on the body will give the side cut at the ridge *when there is no pitch* (at M) as in Fig. 16-43. Placing the square on the jack with the same setting numbers (12, 12) as at S, will give the face cut for the jack when framed to a 45° hip with zero pitch; that is, when all of the timbers lie in the same plane.

OCTAGON RAFTERS

On an octagon or eight-sided roof, the rafters joining the corners are called octagon rafters, and are a little longer than the common rafter and shorter than the hip or valley rafters of a square building of the same span. The relation between the run of an octagon and a common rafter is shown in Fig. 16-44 as being as 13 is to 12. That is, for each foot run of a common rafter, an octagon rafter would have a run of 13 in. Hence, to lay off the top or bottom cut of an octagon rafter, place the square on the timber with the 13 on the tongue and the rise of the common rafter per foot run on the body, as shown graphically in Fig. 16-45. The method of laying out the top and bottom cut with the 13 rise setting is shown in Fig. 16-46.

The length of an octagon rafter may be obtained by scaling the diagonal on the square for 13 on the tongue and the rise in inches per foot run of a common rafter, and multiplying by the number of feet run of a common rafter. The principle involved in determining the amount of backing of an octagon rafter (or rafters of any other polygon) is the same as for hip rafter. The backing is determined by the tangent of the angle whose adjacent side is ½ the rafter thickness and whose angle is equivalent to one-half the central angle.

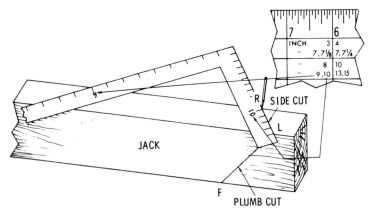

INCH	3	4
"	7,7⅛	7,7¼
"	8	10
"	9,10	13,15

Fig. 16-41. Method of placing framing square on jack to lay off side cut for an 8-in. rise.

PREFABRICATED ROOF TRUSSES

Definite savings in material and labor requirements through the use of preassembled wood roof trusses make truss framing an effective means of cost reduction in small dwelling construction. In a 26' × 32' dwelling, for example, the use of trusses can result in a substantial cost saving and a reduction in use of lumber of almost 30% as compared with conventional rafter and joist construction. In addition to cost savings, roof trusses offer other advantages of increased flexibility for interior planning, and added speed and efficiency in site erection procedures. Today, some 70% of all houses built in the United States incorporate roof trusses.

For many years, trusses were extensively used in commercial and industrial buildings, and were very familiar in bridge construction. In the case of small residential structures, truss construction took a while to catch on, largely because small-house building

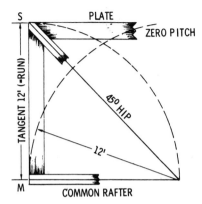

Fig. 16-42. A roof with zero pitch showing the common rafter and the tangent as the same length.

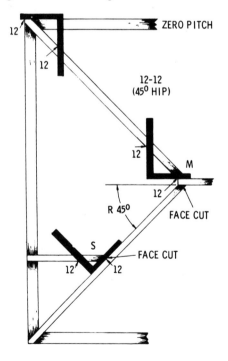

Fig. 16-43. Zero-pitch 45° hip roof showing application of the framing square to give side cuts at ridge.

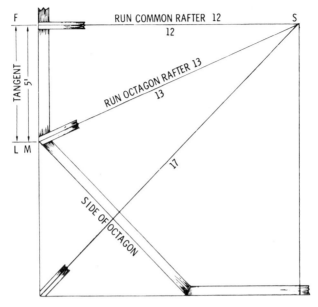

Fig. 16-44. Details of an octagon roof showing relation in length between common and octagon rafters.

has not had the benefit of careful detailing and engineered design that would permit the most efficient use of materials.

During the last several years the truss has been explored and developed for small houses. One of the results of this effort has been the development of light wood trusses which permit substantial savings in the case of lumber. Not only may the framing lumber be smaller in dimension than in conventional framing, but trusses may also be spaced 24 in. on center as compared to the usual 16-in. spacing of rafter and joist construction.

The following figures show percentage savings in the 26′ × 32′ house through the use of trusses 24 in. on center:

Lumber Requirements for Trusses

28.4% less than for conventional framing at 16-in. spacing.

Labor Requirements for Trusses

36.8% less hours than for conventional framing at 16-in. spacing.

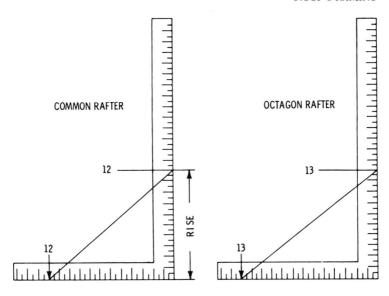

Fig. 16-45. Diagram showing that for equal rise, the run of octagon rafters is 13 in., to 12 in. for common rafters.

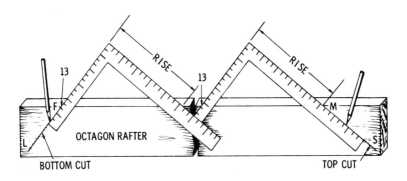

Fig. 16-46. Method of laying off bottom and top cuts of an octagon rafter with a square using the 13 rise setting.

Total Cost of Trusses

29.1% less than conventional framing at 16-in. spacing.

The trusses consisted of 2×4 lumber at top and bottom chords, 1×6 braces, and double 1×6s for struts, with plywood gussets

351

and splices, as shown in Fig. 16-47. The clear span of truss construction permits use of nonbearing partitions so that it is possible to eliminate the extra top plate required for bearing partitions used with conventional framing. It also permits a smaller floor girder to be used for floor construction since the floor does not have to support the bearing partition and help carry the roof load.

Aside from direct benefits of reduced cost and savings in material and labor requirements, roof trusses offer special advantages in helping to speed up site erection and overcome delays due to weather conditions. These advantages are reflected not only in improved construction methods but also in further reductions in cost. With preassembled trusses, a roof can be put over the job quickly to provide protection against the weather.

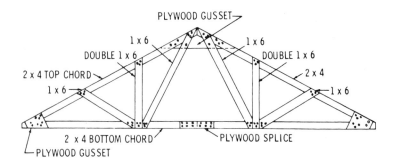

Fig. 16-47. Wood roof truss for small dwellings.

Laboratory tests and field experience show that properly designed roof trusses are definitely acceptable for dwelling construction. The type of truss shown in Fig. 16-47 is suitable for heavy roofing and plaster ceiling finish. In assembling wood trusses, special care should be taken to achieve adequate nailing since the strength of trusses is, to a large extent, dependent on the fastness of the connection between members. Care should also be exercised in selecting materials for trusses. Lumber equal in stress value to No. 2 dimension shortleaf southern pine is suitable; any lower quality is not recommended. Trusses can be assembled with nails and other fasteners, but those put out by such companies as Teco work particularly well.

ROOF VENTILATION

Adequate ventilation is necessary in preventing condensation in buildings. Condensation may occur in the walls, in the crawl space under the structure, in basements, on windows, etc. Condensation is most likely to occur in structures during cold weather when interior humidity is high. Proper ventilation under the roof allows moisture-laden air to escape during the winter heating season, and also allows the hot dry air of the summer season to escape, which will keep the house cooler. The upper areas of a structure are usually ventilated by the use of louvers or ventilators. The various types of ventilators used are as follows:

1. Roof louvers (Fig. 16-48).

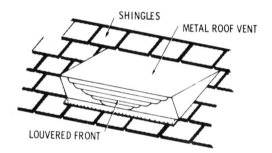

Fig. 16-48. Roof ventilator.

2. Cornice ventilators (Fig. 16-49).

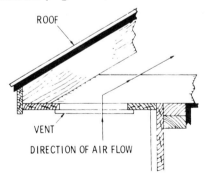

Fig. 16-49. Cornice ventilator.

353

3. Gable louvers (Fig. 16-50).

Fig. 16-50. A typical gable ventilator.

4. Flat-roof ventilators (Fig. 16-51).

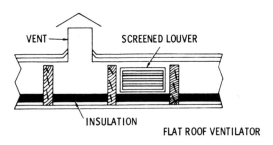

Fig. 16-51. Flat roof vent.

5. Attic fans (Fig. 16-52).

Courtesy Nutone

Fig. 16-52. Roof fan.

6. Ridge ventilators (Fig. 16-53).

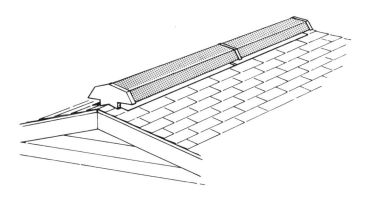

Fig. 16-53. A ridge ventilator.

355

In upper-structure ventilation, one of the most common methods of ventilating is by the use of wood or metal louver frames. The following points should be considered when building or installing the various ventilators.

1. The size and number of ventilators is determined by the size of the area to be ventilated.
2. The minimum net open area should be $\frac{1}{4}$ sq. in. for every square foot of ceiling area.
3. Most wooden louver frames are 5 in. wide and rabbeted out for a screen or door, or both.
4. Three-quarter in. ventilating slots are used and spaced about $1\frac{3}{4}$ in. apart.
5. Sufficient slant or slope to the slots should be provided to prevent rain from driving in.
6. For best operation, upper structure louvers are placed as near the top of the gable roof as possible.

CHAPTER 17

Roofing

A roof includes the roof cover (the upper layer which protects against rain, snow, and wind) or roofing, the sheathing to which it is fastened, and the framing (rafters) which support the other components.

Because of its exposure, roofing usually has a limited life, and so is made to be readily replaceable. It may be made of many widely diversified materials, among which are the following:

1. Wood, usually in the form of shingles (which are uniform, machine-cut) or shakes (which are hand-cut). See Fig. 17-1.
2. Metal or aluminum, which simulates other kinds of roofing.
3. Slate, which may be the natural product or rigid manufactured slabs, often cement asbestos, though these are on the decline since the controversy over asbestos.
4. Tile (Fig. 17-2), which is a burned clay or shale product. Several standard types are available.
5. Built-up covers of asphalt or tar-impregnated felts, with

Fig. 17-1. A wood shingle roof.

    moppings of hot tar or asphalt between the plies and a mopping of tar or asphalt overall. With tar-felt roofs, the top is usually covered with embedded gravel or crushed slag.

6. Roll roofing, which, as the name implies, is marketed in rolls containing approximately 108 sq. ft. Each roll is usually 36 in. wide and may be plain or have a coating of colored mineral granules. The base is a heavy asphalt-impregnated felt.

7. Asphalt shingles (Fig. 17-3), usually in the form of strips with two, three, or four tabs per unit. These shingles are asphalt with the surface exposed to the weather heavily coated with mineral granules. Because of their fire resistance, cost, and durability, asphalt shingles are the most popular roofing material for homes. Asphalt shingles are available in a wide range of colors, including black and white.

8. Glass fiber shingles. These are made partly of a glass fiber mat, which is waterproof, and partly of asphalt. Like asphalt

358

Fig. 17-2. A tile roof. This roof is popular in southwestern states.

shingles, glass fiber shingles come with self-sealing tabs and carry a Class-A fire-resistance warranty. For the do-it-yourselfer they may be of special interest because they are lightweight, about 220 lbs. per square (100 sq. ft. of roofing).

SLOPE OF ROOFS

The slope of the roof is frequently a factor in the choice of roofing materials and method used to put them in place. The lower the pitch of the roof, the greater the chance of wind getting under the shingles and tearing them out. Interlocking cedar shingles resist this wind prying better than standard asphalt shingles. For roofs with less than a 4-in. slope per foot, do not use standard asphalt. Down to 2 in., use self-sealing asphalt. Roll roofing can be

359

used with pitches down to 2 in. when lapped 2 in. For very low pitched slopes, the manufacturers of asphalt shingles recommend that the roof be planned for some other type of covering.

Aluminum strip roofing virtually eliminates the problem of wind prying, but these strips are noisy. Most homeowners object to the noise during a rainstorm. Even on porches, the noise is often annoying to those inside the house.

Fig. 17-3. These asphalt shingles have a three-dimensional look. These shingles are the most popular.

Spaced roofing boards are sometimes used with cedar shingles as an economy measure and because the cedar shingles themselves add considerably to the strength of the roof. The spaced roofing boards reduce the insulating qualities, however, and it is advisable to use a tightly sheathed roof beneath the shingles if the need for insulation overcomes the need for economy.

For drainage, most roofs should have a certain amount of slope. Roofs covered with tar-and-gravel coverings are theoretically satisfactory when built level, but standing water may ultimately do harm. If you can avoid a flat roof, do so. Level roofs drain very slowly; slightly smaller eave troughs and downspouts are used on these roofs. They are quite common on industrial and commercial buildings.

ROLL ROOFING

Roll roofing (Fig. 17-4) is an economical cover especially suited for roofs with low pitches. It is also sometimes used for valley flashing instead of metal. It has a base of heavy asphalt-impregnated felt with additional coatings of asphalt that are dusted to prevent adhesion in the roll. The weather surface may be plain or covered with fine mineral granules. Many different colors are available. One edge of the sheet is left plain (no granules) where the lap cement is applied. For best results, the sheathing must be tight, preferably 1×6 tongue-and-groove, or plywood. If the sheathing is smooth, with no cupped boards or other protuberance, the slate-surfaced roll roofings will withstand a surprising amount of abrasion from foot traffic, although it is not generally recommended for that purpose. Windstorms are the most relentless enemy of roll roofings. If the wind gets under a loose edge, almost certainly a section will be blown off.

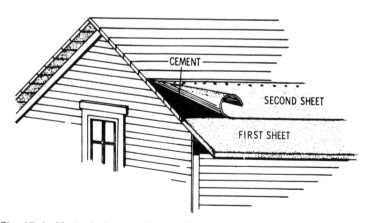

Fig. 17-4. Method of cementing and lapping the first and second strips of roll roofing.

BUILT-UP ROOF

A built-up roof is constructed of: sheathing paper, a bonded base sheet, perforated felt, asphalt, and surface aggregates (Fig. 17-5). The sheathing paper comes in 36-in.-wide rolls and has approximately 500 sq. ft. per roll. It is a rosin-size paper and is used to prevent asphalt leakage to the wood deck. The base sheet is a heavy asphalt saturated felt that is placed over the sheathing paper. It is available in 1, 1½, and 2 square rolls. The perforated felt is one of the primary parts of a built-up roof. It is saturated with asphalt and has tiny perforations throughout the sheet. The perforations prevent air entrapment between the layers of felt. The perforated felt is 36 in. wide and weighs approximately 15 lbs. per square. Asphalt is also one of the basic ingredients of a built-up roof. There are many different grades of asphalt, but the most common are low-melt, medium-melt, high-melt, and extra-high-melt.

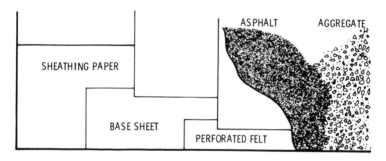

Fig. 17-5. Sectional plan of a built-up roof.

Prior to the application of the built-up roof, the deck should be inspected for soundness. Wood board decks should be constructed of ¾-in. seasoned lumber or plywood. Any knot holes larger than one inch should be covered with sheet metal. If plywood is used as a roof deck it should be placed with the length at right angles to the rafters and be at least ½ in. in thickness.

The first step in the application of a built-up roof is the placing of sheathing paper and base sheet. The sheathing paper should be lapped 2 in. and secured with just enough nails to hold it in place. The base sheet is then placed with 2-in. side laps and 6-in. end laps.

The base sheet should be secured with ½-in. diameter head galvanized roofing nails placed 12 inches on center on the exposed lap. Nails should also be placed down the center of the base sheet. The nails should be placed in two parallel rows 12 in. apart.

The base sheet is then coated with a uniform layer of hot asphalt. While the asphalt is still hot, a layer of roofing felt is placed and mopped with the hot asphalt. Each succeeding layer of roofing felt is placed and mopped in a similar manner with asphalt. Each sheet should be lapped 19 in., leaving 17 in. exposed.

Once the roofing felt is placed, a gravel stop is installed around the deck perimeter, Fig. 17-6. Two coated layers of felt should extend 6 in. past the roof decking where the gravel stop is to be installed. When the other plies are placed, the first two layers are folded over the other layers and mopped in place. The gravel stop is then placed in a ⅛-in.-thick bed of flashing cement and securely nailed every 6 in. The ends of the gravel stop should be lapped 6 in. and packed in flashing cement.

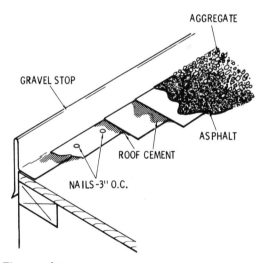

Fig. 17-6. The gravel stop.

After the gravel stop is placed, the roof is flooded with hot asphalt and the surface aggregate is embedded in the flood coat. The aggregates should be hard, dry, opaque, and free of any dust or foreign matter. The size of the aggregates should range from ¼ in. to ⅝ in. When the aggregate is piled on the roof it should be placed on a spot that has been mopped with asphalt. This technique assures proper adhesion in all areas of the roof.

WOOD SHINGLES

The better grades of shingles are made of cypress, cedar, and redwood and are available in lengths of 16 and 18 in. and thicknesses at the butt of ⁵⁄₁₆ and ⁷⁄₁₆ in., respectively. They are packaged in bundles of approximately 200 shingles in random width from 3 to 12 in.

An important requirement in applying wood shingles is that each shingle should lap over the two courses below it, so that there will always be at least three layers of shingles at every point on the roof. This requires that the amount of shingle exposed to the weather (the spacing of the courses) should be less than ⅓ the length of the shingle. Thus in Fig. 17-7, 5½ in. is the maximum amount that 18-in. shingles can be laid to the weather and have an adequate amount of lap. This is further shown in Fig. 17-8.

In case the shingles are laid more than ⅓ of their length to the weather, there will be a space, as shown by MS in Fig. 17-8B, where only two layers of shingles will cover the roof. This is objectionable, because if the top shingle splits above the edge of

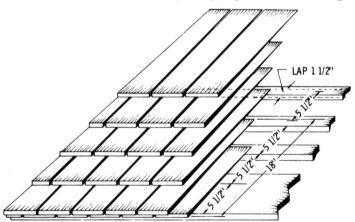

Fig. 17-7. Section of a shingle roof showing the amount of shingle that may be exposed to the weather as governed by the lap.

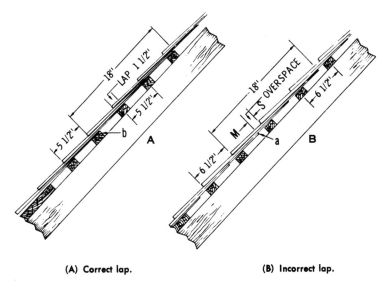

(A) Correct lap.   (B) Incorrect lap.

Fig. 17-8. The amount of lap is an important factor in applying wood shingles.

the shingle below, water will leak through. The maximum spacing to the weather for 16-in. shingles should be 4⅞ in. and for 18-in. shingles should be 5½ in. Strictly speaking, the amount of lap should be governed by the pitch of the roof. The maximum spacing may be followed for roofs of moderate pitch, but for roofs of small pitch, more lap should be allowed, and for a steep pitch the lap may be reduced somewhat, but it is not advisable to do so. Wood shingles should not be used on pitches less than 4 in. per ft.

Table 17-1 shows the number of square feet that 1,000 (five bundles) shingles will cover for various exposures. This table does not allow for waste on hip and valley roofs.

Table 1. Space Covered by 1,000 Shingles

Exposure to weather	4¼	4½	4¾	5	5½	6
Area covered in sq. ft.	118	125	131	138	152	166

Shingles should not be laid too close together, for they will swell when wet, causing them to bulge and split. Seasoned shingles should not be laid with their edges nearer than 3/16 in. when laid by

365

the American method. It is advisable to thoroughly soak the bundles before opening.

Great care must be used in nailing wide shingles. When they are over 8 in. in width, they should be split and laid as two shingles. The nails should be spaced such that the space between them is as small as is practical, thus directing the contraction and expansion of the shingle toward the edges. This lessens the danger of wide shingles splitting in or near the center and over joints beneath. Shingling is always started from the bottom and laid from the eaves or cornice up.

There are various methods of laying shingles, the most common known as:

1. The straight-edge.
2. The chalk-line.
3. The gauge-and-hatchet.

The straight-edge method is one of the oldest. A straight-edge having a width equal to the spacing to the weather or the distance between courses is used. This eliminates measuring, it being necessary only to keep the lower edge flush with the lower edge of the course of shingles just laid; the upper edge of the straight edge is then in line for the next course. This is considered to be the slowest of the three methods.

The chalk-line method consists of snapping a chalk line for each course. To save time, two or three lines may be snapped at the same time, making it possible to carry two or three courses at once. This method is still extensively used. It is faster than the straight-edge method, but not as fast as the gauge-and-hatchet method.

The gauge-and-hatchet method is extensively used in the Western states. The hatchet used is either a lathing or a box-maker's hatchet, as shown in Fig. 17-9. Hatchet gauges to measure the space between courses are shown in Fig. 17-10. The gauge is set on the blade at a distance from the hatchet poll equal to the exposure desired for the shingles.

Nail as close to the butts as possible, if the nails will be well covered by the next course. Only galvanized shingle nails should be used. The 3d shingle nail is slightly larger in diameter than the 3d common nail, and has a slightly larger head.

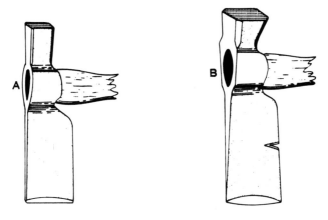

(A) Lathing hatchet.   (B) Box-maker's hatchet.

Fig. 17-9. Hatchets used for shingling.

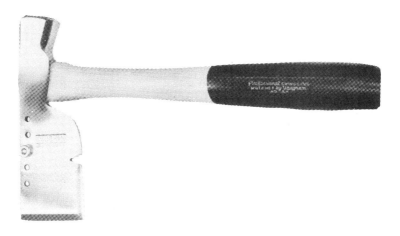

Fig. 17-10. Shingling hatchet.

Hips

The hip is less liable to leak than any other part of the roof as the water runs away from it. However, since it is so prominent, the work should be well done. Fig. 17-11 shows the method of cutting

367

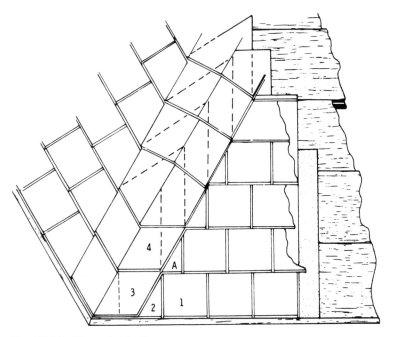

Fig. 17-11. Hip roof shingling.

shingle butts for a hip roof. After the courses 1 and 2 are laid, the top corners over the hip are trimmed off with a sharp shingling hatchet kept keen for that purpose and shingle 3 with the butt cut so as to continue the straight line of courses and again on the dotted line 4, so that shingle A, of the second course squares against it and so on from side to side, each alternately lapping the other at the hip joint. When gables are shingled, this same method may be used up the rake of the roof if the pitch is moderate to steep. It cannot be effectively used with flat pitches. The shingles used should be ripped to uniform width.

For best construction, tin shingles should be laid under the hip shingles, as shown in Fig. 17-12. These tin shingles should correspond in shape to that of hip shingles. They should be at least 7 in. wide and large enough to reach well under the tin shingles of the course above, as at W. At A, the tin shingles are laid so that the lower end will just be covered by the hip shingle of the course above.

368

A variation on the wood shingle is the recently introduced shingle that is part asphalt and part wood composition (Fig. 17-14).

Valleys

In shingling a valley, first a strip of tin, lead, zinc, or copper, ordinarily 20 in. wide, is laid in the valley. Fig. 17-13 illustrates an open type valley. Here the dotted lines show the tin or other material used as flashing under the shingles. If the pitch is above 30°, then a width of 16 in. is sufficient; if flatter, the width should be more. In a long valley, its width between shingles should increase in width from top to bottom about 1 in., and at the top 2 in. is ample width. This is to prevent ice or other objects from wedging when slipping down. The shingles taper to the butt, the reverse of the hip, and need no reinforcing, as the thin edge is held and protected from splitting off by the shingle above it. Care must always be taken to nail the shingle nearest the valley as far from it as practical by placing the nail higher up.

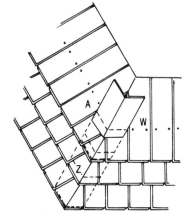

Fig. 17-12. Method of installing metal shingles under wood shingles.

ASPHALT SHINGLES

Asphalt shingles are made in strips of two, three, or four units or tabs joined together, as well as in the form of individual shingles. When laid, strip shingles furnish practically the same pattern as undivided shingles. Both strip and individual types are available in different shapes, sizes, and colors to suit various requirements.

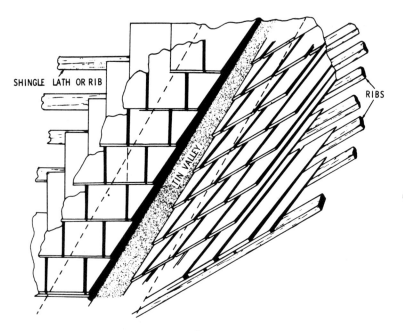

SHINGLE LATH OR RIB

RIBS

TIN VALLEY

Fig. 17-13. Method of shingling a valley.

Asphalt shingles must be applied on slopes having an incline of four inches or more to the foot. Before the shingles are laid, the underlayment should be placed. The underlayment should be 15 lb. asphalt-saturated felt. The underlayment should be placed with 2-in. side laps and 4-in. end laps. Fig. 17-15. The underlayment serves three purposes: (1) it acts as a primary barrier against moisture penetration, (2) it acts as a secondary barrier against moisture penetration, and (3) it acts as a buffer between the resinous areas of the decking and the asphalt shingles. A heavy felt should not be used as underlayment. The heavy felt would act as a vapor barrier and would permit the accumulation of moisture between the underlayment and the roof deck.

The roof deck should be constructed of well-seasoned $1'' \times 6''$ tongue-and-grooved sheathing. The boards should be secured with two 8d nails in each rafter. Plywood sheathing should be

Fig. 17-14. Wood fiber roofing from Masonite is a relatively new product. It is available in fire-rated versions, is bigger than standard shingle.

placed with the long dimension perpendicular to the rafters. The plywood should never be less than ⅜ in. thick.

To efficiently shed water at the roof's edge, a drip edge is usually installed. A drip edge is constructed of corrosion-resistant sheet metal, and extends 3 in. back from the roof edge. To form the drip-edge, the sheet metal is bent down over the roof edges.

The nails used to apply asphalt shingles should be hot-galvanized nails, with large heads, sharp points and barbed shanks. The nails should be long enough to penetrate the roof decking at least ¾ of an inch.

To ensure proper shingle alignment, horizontal and vertical chalk lines should be placed on the underlayment. It is usually recommended that the lines be placed 10 or 20 in. apart. The first course of shingles placed is the starter course. This is used to back up the first regular course of shingles and to fill in the spaces between the tabs. The starter course is placed with the tabs facing up the roof and is allowed to project one inch over the rake and eave, Fig. 17-16. To ensure that all cutouts are covered 3 in. should be cut off the first starter shingle.

371

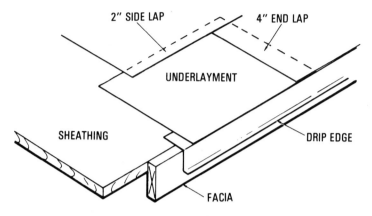

Fig. 17-15. Application of the underlayment.

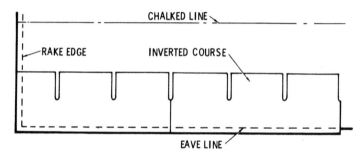

Fig. 17-16. The starter course.

Once the starter course has been placed, the different courses of shingles can be laid. The first regular course of shingles should be started with a full shingle; the second course with a full shingle, minus ½ a tab; the third course is started with a full shingle, Fig. 17-17, and the process is repeated. As the shingles are placed, they should be properly nailed, Fig. 17-18. If a three tab shingle is used, a minimum of 4 nails per strip should be used. The nails should be placed 5⅝ in. from the bottom of the shingle and should be located over the cutouts. The nails on each end of the shingle should be 1 in. from the end. The nails should be driven straight and flush with the surface of the shingle.

If there is a valley in the roof, it must be properly flashed. The two materials that are most often used for valley flashing are 90 lb.

mineral surfaced asphalt roll roofing or galvanized sheet metal. The flashing is 18 inches in width and should extend the full length of the valley. Before the shingles are laid to the valley, chalk lines are placed along the valley. The chalk lines should be 6 in. apart at the top of the valley and should widen ⅛ in. per ft. as they approach the eave line. The shingles are laid up to the chalk lines and trimmed to fit.

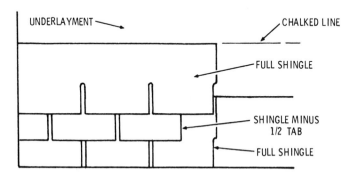

Fig. 17-17. Application of the starter shingles.

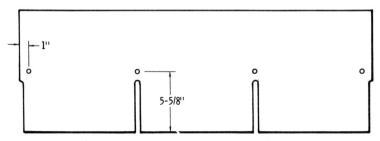

Fig. 17-18. The proper placement of nails.

Hips and ridges are finished by using manufactured hip and ridge units, or hip and ridge units cut from a strip shingle. If the unit is cut from a strip shingle, the two cut lines should be cut at an angle, Fig. 17-19. This will prevent the projection of the shingle past the overlaid shingle. Each shingle should be bent down the center so that there is an equal distance on each side. In cold weather the shingles should be warmed before they are bent. Starting at the bottom of the hip or at the end of a ridge the shingles are placed with a 5 inch exposure. To secure the shingles, a nail is placed on each side of the shingle. The nails should be placed 5½ inches back from the exposed edge and one inch up from the side.

If the roof slope is particularly steep, specifically if it exceeds 60° or 21 in. per ft., then special procedures are required for securing the shingles. This is shown in Fig. 17-20.

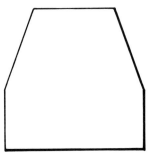

Fig. 17-19. Hip shingles.

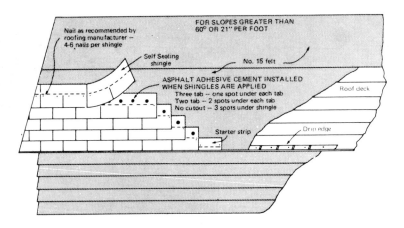

Fig. 17-20. When roof slope exceeds 60°, you have to take special steps in application.

Two other details. For neatness when installing asphalt shingles, the courses should meet in a line above any dormer (Fig. 17-21). And, of course, ventilation must be provided for an asphalt, or really, for any kind of roof, as indicated in Fig. 17-22.

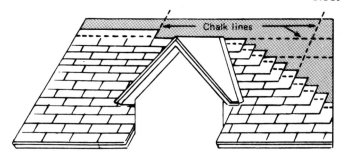

Fig. 17-21. For neatness, shingle courses should meet in a line above dormer.

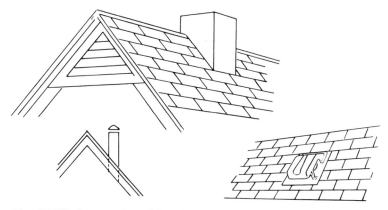

Fig. 17-22. Louvers in gables should be located at opposite ends of the house, as near ridge as possible.

SLATE

Slate is an ideal roofing material and is used on permanent buildings with pitched roofs. The process of manufacture is to split the quarried slate blocks horizontally to a suitable thickness, and to cut vertically to the approximate sizes required. The slates are then passed through planers, and after the operation are ready to be reduced to the exact dimensions on rubbing beds or through the use of air tools and other special machinery.

Roofing slate is usually available in various colors and in standard sizes suitable for the most exacting requirements. On all boarding to be covered with slate, asphalt-saturated rag felt of certain

375

specified thickness is required. This felt should be laid in a horizontal layer with joints lapped toward the eaves and at the ends at least 2 in. A well-secured lap at the end is necessary to properly hold the felt in place, and to protect the structure until covered by the slate. In laying the slate, the entire surface of all main and porch roofs should be covered with slate in a proper and watertight manner.

The slate should project 2 in. at the eaves and 1 in. at all gable ends, and must be laid in horizontal courses with the standard 3-in. headlap. Each course breaks joints with the preceding one. Slates at the eaves or cornice line are doubled and canted $\frac{1}{4}$ in. by a wooden cant strip. Slates overlapping sheet-metal work should have the nails so placed as to avoid puncturing the sheet metal. Exposed nails are.permissible only in courses where unavoidable. Neatly fit the slate around any pipes, ventilators, or other rooftop protuberances.

Nails should not be driven in so far as to produce a strain on the slate. Cover all exposed nail heads with elastic cement. Hip slates and ridge slates are to be laid in elastic cement spread thickly over unexposed surfaces. Build in and place all flashing pieces furnished by the sheeting contractor and cooperate with him in doing the work of flashing. On completion, all slate must be sound, whole, and clean, and the roof left in every respect tight and a neat example of workmanship.

The most frequently needed repair of slate roofs is the replacement of broken slates. When such replacements are necessary, supports similar to those shown in Fig. 17-23 should be placed on the roof to distribute the weight of the roofers while they are working. Broken slates should be removed by cutting or drawing out the nails with a ripper tool. A new slate shingle of the same color and size as the old should be inserted and fastened by nailing through the vertical joint of the slates in the overlying course approximately 2 in. below the butt of the slate in the second course, as shown in Fig. 17-24.

A piece of sheet copper or terneplate about 3″ × 8″ should be inserted over the nail head to extend about 2 in. under the second course above the replaced shingle. The metal strip should be bent slightly before being inserted so that it will stay securely in place. Very old slate roofs sometimes fail because the nails used to fasten the slates have rusted. In such cases, the entire roof covering should be removed and replaced, including the felt underlay materials. The sheathing and rafters should be examined and any broken boards replaced with new material. All loose boards should be nailed in place and, before laying the felt, the sheathing should be swept clean, protruding nails driven in, and any rough edges trimmed smooth.

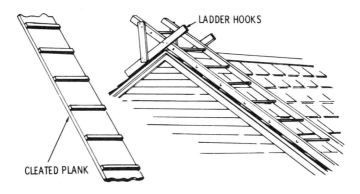

Fig. 17-23. Two types of supports used in repairs of roof.

Fig. 17-24. Method of inserting new pieces of slate shingles.

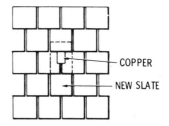

If the former roof was slate, all slates that are still in good condition may be salvaged and relaid. New slates should be the same size as the old ones and should match the original slates as nearly as possible in color and texture. The area to be covered should govern the size of slates to be used and whatever the size, the slates may be of random widths, but they should be of uniform length and punched for a head lap of not less than 3 in. The roof slates should be laid with a 3-in. headlap and fastened with two large-head slating nails. Nails should not be driven too tightly, for

377

the nail heads should barely touch the slate. All slates within 1 ft. of the top and along the gable rakes of the roof should be bedded in flashing cement.

GUTTERS AND DOWNSPOUTS

Most roofs require gutters and downspouts (Fig. 17-25) in order to convey the water to the sewer or outlet. They are usually built of metal. In regions of heavy snow fall, the outer edge of the gutter should be ½ in. below the extended slope of the roof to prevent snow banking on the edge of the roof and causing leaks. The hanging gutter is best adapted to such construction.

Downspouts should be large enough to remove the water from the gutters. A common fault is to make the gutter outlet the same size as the downspout. At 18 in. below the gutter, a downspout has

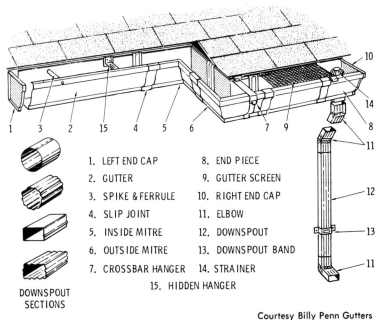

1. LEFT END CAP
2. GUTTER
3. SPIKE & FERRULE
4. SLIP JOINT
5. INSIDE MITRE
6. OUTSIDE MITRE
7. CROSSBAR HANGER

8. END PIECE
9. GUTTER SCREEN
10. RIGHT END CAP
11. ELBOW
12. DOWNSPOUT
13. DOWNSPOUT BAND
14. STRAINER

15. HIDDEN HANGER

DOWNSPOUT SECTIONS

Courtesy Billy Penn Gutters

Fig. 17-25. Various downspouts and fittings.

nearly four times the water-carrying capacity of the inlet at the gutter. Therefore, a good-size ending spout should be provided. Wire baskets or guards should be placed at gutter outlets to prevent leaves and trash from collecting in the downspouts and causing damage during freezing weather.

Gutters come in a variety of materials including wood and metal. Most people favor metal gutters. You can get them in enameled steel or aluminum, with the latter the favorite.

Though it comes in sections, the so-called seamless gutter is easiest to install. A specialist cuts the gutter to the exact lengths needed; no joining of lengths is necessary—and therefore there are no possible leaks.

It should be noted that aluminum gutter is available in various gauges. The .027 size is standard, but .032 is standard in seamless types; .014 is also available, but this should be avoided because it is too flimsy.

SELECTING ROOFING MATERIALS

Roofing materials are commonly sold by dealers or manufacturers on the basis of quantities to cover 100 sq. ft. This quantity is commonly termed "one square" by roofers and in trade literature. When ordering roofing material, it will be well to make allowance for waste such as in hips, valleys, and starter courses. This applies in general to all types of roofing.

The slope of the roof and the strength of the framing are the first determining factors in choosing a suitable covering. If the slope is slight, there will be a danger of leaks with a wrong kind of covering, and excessive weight may cause sagging that is unsightly and adds to the difficulty of keeping the roof in repair. The cost of roofing depends to a great extent on the type of roof to be covered. A roof having ridges, valleys, dormers, or chimneys will cost considerably more to cover than one having a plain surface. Very steep roofs are also more expensive than those with a flatter slope, but most roofing materials last longer on steep grades than on low-pitched roofs. Frequently, nearness to supply centers permits the use, at lower cost, of the more durable materials instead of the commonly lower-priced, shorter-lived ones.

In considering cost, one should keep in mind maintenance and repair and the length of service expected from the building. A permanent structure warrants a good roof, even though the first cost is somewhat high. When the cost of applying the covering is high in comparison with the cost of the material, or when access to the roof is hazardous, the use of long-lived material is warranted. Unless insulation is required, semipermanent buildings and sheds are often covered with low-grade roofing.

Frequently, the importance of fire resistance is not recognized, and sometimes it is wrongly stressed. It is essential to have a covering that will not readily ignite from glowing embers. The building regulations of many cities prohibit the use of certain types of roofings in congested areas where fires may spread rapidly. The Underwriters Laboratories has grouped many of the different kinds and brands of roofing in classes from A to C according to the protection afforded against the spread of fire. Class A is best.

The appearance of a building can be changed materially by using the various coverings in different ways. Wood shingles and slate are often used to produce architectural effects. The roofs of buildings in a farm group should harmonize in color, even though similarity in contour is not always feasible.

The action of the atmosphere in localities where the air is polluted with fumes from industrial works or saturated with salt (as along the seacoast) shortens the life of roofing made from certain metals. Sheet aluminum is particularly vulnerable to acid fumes.

All coal-tar pitch roofs should be covered with slag or a mineral coating, because when fully exposed to the sun, they deteriorate. Observation has shown that, in general, roofings with light-colored surfaces absorb less heat than those with dark surfaces. Considerable attention should be given to the comfort derived from a properly insulated roof. A thin, uninsulated roof gives the interior little protection from heat in summer and cold in winter. Discomfort from summer heat can be lessened to some extent by ventilating the space under the roof. None of the usual roof coverings have any appreciable insulating value. Installing insulation and providing for ventilation are other things considered elsewhere in this book.

DETECTION OF ROOF LEAKS

A well-constructed roof should be properly maintained. Periodic inspections should be made to detect breaks, missing shingles, choked gutters, damaged flashings, and also defective mortar joints of chimneys, parapets, coping, and such. At the first appearance of damp spots on the ceilings or walls, a careful examination of the roof should be made to determine the cause, and the defect should be promptly repaired. When repairs are delayed, small defects extend rapidly and involve not only the roof covering, but also the sheathing, framing, and interior.

Many of these defects can be readily repaired to keep water from the interior and to extend the life of the roof. Large defects or failures should be repaired by people familiar with the work. On many types of roofs, an inexperienced person can do more damage than good. Leaks are sometimes difficult to find, but an examination of the wet spots on a ceiling furnishes a clue to the probable location. In some cases, the actual leak may be some distance up the slope. If near a chimney or exterior wall, the leaks are probably caused by a defective or narrow flashing, loose mortar joints, or dislodged coping. On flat roofs, the trouble may be the result of choked downspouts or an accumulation of water or snow on the roof higher than the flashing. Defective and loose flashing is not uncommon around scuttles, cupolas, and plumbing vent pipes. Roofing deteriorates more rapidly on a south exposure than on a north exposure, which is especially noticeable when wood or composition shingles are used.

Wet spots under plain roof areas are generally caused by holes in the covering. Frequently, the drip may occur much lower down the slope than the hole. Where attics are unsealed and roofing strips have been used, holes can be detected from the inside by light shining through. If a piece of wire is stuck through the hole it can be located from the outside.

Sometimes gutters are so arranged that when choked, they overflow into the house, or ice accumulating on the eaves will form a ridge that backs up melting snow under the shingles. This is a common trouble if roofs are flat and the eaves wide. Leaky downspouts permit water to splash against the wall and the wind-driven water may find its way through a defect into the interior. The exact method to use in repairing depends on the kind of roofing and the nature and extent of the defect.

CHAPTER 18

Cornice Construction

The cornice is the projection of the roof at the eaves that forms a connection between the roof and the side walls. The three general types of cornice construction are the *box*, the *closed*, and the *open*.

BOX CORNICES

The typical box cornice shown in Fig. 18-1 utilizes the rafter projection for nailing surfaces for the facia and soffit boards. The soffit provides a desirable area for inlet ventilators. A frieze board is often used at the wall to receive the siding. In climates where snow and ice dams may occur on overhanging eaves, the soffit of the cornice may be sloped outward and left open ¼ in. at the facia board for drainage.

CLOSED CORNICES

The closed cornice shown in Fig. 18-2 has no rafter projection. The overhang consists only of a frieze board and a shingle or crown moulding. This type is not so desirable as a cornice with a projection, because it gives less protection to the side walls.

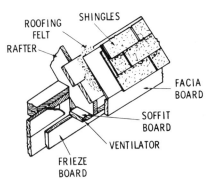

Fig. 18-1. Box cornice construction.

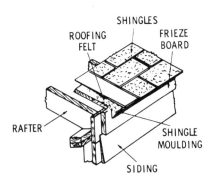

Fig. 18-2. Closed cornice construction.

WIDE BOX CORNICES

The wide box cornice in Fig. 18-3 requires forming members called lookouts, which serve as nailing surfaces and supports for the soffit board. The lookouts are nailed at the rafter ends and are also toenailed to the wall sheathing and directly to the stud. The soffit can be of various materials, such as beaded ceiling, plywood, or aluminum, either ventilated or plain. A bed moulding may be used at the juncture of the soffit and frieze. This type of cornice is often used in hip roof houses, and the facia board usually carries around the entire perimeter of the house.

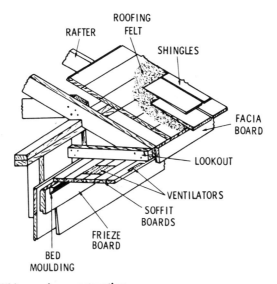

Fig. 18-3. Wide cornice construction.

OPEN CORNICES

The open cornice shown in Fig. 18-4 may consist of a facia board nailed to the rafter ends. The frieze is either notched or cut out to fit between the rafters and is then nailed to the wall. The open cornice is often used for a garage. When it is used on a house, the roof boards are visible from below from the rafter ends to the wall line, and should consist of finished material. Dressed or matched V-beaded boards are often used.

CORNICE RETURNS

The cornice return is the end finish of the cornice on a gable roof. The design of the cornice return depends to a large degree on

385

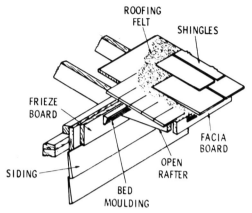

Fig. 18-4. Open cornice construction.

the rake or gable projection, and on the type of cornice used. In a close rake (a gable end with very little projection), it is necessary to use a frieze or rake board as a finish for siding ends, as shown in Fig. 18-5. This board is usually 1 ⅛ in. thick and follows the roof slope to meet the return of the cornice facia. Crown moulding or other type of finish is used at the edge of the shingles.

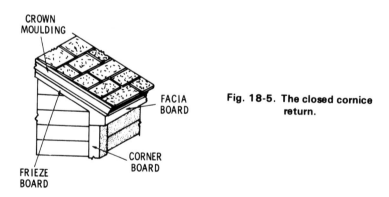

Fig. 18-5. The closed cornice return.

When the gable end and the cornice have some projection as shown in Fig. 18-6, a box return may be used. Trim on the rake projection is finished at the cornice return. A wide cornice with a small gable projection may be finished as shown in Fig. 18-7. Many variations of this trim detail are possible. For example, the frieze board at the gable end might be carried to the rake line and mitered with a facia board of the cornice. This siding is then carried across the cornice end to form a return.

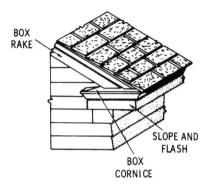

Fig. 18-6. The box cornice return.

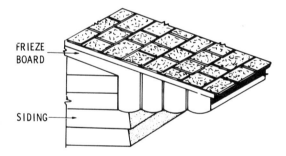

Fig. 18-7. The wide cornice return.

RAKE OR GABLE-END FINISH

The rake section is that trim used along the gable end of a house. There are three general types commonly used: the *closed*, the *box with a projection*, and the *open*. The closed rake, as shown in Fig. 18-8, often consists of a frieze or rake board with a crown moulding as the finish. A 1″ × 2″ square edge moulding is sometimes used instead of the crown moulding. When fiber board sheathing is used, it is necessary to use a narrow frieze board that will leave a surface for nailing the siding into the end rafters.

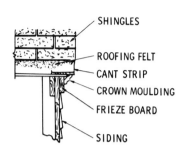

SHINGLES
ROOFING FELT
CANT STRIP
CROWN MOULDING
FRIEZE BOARD
SIDING

Fig. 18-8. The closed end finish.

If a wide frieze is used, nailing blocks must be provided between the studs. Wood sheathing does not require nailing blocks. The trim used for a box rake section requires the support of the projected roof boards, as shown in Fig. 18-9. In addition, lookouts or nailing blocks are fastened to the side wall and to the roof sheathing. These lookouts serve as a nailing surface for both

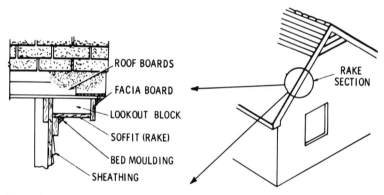

ROOF BOARDS
FACIA BOARD
LOOKOUT BLOCK
SOFFIT (RAKE)
BED MOULDING
SHEATHING

RAKE SECTION

Fig. 18-9. The box end finish.

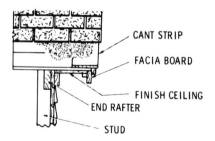

CANT STRIP
FACIA BOARD
FINISH CEILING
END RAFTER
STUD

Fig. 18-10. The open end finish.

the soffit and the facia boards. The ends of the roof boards are nailed to the facia. The frieze board is nailed to the side wall studs, and the crown and bed mouldings complete the trim. The underside of the roof sheathing of the open projected rake as shown in Fig. 18-10, is generally covered with liner boards such as ⅝-in. beaded ceiling. The facia is held in place by nails through the roof sheathing.

CHAPTER 19

Sheathing and Siding

Sheathing is nailed directly to the framework of the building. Its purpose is to strengthen the building, to provide a base wall to which the finish siding can be nailed, to act as insulation, and in some cases to be a base for further insulation. Some of the common types of sheathing include fiberboard, wood, and plywood.

FIBERBOARD SHEATHING

Fiberboard usually comes in 2×8 or 4×8 sheets which are tongue-and-grooved, and generally coated or impregnated with an asphalt material which increases water resistance. Thickness is normally $\frac{1}{2}$ and $\frac{25}{32}$ in. Fiberboard sheathing may be used where the stud spacing does not exceed 16 in., and it should be nailed with 2-in. galvanized roofing nails or other type noncorrosive nails. If the fiberboard is used as sheathing, most builders will use plywood at all corners (the thickness of the sheathing), to strengthen the walls, as shown in Fig. 19-1.

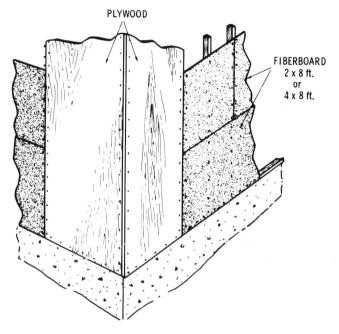

PLYWOOD

FIBERBOARD
2 x 8 ft.
or
4 x 8 ft.

Fig. 19-1. Method of using plywood on all corners as bracing when using
fiberboard as exterior sheathing.

SOLID WOOD SHEATHING

Wood wall sheathing can be obtained in almost all widths, lengths, and grades. Generally, widths are from 6 to 12 in., with lengths selected for economical use. Almost all solid wood wall sheathing used is $^{25}/_{32}$ to 1 in. in thickness. This material may be nailed on horizontally or diagonally, as shown in Fig. 19-2. Wood sheathing is laid on tight, with all joints made over the studs. If the sheathing is to be put on horizontally, it should be started at the foundation and worked toward the top. If the sheathing is installed diagonally, it should be started at the corners of the building and worked toward the center or middle.

Diagonal sheathing should be applied at a 45° angle. This method of sheathing adds greatly to the rigidity of the wall and

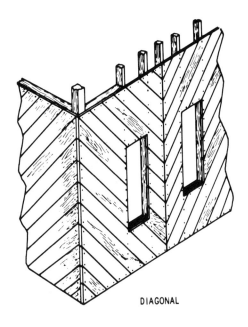

DIAGONAL

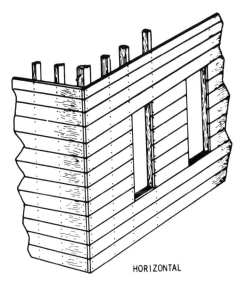

HORIZONTAL

Fig. 19-2. Two methods of nailing on wood sheathing.

eliminates the need for the corner bracing. It also provides an excellent tie to the sill plate when it is installed diagonally. There is more lumber waste than with horizontal sheathing because of the angle cut, and the application is somewhat more difficult. Fig. 19-3 shows the wrong way and the correct way of laying diagonal sheathing.

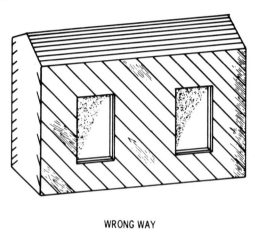

WRONG WAY

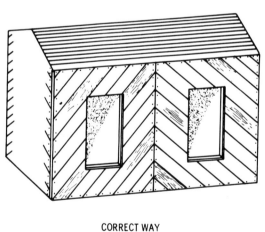

CORRECT WAY

Fig. 19-3. The wrong and correct way of laying sheathing.

394

PLYWOOD SHEATHING

Plywood as a wall sheathing is highly recommended because of its size, weight, and stability, plus the ease and rapidity of installation (Fig. 19-4). It adds considerably more strength to the frame structure than the conventional horizontal or diagonal sheathing.

Courtesy American Plywood Assn.

Fig. 19-4. Plywood is a popular sheathing. Here it is used at corners with fiberboard.

When plywood sheathing is used, corner bracing can also be omitted. Large size panels effect a major saving in the time required for application and still provide a tight, draft-free installation that contributes a high insulation value to the walls. Minimum thickness of plywood wall sheathing is $5/16$ in. for 16-in. stud spacing, and $3/8$ in. for 24-in. stud spacing. The panels should be installed with the face grain parallel to the studs. However, a little more stiffness can be obtained by installing them across the studs, but this requires more cutting and fitting. Nail spacing should not

395

be more than 6 in. on the center at the edges of the panels and not more than 12 in. on center elsewhere. Joints should meet on the centerline of framing members.

URETHANE AND FIBERGLASS

With the accent in recent years on saving energy, a number of other insulations have been developed that have fairly high insulating value. For example, there is urethane, 1¼-in.-thick material that, when combined with regular insulation, yields an R factor of 22 (Fig. 19-5). There is also fiberglass insulation with an R-4.8. Such insulations are particularly good on masonry construction because brick itself has very little insulating value and requires whatever insulation can be built in.

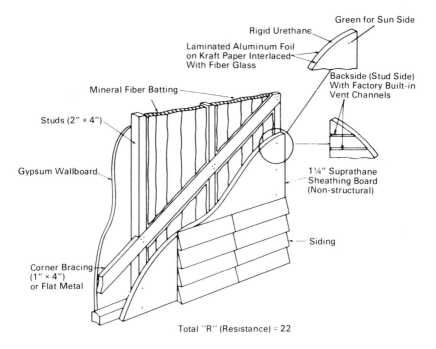

Fig. 19-5. Urethane in combination with batt insulation here produces an R-22.

SHEATHING PAPER

Sheathing paper should be used on frame structure when wood or plywood sheathing is used. It should be water resistant but not vapor resistant. It should be applied horizontally, starting at the bottom of the wall. Succeeding layers should lap about 4 in. and lap over strips around openings. Strips about 6 in. wide should be installed behind all exterior trim on exterior openings.

WOOD SIDING

One of the materials most characteristic of the exteriors of American houses is wood siding. The essential properties required for wood siding are good painting characteristics, easy working qualities, and freedom from warp. These properties are present to a high degree in the cedars, eastern white pine, sugar pine, western white pine, cypress, and redwood.

Material used for exterior siding should preferably be of a select grade, and should be free from knots, pitch pockets, and wavy edges. The moisture content at the time of application should be that which it would attain in service. This would be approximately 12 percent, except in the dry southwestern states, where the moisture content should average about 9 percent.

Bevel Siding

Plain bevel siding, as shown in Fig. 19-6, is made in nominal 4-, 5-, and 6-in. widths from $7/16$ in. butts, 6-, 8-, and 10-in. widths with $9/16$ and $11/16$ in. butts. Bevel siding is generally furnished in random lengths varying from 4 to 20 ft.

Drop Siding

Drop siding is generally $3/4$ in. thick, and is made in a variety of patterns with either matched or shiplap edges. Figs. 19-7 shows three common patterns of drop siding that are applied horizontally. Fig. 19-7A may be applied vertically, for example at the gable ends of a house. Drop siding was designed to be applied directly to the studs, and it thereby serves as sheathing and exterior

397

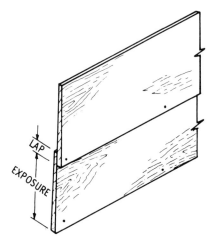

Fig. 19-6. Bevel siding.

wall covering. It is widely used in this manner in farm structures, such as sheds and garages in all parts of the country. When used over or when in contact with other material, such as sheathing or sheathing paper, water may work through the joints and be held between the sheathing and the siding. This sets up a condition conducive to paint failure and decay. Such problems can be avoided when the side walls are protected by a good roof overhang.

Square-Edge Siding

Square-edge or clapboard siding made of $^{25}/_{32}$ -in. board is occasionally selected for architectural effects. In this case, wide boards are generally used. Some of this siding is also beveled on the back at the top to allow the boards to lie rather close to the sheathing, thus providing a solid nailing surface.

Vertical Siding

Vertical siding is commonly used on the gable ends of a house, over entrances, and sometimes for large wall areas. The type used may be plain-surfaced matched boards, patterned matched boards or square-edge boards covered at the joint with a batten strip. Matched vertical siding should preferably not be more than 8 in. wide and should have 2 eight-penny nails not more than 4 ft. apart. Backer blocks should be placed between studs to provide a good nailing base. The bottom of the boards should be undercut to form a water drip.

Batten-type siding is often used with wide square-edged boards which, because of their width, are subjected to considerable expansion and contraction. The batten strips used to cover the

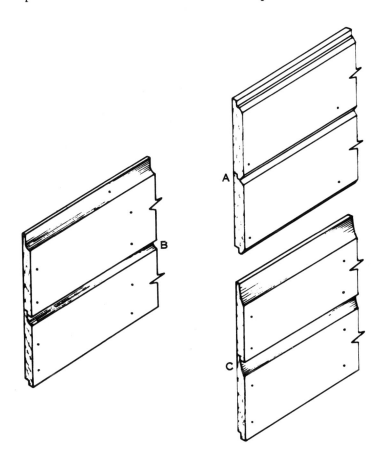

Fig. 19-7. Types of drop siding: (A) V-rustic, (B) drop, (C) rustic drop.

joints should be nailed to only one siding board so the adjacent board can swell and shrink without splitting the boards or the batten strip.

Plywood Siding

Plywood is often used in gable ends, sometimes around windows and porches, and occasionally as an overall exterior wall covering. The sheets are made either plain or with irregularly cut striations. It can be applied horizontally or vertically. The joints can be moulded batten, V-grooves, or flush. Sometimes it is installed as lap siding. Plywood siding should be of exterior grade, since houses are often built with little overhang of the roof, particularly on the gable end. This permits rainwater to run down freely over the face of the siding. For unsheathed walls, the following thicknesses are suggested:

Minimum thickness	Maximum stud space
$\frac{3}{8}$ inch	16 inches on center
$\frac{1}{2}$ inch	20 inches on center
$\frac{5}{8}$ inch	24 inches on center

Treated Siding

In a construction situation which allows rainwater to flow down freely over the face of the siding (such as when there is little roof overhang along the sides or gable ends of the house), water may work up under the laps in bevel siding or through joints in drop siding by capillary action, and provide a source of moisture that may cause paint blisters or peeling.

A generous application of a water repellent preservative to the back of the siding will be quite effective in reducing capillary action with bevel siding. In drop siding, the treatment would be applied to the matching edges. Dipping the siding in the water repellent would be still more effective. The water repellent should be applied to all end cuts, at butt points, and where the siding meets door and window trim.

WOOD SHINGLES AND SHAKES

Cedar shingles and shakes (Fig. 19-9) are also available. They come in a variety of grades and may be applied in several ways. You can get them in random widths 18 to 24 in. long or in a uniform 18 in. The shingles may be installed on regular sheathing or on an undercourse of shingles, which produces a shadowed effect. Cedar, of course, stands up to weather well and does not have to be painted.

Asbestos Shingles

Asbestos shingles offer good economy, though they are brittle and tend to break easily if hit. They come in various colors and are

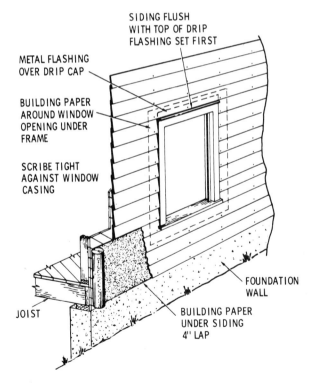

SIDING FLUSH
WITH TOP OF DRIP
FLASHING SET FIRST

METAL FLASHING
OVER DRIP CAP

BUILDING PAPER
AROUND WINDOW
OPENING UNDER
FRAME

SCRIBE TIGHT
AGAINST WINDOW
CASING

FOUNDATION
WALL

JOIST

BUILDING PAPER
UNDER SIDING
4" LAP

Fig. 19-8. Installation of bevel siding.

401

installed by driving nails through predrilled holes on a 15-lb. felt base. It is best to use a shingle cutter when installing asbestos shingles.

INSTALLATION OF SIDING

The spacing for siding should be carefully laid out before the first board is applied. The bottom of the board that passes over the top of the first-floor windows should coincide with the top of the window cap, as shown in Fig. 19-10. To determine the maximum board spacing or exposure, deduct the minimum lap from the overall width of the siding. The number of board spaces between

Fig. 19-9. Wood shingles blend well with stone veneer on this home.

the top of the window and the bottom of the first course at the foundation wall should be such that the maximum exposure will not be exceeded. This may mean that the boards will have less than the maximum exposure.

Siding starts with the bottom course of boards at the foundation, as shown in Fig. 19-11. Sometimes the siding is started on a water

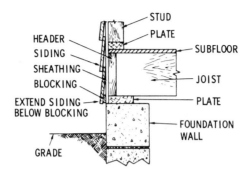

Fig. 19-10. Installation of the first or bottom course.

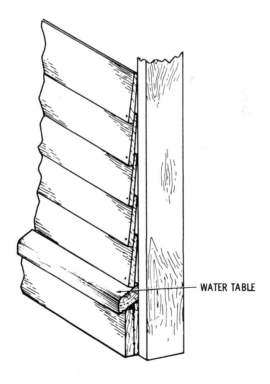

Fig. 19-11. Water table is sometimes used.

403

table, which is a projecting member at the top of the foundation to throw off water, as shown in Fig. 19-12. Each succeeding course overlaps the upper edge of the lower course. The minimum head lap is 1 in. for 4- and 6-in. widths, and 1 ¼ -in. lap for widths over 6 in. The joints between boards in adjacent courses should be staggered as much as possible. Butt joints should always be made on a stud, or where boards butt against window and door casings and corner boards. The siding should be carefully fitted and be in close contact with the member or adjacent pieces. Some carpenters fit the boards so tight that they have to spring the boards in place, which assures a tight joint. Loose-fitting joints allow water to get behind the siding and thereby causes paint deterioration around the joints, and also sets up conditions conducive to decay at the ends of the siding.

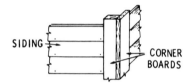

Fig. 19-12. Corner treatment for bevel siding using the corner board.

Types of Nails

Nails cost very little compared to the cost of siding and labor, but the use of good nails is important. It is poor economy to buy siding that will last for years and then use nails that will rust badly within a few years. Rust-resistant nails will hold the siding permanently and will not disfigure light-colored paint surfaces.

There are two types of nails commonly used with siding, one having a small head and the other a slightly larger head. The small-head casing nail is set (driven with a nailset) about ⅟₁₆ in. below the surface of the siding. The hole is filled with putty after the prime coat of paint is applied. The large-head nail is driven flush with the face of the siding, with the head being later covered with paint. Ordinary steel wire nails tend to rust in a short time and cause a disfiguring stain on the face of the siding. In some cases, the small-head nail will show rust spots through the putty and paint. Noncorrosive-type nails (galvanized, aluminum, and stainless steel) that will not cause rust stains are readily available.

Bevel siding should be face nailed to each stud with noncorrosive nails, the size depending upon the thickness of the siding and the type of sheathing used. The nails are generally placed about ½ in. above the butt edge, in which case it passes through the upper edge of the lower course of siding. Another method recommended for bevel siding by most associations representing siding manufacturers, is to drive the nails through the siding just above the lap so that the nail misses the thin edge of the piece of siding underneath. The latter method permits expansion and contraction of the siding board with seasonal changes in moisture content.

Corner Treatment

The method of finishing the wood siding at the exterior corners is influenced somewhat by the overall house design. Corner boards are appropriate to some designs, and mitered joints to others. Wood siding is commonly joined at the exterior corners by corner boards, mitered corners, or by metal corners.

Corner Boards—Corner boards, as shown in Fig. 19-12, are used with bevel or drop siding and are generally made of nominal 1- or 1¼ -in. material, depending upon the thickness of the siding. It may be either plain or moulded, depending on the architectural treatment of the house. The corner boards may be applied vertically against the sheathing, with the siding fitting tightly against the narrow edge of the corner board. The joints between the siding and the corner boards and trim should be caulked or treated with a water repellent. Corner boards, and trim around windows and doors, are sometimes applied over the siding, a method that minimizes the entrance of water into the ends of the siding.

Mitered Corners—Mitered corners, such as shown in Fig. 19-13, must fit tightly and smoothly for the full depth of the miter. To

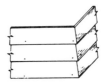

Fig. 19-13. The mitered corner treatment.

405

maintain a tight fit at the miter, it is important that the siding is properly seasoned before delivery, and is stored at the site so as to be protected from rain. The ends should be set in white lead when the siding is applied, and the exposed faces should be primed immediately after it is applied. At interior corners, shown in Fig. 19-14, the siding is butted against a corner strip of nominal 1- or 1¼-in. material, depending upon the thickness of the siding.

Metal Corners—Metal corners, as shown in Fig. 19-15, are made of 8-gauge metals, such as aluminum and galvanized iron. They

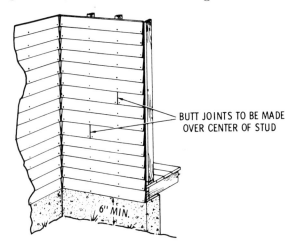

BUTT JOINTS TO BE MADE
OVER CENTER OF STUD

6" MIN.

Fig. 19-14. Construction of an interior corner using bevel siding.

are used with bevel siding as a substitute for mitered corners, and can be purchased at most lumber yards. The application of metal corners takes less skill than is required to make good mitered corners, or to fit the siding to a corner board. Metal corners should always be set in white lead paint.

METAL SIDING

The most popular metal siding is aluminum. It is installed over most types of sheathing with an aluminum building paper (for insulation) nailed on between the sheathing and siding. Its most attractive characteristic is the long-lasting finish obtained on the

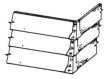

Fig. 19-15. Corner treatment for bevel siding using the corner metal caps.

prefinished product. The cost of painting and maintenance has made this type of siding doubly attractive. Aluminum siding can be installed over old siding that has cracked and weathered, or where paint will not hold up. Installation instructions are furnished with the siding, which is available with insulation built on and in various gauges.

Courtesy Vinyl Siding Institute

Fig. 19-16. Vinyl siding comes in a variety of colors and is easy to keep clean.

407

Vinyl Siding

Also popular is vinyl siding (Fig. 19-16). This comes in a wide variety of colors, textures, and styles. As with aluminum siding, the big advantage of vinyl siding is that it does not need to be painted and will not corrode, dent, or pit. When it is very cold, it is relatively susceptible to cracking if hit.

CHAPTER 20

Windows

The three main window types are gliding, double-hung, and casement, but there are also awning, bow, and bay windows (Fig. 20-1). Basic windows consist essentially of two parts, the frame and the sash. The frame is made up of four basic parts: the head, two jambs, and the sill. Good construction around the window frame is essential to good building. Where openings are to be provided, studding must be cut away and its equivalent in strength replaced by doubling the studs on each side of the opening to form trimmers and inserting a header at the top. If the opening is wide, the header should be doubled and trussed. At the bottom of the opening, a header or rough sill is inserted.

WINDOW FRAMING

This is the frame into which the window sash fits. It is set into a rough opening in the wall framing, and is intended to hold the sash in place. (See Fig. 20-2).

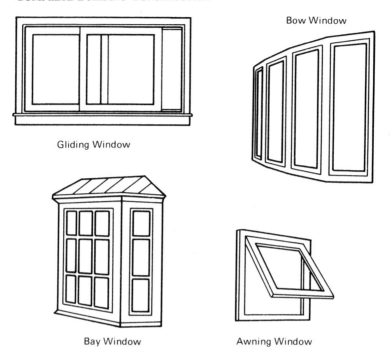

Gliding Window

Bow Window

Bay Window

Awning Window

Fig. 20-1. Various kinds of windows.

DOUBLE-HUNG WINDOWS

The double-hung window is the most common kind of window. It is made up of two parts—an upper and lower sash, which slide vertically past each other. An illustration of this type of window, made of wood, is shown in Fig. 20-3. It has some advantages and some disadvantages. Screens can be installed on the outside of the window without interfering with its operation. For full ventilation of a room, only one-half the area of the window can be utilized, and any current of air passing across its face is, to some extent, lost in the room. Double-hung windows are sometimes more involved in their frame construction and operation than the casement window. Ventilation fans and air conditioners can be placed in the window with it partly closed.

HINGED OR CASEMENT WINDOWS

There are basically two types of casement windows—the outswinging and the inswinging. These windows may be hinged at the side, top, or bottom. The casement window which opens out

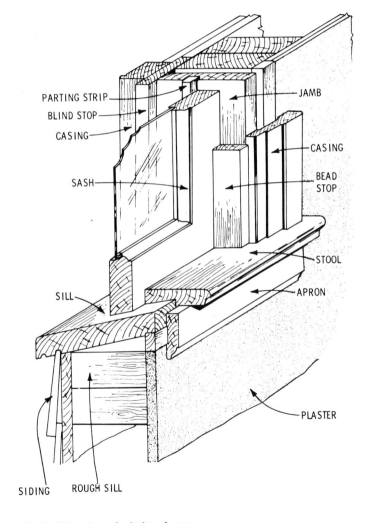

Fig. 20-2. Side view of window frame.

411

Fig. 20-3. The popular double-hung window.

requires the screen to be located on the inside. This type of window, when closed, is most efficient as far as waterproofing. The inswinging, like double-hung windows, are clear of screens, but they are extremely difficult to make watertight. Casement windows have the advantage of their entire area being opened to air currents, thus catching a parallel breeze and slanting it into a

412

room. Casement windows are considerably less complicated in their construction than double-hung units. Sill construction is very much like that for a double-hung window, however, but with the stool much wider and forming a stop for the bottom rail. When there are two casement windows in a row in one frame, they are separated by a vertical double jamb called a mullion, or the stiles may come together in pairs like a french door. The edges of the stiles may be a reverse rabbet, a beveled reverse rabbet with battens, or beveled astragals. The battens and astragals ensure better weathertightness. Fig. 20-4 shows a typical casement window with a mullion.

Fig. 20-4. A casement window.

GLIDING, BOW, BAY, AND AWNING WINDOWS

Gliding windows consist of two sashes that slide horizontally right or left. They are often installed high up in a home to provide light and ventilation without sacrificing privacy.

Awning windows have a single sash hinged at the top and open outward from the bottom. They are often used at the bottom of a fixed picture window to provide ventilation without obstructing the view. They are popular in ranch homes.

Bow and bay windows add architectural interest to a home. Bow windows curve gracefully, while bay windows are straight across the middle and angled at the ends. They are particularly popular in Georgian and Colonial homes.

Experts agree that wood windows are better than metal ones for insulation purposes, simply because metal conducts heat better than wood. But even more important is double glazing, which contains dead air space that inhibits heat escaping—or getting in, should you have air-conditioning. The second pane can be incorporated in the window, as shown in Fig. 20-5, or it can be removable. If you live in an area where heating costs are very high,

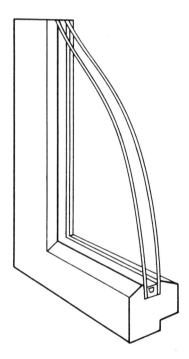

Fig. 20-5. A double-insulated window. The dead air space between the sandwich of glass helps save on fuel.

414

consider triple glazing—three panes of glass with air spaces between. Tinted or reflective glass is good for warding off the sun's rays in warmer climates.

WINDOW SASH

Most frames and sash are made from steel, aluminum or wood, and the sash are either fixed in place or movable. Aluminum windows require very little maintenance except for cleaning, while steel and wood must be kept painted (some wooden windows, however, are vinyl-clad and do not require paint). Glazing compound must be kept in good shape on steel and wooden windows, whereas most aluminum windows do not use it.

Fixed window sash are removable only through a knowledge of carpentry. Movable sash may slide up and down or from side to side; they may be hinged at the top, either side, or bottom in such a manner as to swing in or out. Any of the movable sash are easily removed for cleaning or repair.

Vertically sliding sash in older wooden windows are counterbalanced by two sash weights per sash, the combined weight of which exactly equals the weight of the sash. Ropes (which usually rot and break in time) or thin chain attaches the weight to each sash. Newer counterbalance methods for wooden and aluminum windows of this type make use of spring pressure or tension instead of the weights.

Installing Double-Hung Sash

Installing the typical double-hung window is not difficult. To do it, first place the upper double-hung sash in position and trim off a slight portion of the top rail to ensure a good fit, and tack the upper sash in position. Fit the lower sash in position by trimming off the sides. Place the lower sash in position, and trim off a sufficient amount from the bottom rail to permit the meeting rails to meet on a level. In most cases, the bottom rail will be trimmed on an angle to permit the rail and sill to match both inside and outside, as shown in Fig. 20-6.

415

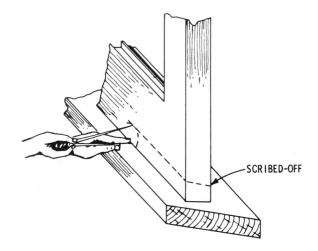

SCRIBED-OFF

Fig. 20-6. Bottom rail trim to match sill plate.

Sash Weights

If sash weights are used, remove each sash after it has been properly cut and sized. Select sash weights equal to one-half the weight of each sash and place in position in the weight pockets. Measure the proper length of sash cord for the lower sash and attach it to the stiles and weights on both sides. Adjust the length of the cord so that the weight will not strike the pulley or bottom of the frame when window is moved up and down. Install the cords and weights for the upper sash and adjust the cord so that the weights run smoothly. Close the pockets in the frame and install the blind stop, parting strip, and bead stop.

There are many other types of window lifts, such as spring-loaded steel tapes, spring-tension metal guides, and full-length coil springs.

GLAZING SASH

The panes or *lights*, as they are called, are generally cut ⅛ in. smaller than the rabbet to allow for irregularities in cutting and in

416

the sash. This leaves an approximate margin of $\frac{1}{16}$ in. between the edge of the glass and the sides of the rabbet. Fig. 20-7 shows two lights or panes of glass in position for glazing. To install the window glass properly, first spread a film of soft glazing compound close to the edge on the inside portion of the glass. After the glass has been inserted, drive or press in at least two glazier points on each side. This is illustrated in Fig. 20-8.

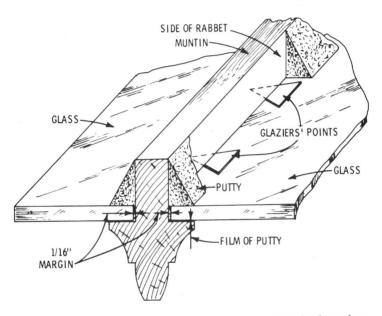

Fig. 20-7. Glazier points that are removed to replace broken glass.

After the glass is firmly secured with the glazier points, the glazing compound (which is soft), is put on around the glass with a putty knife. Do not project the compound beyond the edge of the rabbet so that it will be visible from the other side.

Glazing compound is usually purchased in a can with a lid which provides an airtight seal. The compound should be soft and pliable to work properly with a putty knife. This can be accomplished by working the glazing compound in your hands to warm it. After the compound has hardened, it should be painted to match the window sash.

417

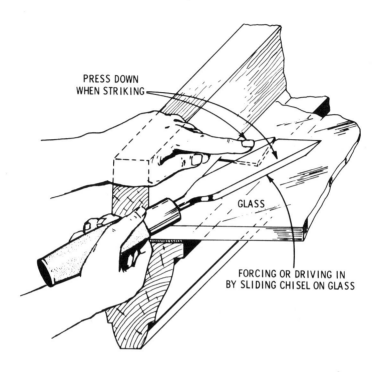

PRESS DOWN
WHEN STRIKING

GLASS

FORCING OR DRIVING IN
BY SLIDING CHISEL ON GLASS

Fig. 20-8. The proper way to install glazier points.

WOODEN WINDOW SCREENS

Wooden window screen frames are usually $1\frac{3}{4}$ or $2\frac{1}{4}$ in. wide. The screen may be attached by stapling or tacking. The frame corners may be constructed with an open mortise, with the rails tenoned into the stiles, with half-lap corners, or with butt joints with corrugated fasteners. In either case, the joints may be nailed and glued. When attaching the screen material, start at one end and tack or staple it with copper staples, holding the material tight as you nail. Hand stretch the screen along the side, working toward the other end and attach, making sure that the weave is parallel to the ends and sides. Tack the sides and then apply the

418

moulding. Copper staples should be used for bronze or copper screen; cadmium staples should be used for aluminum screens.

In most cases, factory built combination aluminum storm windows are installed. The combination storm windows are attractive and can be made to fit most size openings.

Fig. 20-9. Movable-type shutters.

SHUTTERS

In coastal areas where damaging high winds occur frequently, shutters are necessary to protect large plate-glass windows from being broken. The shutters are mounted on hinges, and can be closed at a moment's notice. Throughout te midwest, shutters are generally installed for decoration only, and are mounted stationary to the outside wall. There are generally two types of shutters—the solid panel, and the slat or louver type. Louver shutters can have stationary or movable slats, as shown in Fig. 20-9.

CHAPTER 21

Insulation

Over the past few years, ever since the energy crunch began, there has been a tremendous interest in energy saving, both on new work and existing structures. Carpenters, contractors, builders, and do-it-yourselfers have found that, over time, using certain materials and techniques can result in big savings.

What follows is a round-up of energy savers for new and existing buildings. In most cases the suggestions can be readily adopted. But it is always a good idea to have discussions with knowledgeable people, such as experts at local utility companies, to see what works best. For example, 12-in.-thick insulation may be just the thing for frigid New Hampshire winters, but it would hardly be practical, in terms of original cost, for insulating a stucco home south of Los Angeles.

INSULATION—R FACTORS

A key concept to remember when selecting insulation is the R factor of the material. "R" stands for Resistance: the ability of the material (technically the stratified air spaces) to resist or retard heat. In the winter this resistance works by keeping warm air inside the home. In the summer it keeps hot air out. The critical time is winter, when the expensive fuel used to heat that warm air may be traveling out of the house by conduction, making the furnace work harder—and using up more fuel.

Cold air infiltration is also an energy factor. If the house has cracks and openings where cold air can sneak in (and warm air escape), this will also result in higher fuel costs. Such openings must be plugged up.

Insulation comes in a variety of types, generally ranging from about R-4 to R-38. (All insulation has its R factor stamped on it.) The R-4 material could be urethane board applied to basement walls; the R-38, 12-in.-thick fiberglass batts. The following are the materials available:

Rigid Board Insulation—This is available in various sizes in board form. It may be urethane, fiberglass, or styrofoam, or composition (Figs. 21-1 and 21-2). It is easily worked with standard hand tools, and some contractors favor it as sheathing. While the R values are not high compared with other kinds of insulation, they are better than those of board or plywood. Rigid board insulation is also handy to use in the basement, where it can be glued directly to walls. Before using it, however, you should check for fire restrictions. Some rigid board insulation is flammable and must be covered by a nonflammable material before any other flammable material, such as wood paneling, can be installed.

Siding with Insulation—Some siding, chiefly aluminum, has built-in insulation. The R factors vary, however, and should be checked out before any material is purchased.

Batts and Blankets—Batts are flexible insulation sections cut to fit between standard studs (on 16-in. centers) that are about 8 ft. long. Blankets are the same width as batts but come in virtually infinite lengths (up to 100 ft.). Batts and blankets may be faced on

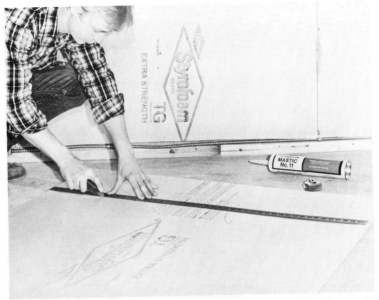

Courtesy Dow Chemical

Fig. 21-1. Styrofoam can be cut easily.

both sides with a paperlike material, and one side will have a vapor barrier—either aluminum or Kraft paper (Fig. 21-3) or be unfaced (Fig. 21-4). Mineral fiber is the filler for such insulation—fiberglass such as that made by Owens-Corning, or mineral wool. On the edges of batts or blankets are strips, which are stapled to framing members.

Batts and blankets are favored by contractors and do-it-yourselfers alike. This insulation is used on exposed stud walls, on attic floors (between joists), and on open ceilings. It ranges in R factor from around R-4 to R-38.

Fill-Type Insulation—Another insulation is the fill type (Fig. 21-5), which comes as a granular material (Vermiculite) or a macerated paper (cellulose). This material may be poured or blown in place, such as between attic joists, but its primary use is to be blown into areas inaccessible for installing other insulation, mainly walls in finished homes. Sections of siding are removed, holes are

423

Courtesy Dow Chemical

Fig. 21-2. It is good for basement walls.

drilled between pairs of studs, a hose is inserted, and the material is blown into place. It fills up the cavities and insulates (fig. 21-6).

Foam—Like fill-type insulation, foam is designed to be used on finished work that is inaccessible to the batts or blankets. The foam is mixed and is applied as a liquid. It foams up, filling the spaces, then becomes hard. There has been controversy about foam insulation. Complaints have been voiced—and indeed the government has investigated them—about medical problems caused by the fumes of foam, and it has been banned in many places. In fairness to foam makers and installers, a wise precaution is to know the type of foam an installer plans to use. Some foams are considered harmless, but may suffer from a blanket indictment of all foam. Foam is practical, too; nothing has a higher R value, inch for inch, than foam.

Courtesy Owens-Corning

Fig. 21-3. Some insulation is faced with Kraft paper. Here it is being applied with Kraft paper facing wrong direction. Aluminum or Kraft paper, which are vapor barriers, always face warm side of house.

REDUCING LEAKAGE AROUND DOORS AND WINDOWS

A large amount of the heat delivered to the rooms is lost by air leakage around doors and windows. Another large factor in heat loss is the heat leakage through the window glass.

These various heat losses may be greatly prevented by:

1. Caulking.
2. Weatherstrip.
3. Storm or double-glass windows.
4. Storm Doors.

Though the house may have been tight when built, a few years of usage will sometimes permit a considerable amount of leakage. In addition, small cracks often develop in masonry walls which, in

425

Courtesy Owens-Corning

Fig. 21-4. Batt insulation is the most popular kind. Here it is being used to add to existing attic insulation.

time, may become sufficiently large to permit a considerable air leakage.

Caulking

The most common remedy against air leakage is caulking—filling the cracks or seams of the house: between windows and siding, doors and siding, foundations and siding, etc. A caulking gun makes the job simple (Fig. 21-7).

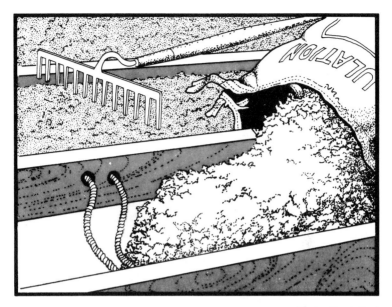

Courtesy Certain-Teed

Fig. 21-5. Insulation in chopped-up form can be applied to open areas, most commonly the attic.

Weatherstripping

Weatherstripping is another excellent idea. Weatherstrip comes in a variety of types and is designed to be used where there is a natural opening in the house, such as beneath the garage or exterior doors, or where windows meet framework. You can use felt material, which can be tacked in place to V-shaped strips that have one adhesive edge. Also available is a moldable weatherstrip and adhesive-backed felt. Whichever kind you use, weatherstripping is an energy saver.

Window and Door Glazing

A large amount of heat can be lost through window glass, and so storm windows and storm doors are an excellent idea. Indeed,

427

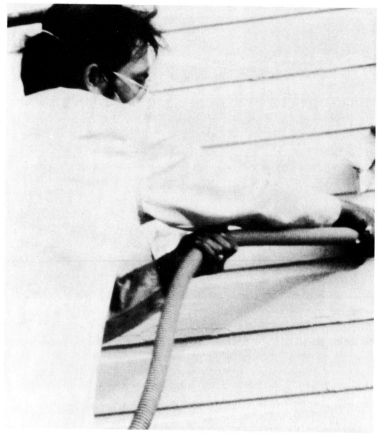

Courtesy United Gilsonite

Fig. 21-6. Insulation being blown through siding into wall cavity. It is generally a bad idea to drill through siding, but in some cases it cannot be helped.

they are just as important as insulation. If you find that you cannot afford storm windows this winter, you can use polyethylene sheeting and make your own. Polyethylene is as effective as glass, though it is only temporary, of course.

An even better solution for windows is double- or even triple-glazed windows. A dead air space is trapped between the layers of glass and retards heat transmission. Such glazing is initially expensive, but it pays for itself over time.

Fig. 21-7. Caulking being applied to window-siding seam.

WHERE TO INSULATE

A basic question is where to insulate. Generally, insulation should be placed on all outside walls and in the ceiling, as shown in Fig. 21-8. In houses with unheated crawl spaces, insulation should be placed between the floor joists. If a blanket type of insulation is used, it should be well supported by slats with galvanized mesh wire or by a rigid board. The next important area to insulate is the attic, because heat rises.

In 1½-storey houses, insulation should be placed along all areas that are adjacent to unheated areas, as shown in Fig. 21-9. These include stairways, dwarf walls, and dormers. Provisions should be made for ventilation of the unheated areas. Where attic storage space is unheated and a stairway is included, insulation should be used around the stairway as well as in the first floor ceiling, as shown in Fig. 21-10. The door leading to the attic should be weatherstripped to prevent loss of heat. Walls adjoining an unheated garage or porch should be well insulated.

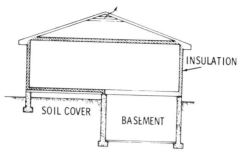

Fig. 21-8. The proper way to insulate walls, floors, and ceilings.

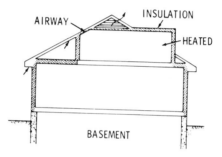

Fig. 21-9. A method of insulating 1½-storey houses.

In houses with flat or low-pitched roofs, as shown in Fig. 21-11, insulation should be used in the ceiling area with sufficient space allowed above for a clear ventilating area between the joists. Insulation should be used along the perimeter of a house built on a slab. A vapor barrier should be included under the slab. Insulation can be used effectively to improve comfort conditions within the house during hot weather. These surfaces exposed to the direct rays of the sun may attain temperatures of 50°F or more above shade temperature and, of course, tend to transfer this heat toward the inside of the house. Insulation in the roof and walls retards the flow of heat and, consequently, less heat is transmitted through such surfaces.

Where any system of cooling hot-weather air is used, insulation should be used in all exposed ceilings and walls in the same manner as for preventing heat loss in cold weather. Of course, where cooling is used, the windows and doors should be kept closed

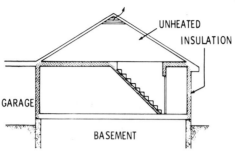

Fig. 21-10. A method of shielding a heated area from an unheated area by insulating.

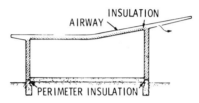

Fig. 21-11. Insulation of a typical flat-roof house.

during periods when outdoor temperatures are above inside temperatures. Windows exposed to the sun should be shaded with awnings, blinds, or draperies.

Ventilation of attic and roof space is an important adjunct to insulation. Without ventilation, an attic space may become very hot and hold the heat for many hours. Obviously, more heat will be transmitted through the ceiling when the attic temperature is 150°F than if it is 120°F. Ventilation methods suggested for protection against cold-weather condensation apply equally well to protection against excessive hot-weather roof temperature. (More details on ventilation are given later.)

INSTALLATION OF INSULATION

Blanket or batt insulation should be installed between framing members so that the tabs lap all the face of the framing members or on the sides. The vapor barrier should face the inside of the house. Where there is no vapor barrier on the insulation, a vapor barrier should be used over these unprotected areas. A hand stapling machine is usually used to fasten insulation tabs and vapor barriers in place. A vapor barrier should be used on the warm side (the bottom, in case of ceiling joists) before insulation is placed. Figs. 21-12 through 21-18 show a number of other installation tips.

Fig. 21-12. Investigate and measure attic for material before you start.

VAPOR BARRIERS

Most building materials are permeable to water vapor. In cold climates during cold weather such vapor, generated in the house from cooking, dishwashing, laundering, bathing, humidifiers, and other sources, may pass through wall and ceiling materials and condense in the wall or attic space, where it may subsequently do

Courtesy Owens-Corning

Fig. 21-13. Batt, sized to fit neatly between framing members, being installed in attic.

Courtesy Certain-Teed

Fig. 21-14. Shown is R-38, 12-in., the largest insulation available.

433

Courtesy Owens-Corning

Fig. 21-15. When you insulate, you must make provision for electrical devices.

damage to exterior paint and interior finish, or may even cause decay in structural members. As a protection, a material highly resistive to vapor transmission, called a *vapor barrier*, should be used on the warm side of a wall or below the insulation in a roof. Among the effective and more common vapor barrier materials is polyethelene sheeting. This is sold by the foot off rolls and can be attached easily.

Some batt and blanket insulation has a barrier material on one side of the blanket. Such material should be attached with the tabs at their sides fastened on the faces of the studs. When polyethene is used, it should be applied over the stud faces and stapled on. The wall material is then applied. Vapor barriers should be cut tightly to fit around wall outlets and switches.

434

Courtesy Owens-Corning

Fig. 21-16. Insulation is always applied with vapor barrier facing the warm side of the house.

Fig. 21-17. If insulation comes without vapor barrier, one must be provided. Polyethelene film works well.

Fig. 21-18. In some places, such as above sill, insulation should be stuffed in place.

VENTILATION

It is an excellent idea to button up a house with insulation and other energy-saving steps. But it is also important to ventilate a house properly to avoid condensation and high temperatures. The attic and crawl spaces, in particular, deserve attention. Vents—power types and standard vents—let moisture escape, and in summer the moving air in an attic tends to keep it cooler. Indeed, an attic can reach an almost unbelievable 140°F in summer.

Attics

At least two vent openings should always be provided. They should be located so that air can flow in one vent and out the other. A combination of vents at the eaves and gable vents is better than gable vents alone. A combination of eave vents and continuous ridge venting is best.

An attic should have a minimum amount of vent area, as follows:

Combinations of eave vents and gable vents without a vapor barrier: 1 sq. ft. inlet and 1 sq. ft. outlet for each 600 sq. ft. of ceiling area with at least half the vent area at the tops of the gables and the balance at the eaves.

Gable vents only without a vapor barrier: 1 sq. ft. inlet and 1 sq. ft. outlet for each 300 sq. ft. of ceiling area.

Gable vents only with a vapor barrier: 1 sq. ft. inlet and 1 sq. ft. outlet for each 600 sq. ft. of ceiling area.

Crawl Spaces

For a crawl space that is unheated, there should be at least two vents opposite one another. The minimum opening size, with moisture seal (4-mil or thicker polyethylene sheeting or 55-lb. asphalt roll roofing lapped 3 in.) on the ground should be 1 sq. ft. for every 1,500 ft. of crawl space. If there is no moisture seal, then there should be an opening of 1 sq. ft. for every 150 sq. ft.

Covered Spaces

When attic or crawl space vents are covered by screening or other materials, this results in a reduction in the size of the opening. Openings sizes should be increased, as indicated by the following:

Covering	Size of Opening
¼″ Hardware Cloth	1 × Net Vent Area
¼″ Hardware Cloth and Louvers	2 × Net Vent Area
⅛″ Screen	1 ¼ × Net Vent Area
⅛″ Screen and Rain Louvers	2 ¼ × Net Vent Area
¹⁄₁₆″ Screen	2 × Net Vent Area
¹⁄₁₆″ Screen and Rain Louvers	3 × Net Vent Area

CHAPTER 22

Plaster Walls
and Ceilings

Interior walls are usually (80% in residential construction) made of plasterboard, also known as sheet rock and drywall, but plaster is still used. Here, we consider the latter. Plaster walls generally have a hard smooth finish that is easily painted or papered. It is (to some extent) fire retarding and has some sound deadening qualities. The plaster is applied to a base which is attached to studs in residences or to metal studs in many industrial buildings, as shown in Fig. 22-1. Plaster may be applied directly to brick or masonry walls.

Two methods are used to apply plaster to a wall. The older method is one in which plaster is applied to properly spaced horizontally mounted wood lath. When applied in a thick coat, the plaster is forced into the spaces between the lath, locking it firmly to the wall. Wood lath has been replaced in later construction by gypsum lath and expanded metal lath.

Plastering requires a tremendous amount of labor and considerable time because of the two or three coats applied and the drying

439

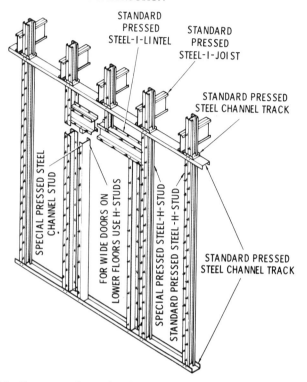

STANDARD
PRESSED
STEEL-I-LINTEL

STANDARD
PRESSED
STEEL-I-JOIST

STANDARD PRESSED
STEEL CHANNEL TRACK

SPECIAL PRESSED STEEL
CHANNEL STUD

FOR WIDE DOORS ON
LOWER FLOORS USE H-STUDS

SPECIAL PRESSED STEEL-H-STUD

STANDARD PRESSED STEEL-H-STUD

STANDARD PRESSED
STEEL CHANNEL TRACK

Fig. 22-1. One type of metal wall and ceiling structure designed to hold plasterboard. In homes, framing members are wood.

time between coats. To save time and expense, plastering has steadily given way to plasterboard, in which 4′ × 8′, 10′, or 12′ plasterboard sheets are nailed or screwed to the wall studs. Plasterboard has a number of advantages: it eliminates the need to mix plaster on the job site which saves considerably on the amount of labor involved and its installation is easier to learn than plastering. It is made in thicknesses from ¼″ to ¾″ and usually in 4′ × 8′.

INSTALLING PLASTER: WALL PREPARATION

Old-fashioned wooden lath is rough-sawed hardwood stripping cut 4 ft. long, 1½ in. wide, and ¼ in. thick. It is nailed horizontally to the studs, with each lath spaced ³⁄₈ in. vertically from its neighbor (Fig. 22-2) in order to provide a space for wet plaster to ooze through and form a "key" to lock itself to the lath. Although wood lath is still available at some lumberyards, its use in new plasterwork is, as far as we know, nonexistent except in absolutely faithful restorations.

Plaster is now applied primarily over gypsum lath (Fig. 22-3) and/or metal lath (Fig. 22-4) fastened to wooden or metal studding in frame walls, or directly to the brick or block in masonry walls.

Gypsum and Fiberwood Lath

One of the popular types of plaster base which may be used on the side walls and ceilings is gypsum-board lath. Such lath is generally 16″ × 48″, and is applied horizontally to the frame members. This type of board has a proper face with a gypsum filler. For studs and joists with a spacing of 16 in. on center, ³⁄₈-in. thickness is used, and for 24 in. on center spacing, ¹⁄₂-in. thickness

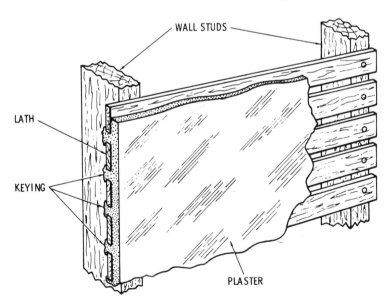

WALL STUDS

LATH

KEYING

PLASTER

Fig. 22-2. Lath and plaster wall construction. Wood furring strips are nailed to the wall studs and plaster is applied in several coats.

441

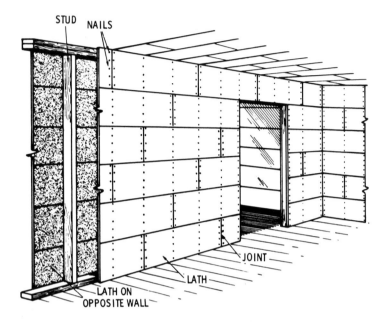

Fig. 22-3. Application of gypsum plaster base.

is used. This material can be obtained with a foil backing that serves as a vapor barrier, and if it faces an air-space, it has some insulating value. It is also available with perforations, which improves the bonding strength of the plaster base.

Insulating fiberboard lath may also be used as a plaster base. It is usually ½ in. thick and generally comes in strips of 18″ × 48″. It often has a shiplap edge and may be used with metal clips that are located between studs and joists to stiffen the horizontal joints. Fiberboard lath has a value as insulation and may be used on the walls or ceilings adjoining exterior or·unheated areas.

Gypsum lath should be applied horizontally, with the joints broken as shown in Fig. 22-3. Vertical joints should be made over the center of the studs or joists, and should be nailed with 13-gauge gypsum lathing nails 1⅛ in. long and having a ⅜-in. flat head. Nails should be spaced 4 in. on center, and should be nailed at each stud or joist crossing. Lath joints over the heads of door and window openings should not occur at the jamb lines. Insulating

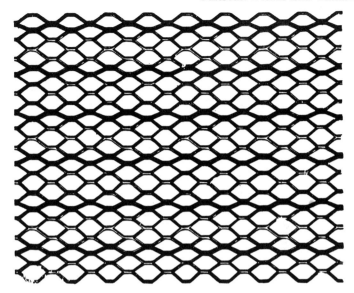

Fig. 22-4. Expanded-metal screening, instead of wood furring strips, is applied to the wall.

lath should be installed much like gypsum lath, except that 13-gauge $1\frac{1}{4}$-in. blued nails should be used.

Metal Lath

Metal lath comes in sheets, usually measuring 27 by 96 in., which have been die cut with numerous small slits and then "expanded,"or pulled apart, to form openings for the keying of the plaster. The sheets come prepainted or galvanized to prevent rust and thus assure a tight plaster bond.

Expanded metal lath can be used by itself as a plaster base in high-moisture applications (Fig. 22-5) or on furring strips installed over masonry unsuitable for direct plaster application (Fig. 22-6). It can also be used to reinforce gypsum lath over window and door openings (Fig. 22-7) when cut into strips about 8 by 20 in. and lightly nailed in place.

Inside corners at the juncture of walls and ceilings should be reinforced with corners of metal lath or wire fabric, as shown in Fig. 22-8, except where special clip systems are used for installing

443

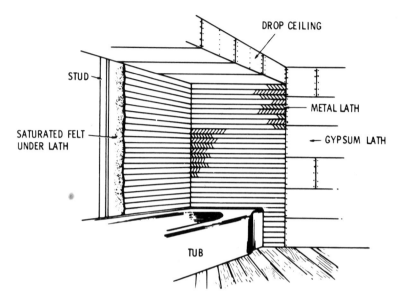

Fig. 22-5. Application of expanded-metal lath around bathtub for the installation of ceramic tile.

the lath. The minimum width of the lath in the corners should be 5 in., or 2½ in. on each surface or internal angle, and should also be lightly nailed in place. Corner beads, as shown in Fig. 22-9, of expanded metal lath or perforated metal, should be installed on all exterior corners. They should be applied plumb and level. The bead acts as a leveling edge when the walls are plastered and reinforces the corner against mechanical damage.

Metal lath should be used under large flush beams, as shown in Fig. 22-10, and should extend well beyond the edges. Where reinforcing is required over solid wood surfaces, such as drop beams, the metal lath should either be installed on strips or else self-furring nails should be used to set the lath out from the beam. The lath should be lapped on all adjoining gypsum lath surfaces.

Masonry

Concrete masonry provides an excellent base for plaster. The surface of any masonry unit should be rough to provide a good

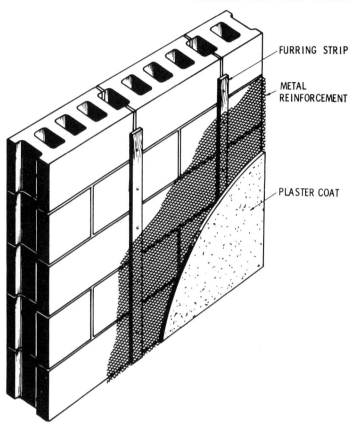

FURRING STRIP

METAL
REINFORCEMENT

PLASTER COAT

Fig. 22-6. Furring strips applied to concrete masonry for the application of plaster.

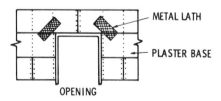

METAL LATH

PLASTER BASE

OPENING

Fig. 22-7. Reinforcement of plaster over openings using expanded-metal lath.

445

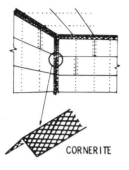

CORNERITE

**Fig. 22-8. Use of cornerites
made from metal lath.**

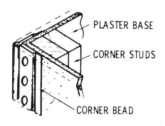

PLASTER BASE

CORNER STUDS

CORNER BEAD

Fig. 22-9. Use of a corner bead.

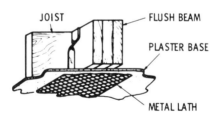

JOIST

FLUSH BEAM

PLASTER BASE

METAL LATH

Fig. 22-10. Use of expanded-metal lath with large flush beams.

mechanical key, and should be free from paint, oil, dust, and dirt, or any other material that might prevent a satisfactory bond. Proper application of plaster requires:

1. That the plaster base material bonds and becomes an integral part of the base to which it is applied, as shown in Fig. 22-11.
2. That it be used as a thin reinforced base for the finished surface.

446

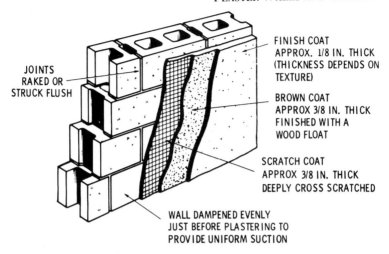

Fig. 22-11. Application of plaster to concrete masonry.

Old masonry walls which have been softened by weathering, or surfaces that cannot be cleaned thoroughly, must be covered with metal lath reinforcement before applying the plaster. Metal lath reinforcement should be applied to wood furring strips, as shown in Fig. 22-6. The metal lath reinforcement should be well braced and rigid to prevent cracking the plaster. This type of construction will give some insulating qualities due to the air space between masonry and plaster.

Plaster Grounds

Plaster grounds are strips of wood the same thickness as the lath and plaster, and are attached to the framing before the plaster is applied. Plaster grounds are used around window and door openings as a plaster stop, and along the floor line for attaching the baseboard. They also serve as a leveling surface when plastering, and as a nailing base for the finish trim, as shown in FIG. 22-12. There are two types of plaster grounds; those that remain in place (Fig. 22-13), and those that are removed after plastering is completed (Fig. 22-14). The grounds that remain in place are usually ⅞ in. thick and may vary in width from 1 inch around openings to 2 in. for those used along the floor line. These grounds are nailed securely in place before the plaster base is installed. Where a

447

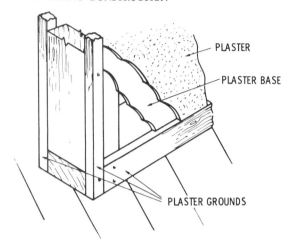

Fig. 22-12. Plaster grounds used at doors and floor line.

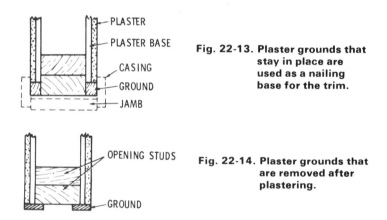

Fig. 22-13. Plaster grounds that stay in place are used as a nailing base for the trim.

Fig. 22-14. Plaster grounds that are removed after plastering.

painted finish is used, finished door jambs are sometimes placed in the rough openings, and the edges of the jambs serve as the grounds during the plastering operation.

Lath Nailers

Lath nailers are horizontal or vertical members to which lath, gypsum boards, or other covering materials are nailed. These

448

members are required at interior corners of the walls and at the juncture of the wall and ceiling. Vertical lath nailers may be composed of studs so arranged as to provide nailing surfaces, as shown in Fig. 22-15. This construction also provides a good tie between walls.

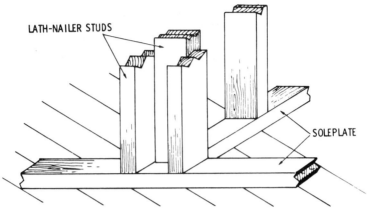

LATH-NAILER STUDS

SOLEPLATE

Fig. 22-15. Lath nailers at wall intersections.

Another vertical nailer construction consists of a 2″ × 6″ lathing board that is nailed to the stud of the intersecting wall, as shown in Fig. 22-16. Lathing headers are used to back up the board. The header should be toe-nailed to the stud. Doubling of the ceiling joist over the wallplates provides a nailing surface for interior finish material, as shown in Fig. 22-17. Walls may be tied to the ceiling framing in this method by toe-nailing through the joists into the wall plates.

Another method of providing nailing surface at the ceiling line is similar to that used on the walls. A 1″ × 6″ lathing board is nailed to the wall plate as shown in Fig. 22-18. Headers are used to back up this board, and the header, in turn, can be tied to the wall by toenailing into the wall plate.

PLASTER CONTENT

Plaster for interior finishing is made from combinations of sand, lime or prepared plaster, and water. Most plasters contain a

449

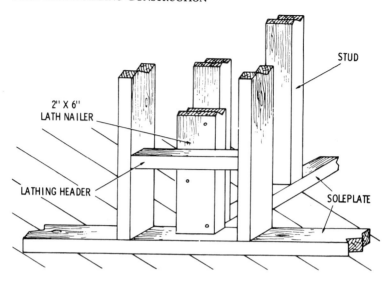

STUD

2" X 6"
LATH NAILER

LATHING HEADER

SOLEPLATE

Fig. 22-16. Another method of installing lath nailers at intersection walls.

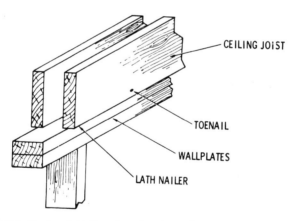

CEILING JOIST

TOENAIL

WALLPLATES

LATH NAILER

Fig. 22-17. Horizontal lath nailers for plaster base formed by ceiling joists.

number of minerals, including gypsum, which alter the mix characteristics slightly to meet certain requirements. Properly mixed plaster is smooth, plastic, and workable with standard plasterer's tools. For application over lath, it is supplied in bags with the

450

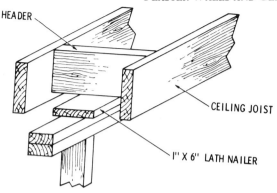

Fig. 22-18. Horizontal lath nailers for ceiling provided by lathing boards.

necessary ingredients already mixed, and requires only the addition of water according to instructions on each bag.

Plaster must be mixed in clean wooden boxes. Mix only as much as can be applied within an hour or hydration will begin to take place and the plaster will become too hard to handle with ease. Each time a new batch is to be mixed, the box must first be thoroughly washed out. Leave the box clean at the end of the day ready for use the next day.

PLASTER APPLICATION

Plaster is carried to the work on a hand-held square wooden platen, and applied to the lathing with a rectangular trowel.

Over metal lath and over masonry, plaster is usually applied in three coats which are termed in order of application as follows:

1. Scratch coat.
2. Second or brown coat.
3. Finish or set coat.

Scratch Coat—This coat should be made approximately ⅜ in. thick measured from the face of the backing and carried to the full length of the wall or the natural breaking points such as doors or windows. Before the scratch coat begins to harden, however, it should be cross-scratched to provide a mechanical key for the second or brown coat.

Brown Coat—The brown coat should be approximately the same thickness as that of the scratch coat. Before applying the brown coat, dampen the surface of the scratch coat evenly by means of a fog spray to obtain uniform suction. This coat may be applied in two thin coats, one immediately following the other. Such a method may prove helpful in applying sufficient pressure to ensure a proper bond with the base coat.

Bring the brown coat to a true even surface then roughen with a wood float or cross-scratch it lightly to provide a bond for the finish coat. Damp-cure the brown coat for at least two days then allow it to dry.

Finish Coat—The brown coat (as mentioned above) should be dampened for at least two days before application of the finish coat. Begin moistening as soon as the brown coat has hardened sufficiently, applying the water in a fine fog spray. Avoid soaking the wall but give it as much water as will readily be absorbed.

"Double-up" work, combining the scratch and brown coat, is used on gypsum or insulating lath, and the leveling and plumbing of walls and ceilings are done during the application of this work. The final or finish coat consists of two general types—the sand-float and the putty finish. In the sand-float finish, lime is mixed with sand and results in a textured finish, with the texture depending on the coarseness of the sand used. Putty finish is used without sand and has a smooth finish. This is commonly used in kitchens and bathrooms, where a gloss paint or enamel finish is often employed, and in other rooms where a smooth finish is desired.

The plastering operation should not be done in freezing weather without the use of constant heat for protection from freezing. In normal construction, the heating unit is in place before plastering is started.

Insulating plaster, consisting of a vermiculite, perlite, or other aggregate used with the plaster mix, may also be used for wall and ceiling finishes. This aggregate properly mixed with the plaster produces small hollow air pockets which act as insulating material. The vermiculite is a material developed from a mica base which is exploded in size, and when mixed with the plaster, reduces the added weight in a conventional plastered wall or ceiling.

trowel is light, so that the tool is easily used. Trowels are classed as *browning* or *finishing*.

The following points in plaster maintenance are worthy of attention:

1. In a newly constructed house, a few small plaster cracks may develop during or after the first heating season. These cracks are usually due to drying and shrinking of the structural members. For this reason, it is advisable to wait until after a heating season before painting the plaster. These cracks can then be filled before painting has begun.

2. Because of the curing period ordinarily required for a plastered wall, it is not advisable to apply oil-base paints until at least 60 days after plastering is completed. Water-mix, or resin-base paints may be applied without the necessity of an aging period.

3. Large plaster cracks often indicate a structural weakness in the framing. One of the common areas that may need correction is around a basement stairs. Framing may not be adequate for the loads of the walls and ceilings. In such cases, the use of an additional post and pedestal may be required to correct this fault. Inadequate framing around fireplaces and chimney openings, and joists that are not doubled under partitions, are other common sources of weakness.

PLASTERING TOOLS

The assortment of tools used by the plasterer is very similar to those employed by the bricklayer. Essential plastering tool are as follows:

Hawk—This tool is usually made of hard pine or cedar and is usually about 13″ or 14″ square (Fig. 22-19). It is held in the left hand forming a small "hand table" which holds a supply of plaster. The plasterer scoops the plaster from the hawk to the work where he spreads it over the surface.

Trowel—As distinguished from the bricklayers' trowel, the plasterers' trowel is rectangular in shape, as shown in Fig. 22-20. The handle is attached to a mounting which stiffens the blade. The

453

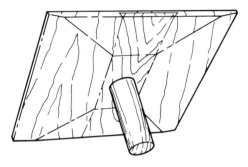

Fig. 22-19. A wood hawk used to carry plaster to the wall or ceiling.

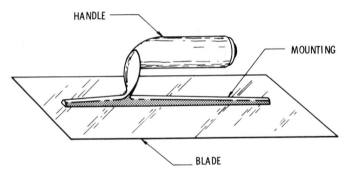

Fig. 22-20. The trowel used to apply plaster is generally made of 24-gauge polished steel.

The browning trowel is used for rough coating and has a heavier blade than that used on the finishing trowel, otherwise the construction of both is the same.

Float—The common form of float consists of a piece of hard pine board 10″ or 12″ × ⅝″ to ⅞″ having a wooden handle, preferably of hard wood screwed to the back. Because of the great friction, the face of a float soon wears off and becomes thin; hence, there is usually an adjustable handle fastened with bolts which can be fixed to new face pieces as they are required. Floats are applied in smoothing and finishing with a rotary motion sometimes reversed as left to right and vice versa. Various types of floats are shown in Fig. 22-21.

454

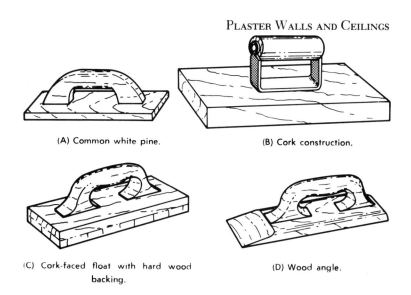

(A) Common white pine.

(B) Cork construction.

(C) Cork-faced float with hard wood backing.

(D) Wood angle.

Fig. 22-21. Several types of wooden floats.

Darby—This tool is simply a flat straight strip of wood (or metal) provided with handles to enable the workman to level up and straighten large surfaces as they are put on.

The tool is held by both hands and moved with a sliding up and down diagonal and horizontal motion to level off by rubbing and pressing any lumps or high spots which may be left after applying the mortar with the trowels. This work is very laborious, especially on ceilings or any job above the line of the shoulders. The tool is also essential in preserving an even thickness of each coat. Fig. 22-22 shows an ordinary wooden darby.

It is important that the first coat of plaster have the right consistency, to establish a firm key onto the lath. If the plaster is too thick it will stand out behind the lathing material but not grab securely. A thinner plaster tends to turn down behind the lathing material and form a key and lock into place. This is illustrated in Fig. 22-23 with wood lathing. The same reasoning applies to metal lathing, although keying to expanded metal lathing is more secure even with stiff plaster material.

Each coat must be thoroughly dry before applying the next coat. While the second coat is fogged with a spray of water before

455

Fig. 22-22. An ordinary wooden darby used to level large areas of plaster.

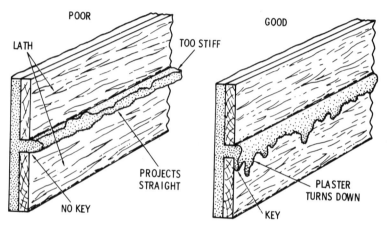

Fig. 22-23. Good and bad embedding of plaster on furring strips. The plaster must form a key behind the strips to hold wall.

applying the finish coat, the second coat must be allowed to dry before fogging.

It is often desirable to obtain a hard glossy surface on the finish coat. This is done by brushing water on the surface, using a large painter's brush, then floating again with a hardwood float. If this is done a couple of times, the surface will take on a very hard finish.

ONE-COAT PLASTERING

In addition to the plastering methods described, there has been one other development in plastering that reduces the plastering process to "one coat" and that gives the process its name. Here, "green" or "blue" ⅝-in. gypsum board is installed like regular gypsum board, and the premixed plaster material, which comes in 5-gal. cans, is applied by the plasterer. That's it. The applied material cures very hard, and the gypsum board has the tooth to keep it in place.

While the description given above for applying plaster is brief, it can readily be seen how much labor is involved. This is why this kind of plastering has given way so much to the use of plasterboard.

PLASTERBOARD

Plasterboard (sometimes referred to as drywall or wallboard) is available in solid sheets made of fireproof material. Standard width is usually 4 ft., with a few types in 2-ft. widths. Lengths are from 6 ft. to 16 ft. and are available with thicknesses from ¼ to ¾ in. Edges are tapered to permit smooth edge finishing, although some boards can be purchased with square edges.

The most common length is 8 ft. to permit wasteless mountings to the wall studs. This particular size will fit standard 8-ft. ceiling heights when vertically mounted. However, recommended mounting for easier handling is horizontal. Two lengths, one above the other, fit the 8-ft. ceiling heights, when mounted horizontally. In addition to regular wallboard, many other types are available for special purposes.

Regular wallboard comes in thicknesses of ¼, ⅜, ½, and ⅝ in. The ¼-in. board is used in two or three layer wallboard construction and generally has square edges. The most commonly used thickness is ⅜ in., which can be obtained in lengths up to 16 ft. All have a smooth cream-colored paper covering that takes any kind of decoration.

Insulating (foil-backed) wallboards have aluminum foil which is laminated to the back surface. The aluminum foil creates a vapor barrier and provides reflected insulation value. It can be used in a single-layer construction or as a base layer in two-layer construction. Thicknesses are ⅜, ½, and ⅝ in., with lengths from 6 to 14 ft. All three thicknesses are available with either square or tapered edges.

Wallboard with a specially formulated core, which provides increased fire-resistance ratings when used in recommended wall and ceiling systems, is made by many companies. National Gypsum Co. calls their product Fire-Shield which achieves a one-hour fire rating in single-layer construction over wood studs. It is manufactured in ½ - and ⅝ -in. thicknesses with lengths from 6 to 14 ft., with tapered edges.

Some suppliers have a lower cost board called backer board. These are used for the base layer in two-layer construction. Thicknesses are ⅜ , ½ , and ⅝ in, with a width of 2 ft. as well as the usual 4 ft. Length is 8 ft. only. The edges are square but the 2-ft. wide, in ½ -, and ⅝ -in. thickness can be purchased with tongue-and-groove edges.

Another type of backer board has a vinyl surface. It is used as a waterproof base for bath and shower areas. The size is 4′ × 11′ which is the right size for enclosing around a standard bathtub. It is available in ½ - and ⅝ -in. thicknesses with square edges.

A moisture-resistant board is specially processed for use as a ceramic tile base. Both core and facing paper are treated to resist moisture and high humidity. Edges are tapered and the regular tile adhesive seals the edges. The boards are 4 ft. wide with lengths of 8 ft. and 12 ft.

There is also vinyl-covered decorator plasterboard, which eliminates the need for painting or wallpapering. The vinyl covering is available in a large number of colors and patterns to fit many decorator needs. Many colors in a fabric like finish plus a number of wood grain appearances can be obtained. Only mild soap and water are needed to keep the finish clean and bright. Usual installation is by means of an adhesive, which eliminates nails or screws except at the top and bottom where decorator nails are generally used. The square edges are butted together (Fig. 22-24). They are available in ½ - and ⅜ -in. thicknesses, 4 ft. wide, and 8 ft., 9 ft., and 10 ft. long.

A similar vinyl covered plasterboard, called monolithic, has an extra width of vinyl along the edges. The boards are installed with adhesive or nails. The edges are brought over the fasteners and pasted down then cut flush at the edges. This method hides the fasteners. Extruded aluminum bead and trim accessories, covered with matching vinyl, can be purchased to finish off inside, outside, and ceiling corners, for vinyl covered plasterboard.

Courtesy National Gypsum Co.

Fig. 22-24. Home interior design utilizes vinyl-finish plasterboard.

Plasterboard Ratings

The principal manufacturers of plasterboard maintain large laboratories for the testing of their products for both sound absorption ability and fire retardation. They are based on established national standards that have been industry accepted and all follow the same procedure. Tests for sound absorption result in figures of merit and those for fire retardation on the length of time a wall will retard a fire. Both vary depending on the material of the wallboard, the layers and thickness, and the method of wall construction.

CONSTRUCTION WITH PLASTERBOARD

Several alternate methods of plasterboard construction may be used, depending on the application and desires of the customer, or

459

yourself if you are the homeowner. Most residential homes use 2″ × 4″ wall studs and single-layer plasterboard walls are easily installed. Custom built homes, commercial buildings, and party walls between apartments should use double-layer plasterboard construction. It provides better sound insulation and fire retardation. For the greatest protection against fire hazards, steel frame partitions, plus plasterboard, provide an all noncombustible system of wall construction. All wood or the all steel construction apply to both walls and ceilings.

Plasterboard may also be applied directly to either insulated or uninsulated masonry walls. Plasterboard may be fastened in place by any of several methods such as, nails, screws, or adhesives. In addition, there are special resilient furrings and spring clips, both providing added sound deadening.

Single Layer on Wood Studs

The simplest method of plasterboard construction is a single layer of plasterboard nailed directly to the wood studs and ceiling joists (Fig. 22-25). For single-layer construction, ⅜-, ½-, or ⅝-in. plasterboard is recommended.

The ceiling panels should be installed first, then the wall material. Install plasterboard perpendicular to the studs for minimum joint treatment and greater strength. This applies to the ceiling as well. Either nails, screws, or adhesive may be used to fasten the plasterboard. Blue wallboard nails have annular rings for better grip than standard nails and the screws have Phillips heads for easy installation with a power driver. In single-nail installations, as shown in the illustration of Fig. 22-26, the nails should be spaced not to exceed 7 in. on the ceilings and 8 in. on the walls. An alternate method, one which reduces nail popping, is the double-nail method. Install the boards in the single-nail method first, but, with nails 12 in. apart. Then, drive a second set of nails about 1½ to 2 in. from the first (Fig. 22-26), from the center out, but not at the perimeter. The first series of nails are then struck again to assure the board being drawn up tight.

Use 4d cooler type nails for ⅜-in. regular foil backed or backer board if used, 5d for ½-in. board, and 6d for ⅝-in. board.

Courtesy National Gypsum Co.

Fig. 22-25. The most common interior finishing uses plasterboard, known as drywall construction.

Screws are even better than nails for fastening plasterboard, since they push the board up tight against the studding and will not loosen. Where studs and joists are 16 in. on center, the screws can be 12 in. apart on ceilings and 16 in. apart on walls. If the studs are 24 in. on center, the screws should not be over 12 in. apart on the walls. Fig. 22-27 shows a power screwdriver in use. Strike the heads of all nails, and screws, to just below the surface of the board. The dimples will be filled in when joints and corners are finished.

A number of adhesives are available for installing plasterboard. Some are quick-drying, others are slower. Adhesives can be purchased in cartridge or bulk form, as shown in Fig. 22-28. The adhesive is applied in a serpentine bead, as shown in Fig. 22-29, to the facing edges of the studs and joists. Place the wallboard in position and nail it temporarily in place. Use double-headed nails

461

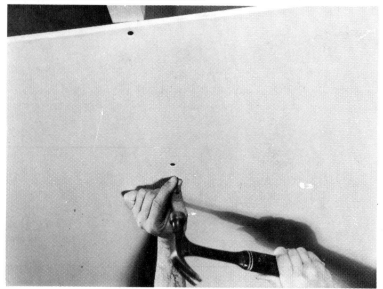

Courtesy National Gypsum Co.

Fig. 22-26. In the double-nail method of framing wallboard, the first set of nails is put in place and a second set is driven in place about 1 1/2 to 2 in. from the first.

or nails through a piece of scrap plasterboard. When the adhesive has dried, the nails can be removed and the holes filled when sealing the joints.

Plasterboard is butted together but not forced into a tight fit. The treatment of joints and corners is covered later. The adhesive method is ideal for prefinished wallboard, although special nails with colored heads can be purchased for the decorator boards. Do not apply more adhesive than can permit installation within 30 minutes.

Two-Layer Construction

As mentioned, two layers of drywall improves sound deadening and fire retardation. The first layer of drywall or plasterboard is applied as described before. Plasterboard may be less expensive used as a backing type board. Foil backed or special sound reducing board may also be used. Nails need only be struck and screws

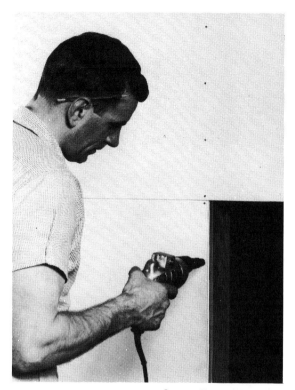

Courtesy National Gypsum Co.

Fig. 22-27. Wallboard fastened to studs with a power screwdriver.

driven with their heads just flush with the surface of the board. The joints will not be given special treatment since they will be covered by the second layer.

The second layer is cemented to the base layer. The facing layer is placed over the base layer and temporarily nailed at the top and bottom. Temporary nailing can be done with either double-headed nails or blocks of scrap wood or plasterboard under the heads. The face layer is left in place until the adhesive is dry.

To make sure of good adhesion, lay bracing boards diagonally from the center of the facing layer to the floor. Prebowing of the boards is another method. Lay the panels finish face down across a 2″ × 4″ for a day or two. Let the ends hang free.

463

Courtesy National Gypsum Co.

Fig. 22-28. Adhesives are available in bulk or cartridge form.

Joints of boards on each layer should not coincide but should be separated by about 10 in. One of the best ways of doing this is to install the base layer horizontally and finish layer vertically. When using adhesive, a uniform temperature must be maintained. If construction is in the winter, the rooms should be heated to somewhere between 55°F and 70°F, and kept well ventilated.

Double-layer construction, with adhesive, is the method used with decorator type vinyl finished panelling except that special matching nails are used at the base and ceiling lines and left permanently in place.

Fig. 22-30 shows the basic double-layer construction, with details for handling corners, and an alternate method of wall construction for improved sound deadening.

Fig. 22-29. Using a cartridge to apply a bead of adhesive to wall studs.

Courtesy National Gypsum Co.

Other Methods of Installation

Sound transmission through a wall can be further reduced if the vibrations of the plasterboard can be isolated from the studs or joists. Two methods are available to accomplish this. One uses metal furring strips whose edges are formed in such a way as to give them flexibility. The other is by means of metal push-on clips with bowed edges also for flexibility, used on ceilings only.

Fig. 22-31 shows the appearance and method of installing resilient furring channels. They are 12 ft. long strips of galvanized steel and include predrilled holes spaced every inch. This permits nailing to studs 16 or 24 in. on center. Phillips head self-tapping screws are power driven through the plasterboard and into the surface of the furring strips.

As shown in Fig. 22-31, strips of $3'' \times \frac{1}{2}''$ plasterboard are fastened to the sole and plate at top and bottom to give the plasterboard a solid base. For best sound isolation, the point of intersection between the wall and floor should be caulked prior to application of the baseboard.

On ceiling joists, use two screws at each joist point to fasten the furring strips. Do not overlap ends of furring strips, leave about a

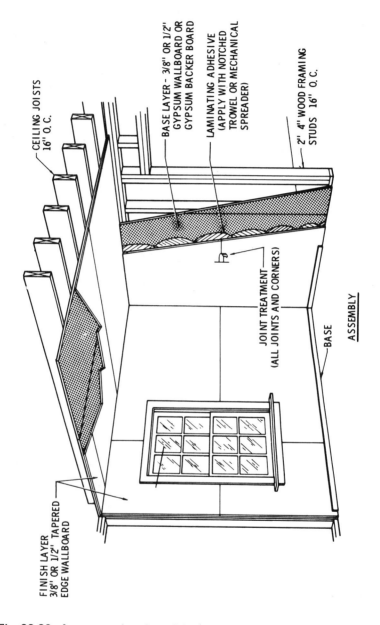

CEILING JOISTS 16" O.C.

BASE LAYER - 3/8" OR 1/2" GYPSUM WALLBOARD OR GYPSUM BACKER BOARD

LAMINATING ADHESIVE (APPLY WITH NOTCHED TROWEL OR MECHANICAL SPREADER)

2" 4" WOOD FRAMING STUDS 16" O.C.

JOINT TREATMENT (ALL JOINTS AND CORNERS)

BASE

ASSEMBLY

FINISH LAYER 3/8" OR 1/2" TAPERED EDGE WALLBOARD

Fig. 22-30. A cross-section view of double-layer wall construction, details

466

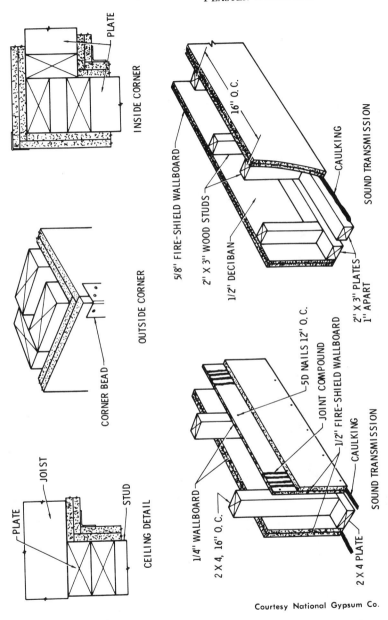

PLATE

INSIDE CORNER

OUTSIDE CORNER

CORNER BEAD

JOIST

PLATE

STUD

CEILING DETAIL

5/8" FIRE-SHIELD WALLBOARD

2" X 3" WOOD STUDS

1/2" DECIBAN

16" O. C.

CAULKING

SOUND TRANSMISSION

2" X 3" PLATES
1" APART

5D NAILS 12" O. C.

JOINT COMPOUND

1/2" FIRE-SHIELD WALLBOARD

1/4" WALLBOARD

2 X 4, 16" O. C.

CAULKING

SOUND TRANSMISSION

2 X 4 PLATE

Courtesy National Gypsum Co.

for handling corners, and special wall construction for sound deadening.

467

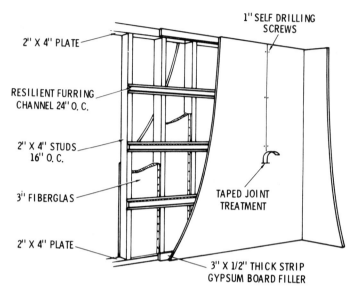

1" SELF DRILLING SCREWS

2" X 4" PLATE

RESILIENT FURRING CHANNEL 24" O. C.

2" X 4" STUDS 16" O. C.

3" FIBERGLAS

TAPED JOINT TREATMENT

2" X 4" PLATE

3" X 1/2" THICK STRIP GYPSUM BOARD FILLER

PERSPECTIVE

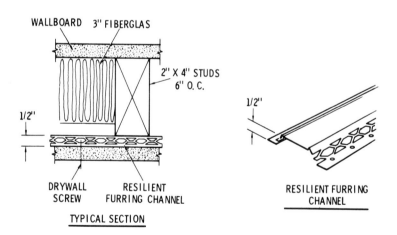

WALLBOARD   3" FIBERGLAS

2" X 4" STUDS 6" O. C.

1/2"

1/2"

DRYWALL SCREW   RESILIENT FURRING CHANNEL

RESILIENT FURRING CHANNEL

TYPICAL SECTION

Courtesy National Gypsum Co.

Fig. 22-31. Resilient furring strips, with expanded edges, may be installed to isolate the plasterboard from the stud.

⅜-in. space between ends. Use only ½- or ⅝-in. wallboard with this system.

Suspending plasterboard ceilings from spring clips is a method of isolating the vibrations from the ceiling joists and floor above. Details are shown in Fig. 22-32.

The push-on clips are placed on 1″ × 2″ wood furring strips. The clips are then nailed to the sides of the ceiling joists with short annular ringed or cooler type nails. With furring strips installed, they will be against the nailing edge of the joists. This provides a solid foundation for nailing the plasterboard to the clips. The weight of the plasterboard after installation will stretch the clips to a spring position.

The furring strips must be of high quality material so they will not buckle, twist, or warp. Nails for attaching the wallboard should be short enough so they will not go through the furring strips and into the joist edges. The wallboard must be held firmly against the strips while it is being nailed. Expansion joints must be provided every 60 ft. or for every 2,400 sq. ft. of surface.

It is essential that the right number of clips be used on a ceiling, depending on the thickness of the wallboard and whether single or double layered. The spring or efficiency of the clips is affected by the weight.

Wall Tile Backing

At least two types of plasterboard are available as backing material for use in tiled areas where moisture protection is important, such as tub enclosures, shower stalls, powder rooms, kitchen-sink splash boards, and locker rooms. One is a vinyl-covered plasterboard and the other is a specially designed board material to prevent moisture penetration.

The vinyl-covered board is 4′ × 11′ the right length for a tub enclosure. By scoring it and snapping it to length, the vinyl covering can be a continuous covering all around the tub and no corner sealing is required. This is shown in one of the sketches in Fig. 22-33. This board has square edges and is nailed or screwed to the studs without the need to countersink the nail or screws. The waterproof tile adhesive is applied and the tile is installed over the board. This board is intended for full tile treatment and is not to extend beyond the edges of the tile.

469

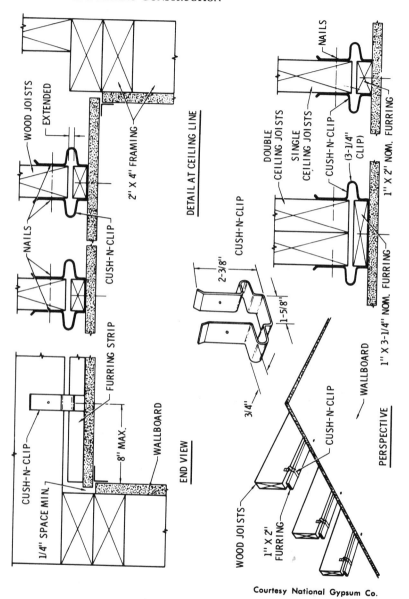

Courtesy National Gypsum Co.

Fig. 22-32. Details for installing spring clips to ceiling joist. Clips suspend plasterboard from the stud.

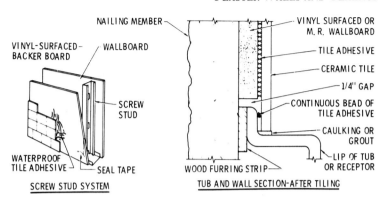

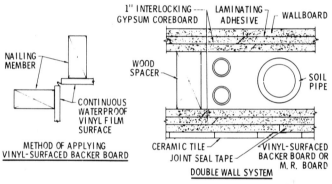

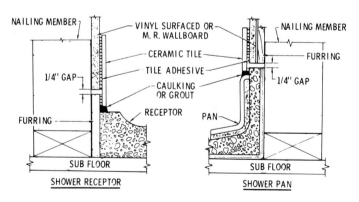

Fig. 22-33. Details for installing vinyl-locked and moisture-resistant wall-board for bath and shower use.

471

Unlike vinyl-surfaced board, the moisture resistant board may be extended beyond the area of the tile. The part extending beyond the tile may be painted with latex, oil-based paint, or papered. It has tapered edges and is installed and treated in the same manner as regular plasterboard. Corners are made waterproof by the tile adhesive. The sketches in Fig. 22-33 show details for installation of either board.

Plasterboard over Masonry

Plasterboard may be applied to inside masonry walls in any one of several methods: directly over the masonry using adhesive cement; over wood or furring strips fastened to the masonry; polystryrene insulated walls by furring strips over the insulation; or by lamination directly to the insulation.

Regular or prefinished drywall may be laminated to unpainted masonry walls such as concrete or block interior partitions and to the interior of exterior walls above or below grade. While nearly all of the adhesive available may be used either above or below grade, the regular joint compounds are recommended for use only above grade. Exterior masonry must be waterproofed below grade and made impervious to water above grade.

The masonry must be clean and free of dust, dirt, grease, oil, loose particles, or water soluble particles. It must be plumb, straight, and in one plane. Fig. 22-34 shows the method of applying the adhesive to masonry walls. Boards may be installed either horizontally or vertically. Ceiling wallboard should normally be applied last to allow nailing temporary bracing to the wood joists. The wallboard should be installed with a clearance of ⅛ in. or more from the floor, to prevent wicking. If there are expansion joints in the masonry, cut the wallboards to include expansion joints to match that of the masonry.

The sketches of Fig. 22-35 show the installation of wallboard to masonry with furring strips. Furring strips may be wood or U-shaped metal as used for foam insulation. The furring strips are fastened to the masonry with concrete nails. They may be mounted vertically or horizontally but there must be one horizontal strip along the base line. The long dimension of the wallboards are to be perpendicular to the furring strips. Furring strips are

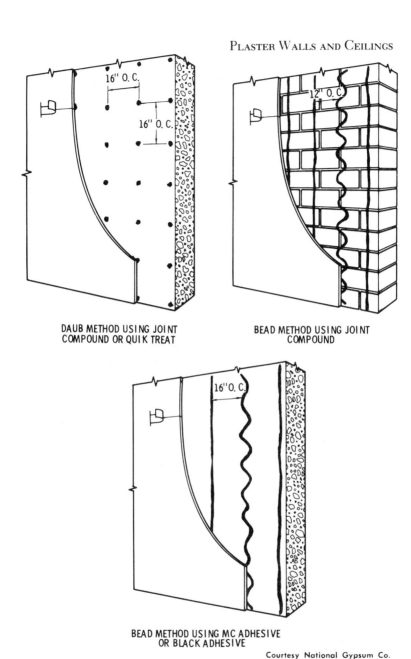

DAUB METHOD USING JOINT
COMPOUND OR QUIK TREAT

BEAD METHOD USING JOINT
COMPOUND

BEAD METHOD USING MC ADHESIVE
OR BLACK ADHESIVE

Courtesy National Gypsum Co.

Fig. 22-34. Method of applying adhesive directly to masonry walls.

473

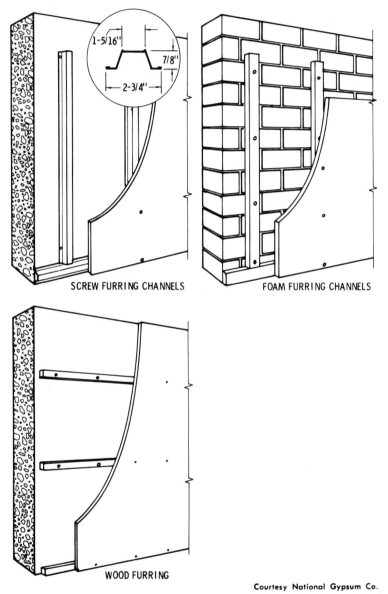

SCREW FURRING CHANNELS

FOAM FURRING CHANNELS

WOOD FURRING

Courtesy National Gypsum Co.

Fig. 22-35. Three types of furring strips used between masonry walls and plasterboard.

fastened 24 in. on center. The wallboard is attached to the furring strips in the usual way. Use self-drilling screws on the metal furring strips and nails or screws on the wood strips.

Insulation may be applied between the plasterboard and the masonry, using urethane foam sheets or extruded polystyrene. Plasterboard is then secured to the wall either with the use of furring strips over the insulation or by laminating directly to the insulation.

Special U-shaped metal furring strips are installed over the insulation. This is shown in Fig. 22-36. The insulation pads may be held against the wall while the strips are installed or they can to be held in place with dabs of plasterboard adhesive. Use a fast drying adhesive. If some time is to elapse before the furring strips and

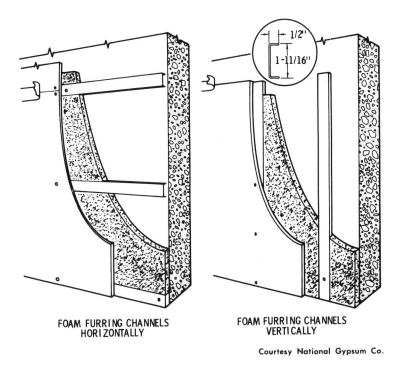

FOAM FURRING CHANNELS
HORIZONTALLY

FOAM FURRING CHANNELS
VERTICALLY

Courtesy National Gypsum Co.

Fig. 22-36. U-shaped furring channels installed horizontally or vertically over foam insulating material.

plasterboard are to be applied, dot the insulation with adhesive every 24 in. in both directions on the back on the insulation pads.

The metal furring strips are fastened in place with concrete nails. Place them a maximum of 24 in. apart, starting about 1 in. to 1 ½ in. from the ends. The strips may be mounted vertically or horizontally, as shown in Fig. 22-36. Fasten the wallboard to the furring strips with a power driver (Fig. 22-37), using self-drilling screws. Screws must not be any longer than the thickness of the wallboard plus the ½-in. furring strip. Space screws 24 in. on center for ½-in. and ⅝ in. board, 16 in. on center for ⅜-in. board.

Courtesy National Gypsum Co.

Fig. 22-37. Special self-drilling screws are power-driven through the plasterboard and into the metal furring strip.

Plasterboard may be laminated directly to the foam insulation. The sketches in Fig. 22-38 show how this is done for both horizontal and vertical mounted boards. Install wood furring strips onto the masonry wall, for the perimeter of the wallboard. The strips should be 2 in. wide and ⅟₃₂-in. thicker than the foam insulation. Include furring where the boards join, as shown in the left hand sketch of Fig. 22-38.

476

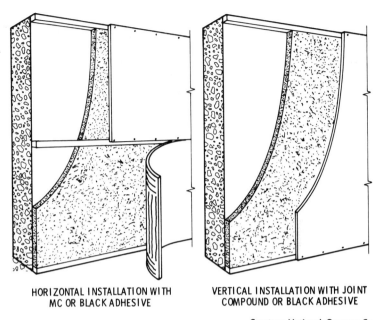

HORIZONTAL INSTALLATION WITH
MC OR BLACK ADHESIVE

VERTICAL INSTALLATION WITH JOINT
COMPOUND OR BLACK ADHESIVE

Courtesy National Gypsum Co.

Fig. 22-38. Plasterboard may be installed against the foam insulation with adhesive.

Apply a ⅜-in. diameter bead over the back of the foam insulation and in a continous strip around the perimeter. Put the bead dots about 16 in. apart. Apply the foam panels with a sliding motion and hand press the entire panel to ensure full contact with the wall surface. For some adhesives, it is necessary to pull the panel off of wall to allow flash-off of the solvent. Then reposition the panel. Read the instructions with the adhesive purchased.

Coat the back of the wallboard and press it against the foam insulation. Nail or screw the board to the furring strips. The nails or screws must not be too long or they will penetrate the furring strips and press against the masonry wall. The panels must clear the floor by ⅛ in. About any type of adhesive may be used with urethane foam but some adhesives cannot be used with polystryrene. Also, in applying prefinished panels, be careful on the choice of adhesive. Some of the quick-drying types are harmful to the finish. Read the instructions on the adhesive before you buy.

FIREPROOF WALLS AND CEILINGS

For industrial applications, walls or ceilings may be made entirely of noncombustible materials, using galvanized steel wall and ceiling construction and faced with plasterboard. Assembly is by self-drilling screws, driven by a power screwdriver with a Phillips bit.

WALL PARTITIONS

Construction consists of U-shaped track which may be fastened to existing ceilings, or the steel member ceilings, and to the floors. U-shaped steel studs are screwed to the tracks. Also available are one-piece or three-piece metal door frames. The three-piece frames are usually preferred by contractors since they permit finishing the entire wall before the door frames are installed.

The sketches in Fig. 22-39 show details of wall construction as well as the cross section of the track and stud channels. For greater sound deadening, double-layer board construction with resilient furring strips between the two layers is recommended. This is shown in a cutaway view in the lower right hand corner of Fig. 22-39. Heavier material and wider studs are required for high wall construction.

Fig. 22-40 and 22-41 show details for intersecting walls, jambs, and other finishing needs. Chase walls are used (Fig. 22-42), where greater interior wall space is needed (between the walls) for equipment.

Fig. 22-43 shows how brackets are attached for heavy loads. Table 22-1 lists the allowable load for bolts installed directly to the plaster board.

STEEL FRAME CEILINGS

Three types of furring members are available for attaching plasterboard to ceilings. They are screw-furring channels, resilient screw-furring channels, and screws studs. Self-drilling screws attach the wallboard to the channels. These are illustrated in Fig.

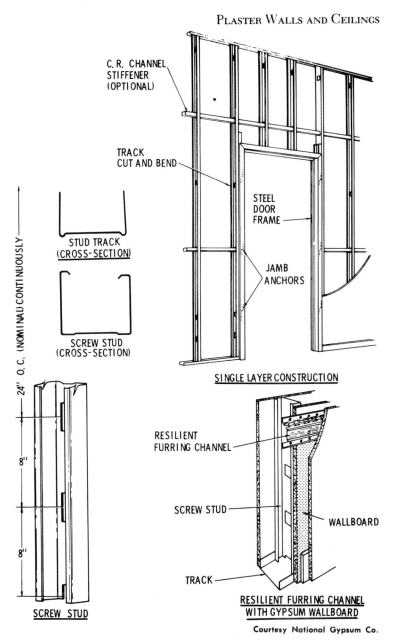

C. R. CHANNEL
STIFFENER
(OPTIONAL)

TRACK
CUT AND BEND

STEEL
DOOR
FRAME

JAMB
ANCHORS

24" O. C. (NOMINAL) CONTINUOUSLY

STUD TRACK
(CROSS-SECTION)

SCREW STUD
(CROSS-SECTION)

SINGLE LAYER CONSTRUCTION

8"

8"

RESILIENT
FURRING CHANNEL

SCREW STUD

WALLBOARD

TRACK

SCREW STUD

RESILIENT FURRING CHANNEL
WITH GYPSUM WALLBOARD

Courtesy National Gypsum Co.

Fig. 22-39. Details of metal wall construction using metal studs and tracks.

479

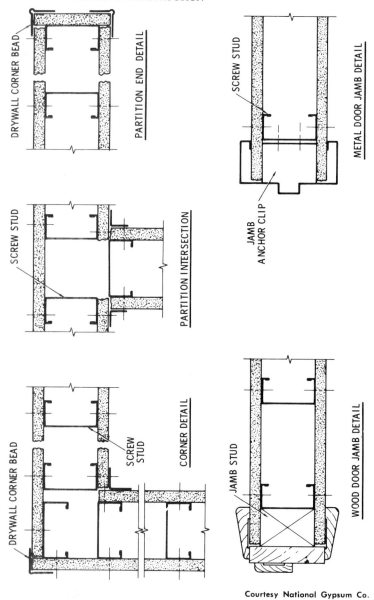

Courtesy National Gypsum Co.

Fig. 22-40. Details for intersections and jambs using metal wall construction.

480

22-44. Any of the three may be attached to the lower chord of steel joists or carrying channels in suspended ceiling construction. Either special clips or wire ties are used to fasten the channels. Fig. 22-45 shows the details in a complete ceiling assembly.

SEMISOLID PARTITIONS

Complete partition walls may be made of all plasterboard, at a considerable savings in material and time over the regular wood and plasterboard types. These walls are thinner than the usual $2'' \times 4''$ stud wall, but they are not load-bearing walls.

Fig. 22-46 shows cross-section views of the all-plasterboard walls, with details for connecting them to the ceiling and floor. Also shown is a door frame and a section of the wall using the baseboard. Partitions may be $2\frac{1}{4}$, $2\frac{5}{8}$, or $2\frac{7}{8}$ in. thick, depending on the thickness of the wallboard used and the number of layers. The center piece is not solid, but a piece of 1- or $1\frac{5}{8}$-in. plasterboard about 6 in. wide acting as a stud element.

Fig. 22-47 shows how the wall is constructed. The boards are vertically mounted, so their length must be the same distance as from the floor to the ceiling, but not exceeding 10 ft. The layers are prelaminated on the job, then raised into position.

Place two pieces of wallboard on a flat surface with face surfaces facing each other. These must be the correct length for the height of the wall. Cut two studs 6 in. wide and a little shorter than the length of the large wallboard. Spread adhesive along the entire length of the two plasterboard studs and put one of them, adhesive face down, in the middle of the top large board. It should be 21 in. from the edges and equidistant from the ends. Set the other stud, adhesive face up flush along one edge of the large plasterboard.

Place two more large panels on top of the studs, with edge of the bundle in line with the outer edge of the uncoated stud. Temporarily place a piece of plasterboard under the opposite edge for support. In this system each pair of boards will not be directly over each other, but alternate bundles will protrude. Continue the procedure until the required number of assemblies are obtained. Let dry, with temporary support under the overhanging edges. Fig. 22-47 shows the lamination process.

481

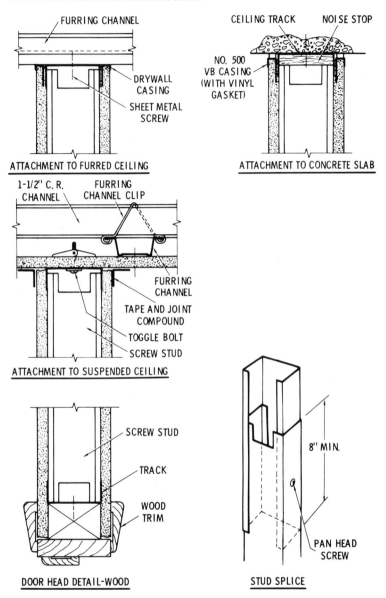

FURRING CHANNEL

DRYWALL
CASING

SHEET METAL
SCREW

ATTACHMENT TO FURRED CEILING

CEILING TRACK   NOISE STOP

NO. 500
VB CASING
(WITH VINYL
GASKET)

ATTACHMENT TO CONCRETE SLAB

1-1/2" C. R.   FURRING
CHANNEL   CHANNEL CLIP

FURRING
CHANNEL

TAPE AND JOINT
COMPOUND

TOGGLE BOLT

SCREW STUD

ATTACHMENT TO SUSPENDED CEILING

SCREW STUD

TRACK

WOOD
TRIM

DOOR HEAD DETAIL-WOOD

8" MIN.

PAN HEAD
SCREW

STUD SPLICE

Fig. 22-41. How to handle

482

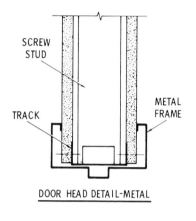

DOOR HEAD DETAIL-METAL

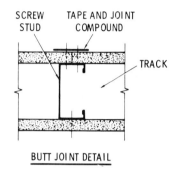

BUTT JOINT DETAIL

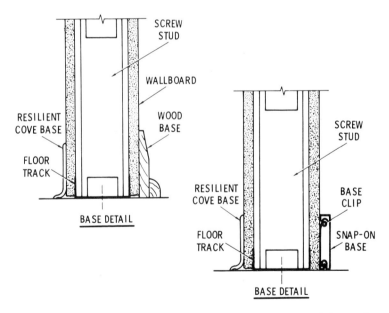

BASE DETAIL

BASE DETAIL

Courtesy National Gypsum Co.

ceiling and base finishing.

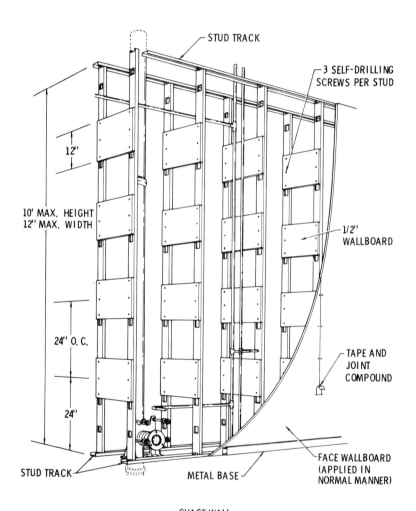

STUD TRACK

3 SELF-DRILLING
SCREWS PER STUD

12"

10' MAX. HEIGHT
12" MAX. WIDTH

1/2"
WALLBOARD

24" O. C.

TAPE AND
JOINT
COMPOUND

24"

STUD TRACK

METAL BASE

FACE WALLBOARD
(APPLIED IN
NORMAL MANNER)

CHASE WALL

Fig. 22-42. Deep-wall construction where space is

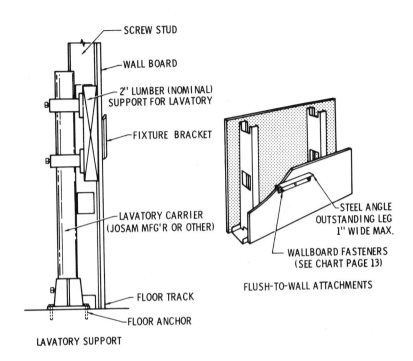

SCREW STUD

WALL BOARD

2" LUMBER (NOMINAL)
SUPPORT FOR LAVATORY

FIXTURE BRACKET

LAVATORY CARRIER
(JOSAM MFG'R OR OTHER)

FLOOR TRACK

FLOOR ANCHOR

LAVATORY SUPPORT

STEEL ANGLE
OUTSTANDING LEG
1" WIDE MAX.

WALLBOARD FASTENERS
(SEE CHART PAGE 13)

FLUSH-TO-WALL ATTACHMENTS

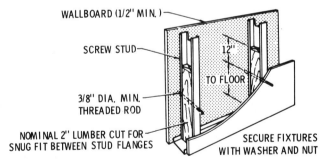

WALLBOARD (1/2" MIN.)

SCREW STUD

3/8" DIA. MIN.
THREADED ROD

NOMINAL 2" LUMBER CUT FOR
SNUG FIT BETWEEN STUD FLANGES

12"

TO FLOOR

SECURE FIXTURES
WITH WASHER AND NUT

FOR WALL HUNG FURNITURE
(BOTH SIDES OF PARTITION)
ALLOWABLE 60 FT. LBS. PER FASTENER- (2'-0" O. C. STUD SPACING)

Courtesy National Gypsum Co.

needed for reasons such as plumbing and heating pipes.

485

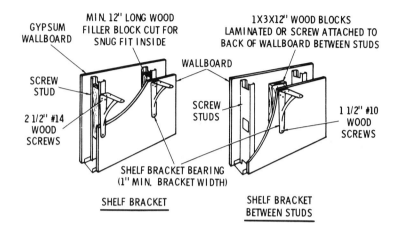

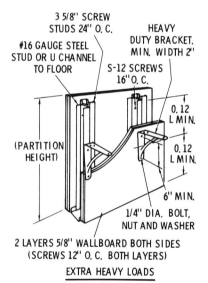

Courtesy National Gypsum Co.

Fig. 22-43. Recommended method for fastening shelf brackets to plaster board.

Table 5. Allowable Carrying Loads for Anchor Bolts

TYPE FASTENER	SIZE	ALLOWABLE LOAD	
		1/2" WALLBOARD	5/8" WALLBOARD
HOLLOW WALL SCREW ANCHORS	1/8" dia. SHORT	50 LBS.	——
	3/16" dia. SHORT	65 LBS.	——
	1/4", 5/16", 3/8" dia. SHORT	65 LBS.	——
	3/16" dia. LONG	——	90 LBS.
	1/4", 5/16", 3/8" dia. LONG	——	95 LBS.
COMMON TOGGLE BOLTS	1/8" dia.	30 LBS.	90 LBS.
	3/16" dia.	60 LBS.	120 LBS.
	1/4", 5/16", 3/8" dia.	80 LBS.	120 LBS.

Courtesy National Gypsum Co.

To raise the wall, first install a 24-in. wide starter section of wallboard with one edge plumbed against the intersecting wall. Spread adhesive along the full length of the plasterboard stud of one wall section and erect it opposite to the 24-in. starter section. The free edge of the starter section should center on the stud piece. Apply adhesive to the plasterboard stud of another section and erect it adjacent to the starter panel. Continue to alternate sides as you put each section in place.

Some plasterboard manufacturers make thick solid plasterboards for solid partition walls. Usually they will laminate two pieces of 1-in. thick plasterboard to make a 2-in. thick board. Chase walls, elevator shafts, and stair wells usually require solid and thick walls which can be of all plasterboard at reduced cost of construction. There is no limit to the thickness that can be obtained, depending on the need.

487

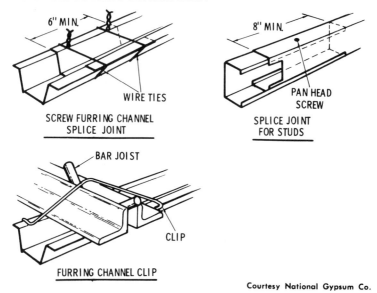

6" MIN.

WIRE TIES

SCREW FURRING CHANNEL
SPLICE JOINT

8" MIN.

PAN HEAD
SCREW

SPLICE JOINT
FOR STUDS

BAR JOIST

CLIP

FURRING CHANNEL CLIP

Courtesy National Gypsum Co.

Fig. 22-44. Metal ceiling channels may be fastened to metal joists by wire ties or special clips.

JOINT FINISHING

Flat and inside corner joints are sealed with perforated tape embedded in joint compound and with finishing coats of the same compound. Outside corners are protected with a metal bead, nailed into place and finished with jointing compound, or a special metal-backed tape, which is also used on inside corners.

Two types of jointing compound are generally available— regular, which takes about 24 hours to dry, and quick-setting, which takes about 2½ hours to dry. The installer must make the decision on which type to use, depending on the amount of jointing work he has to do and when he can get around to subsequent coats.

Begin by spotting nail heads with jointing compound. Use the broad knife to smooth out the compound. Apply compound over the joint (Fig. 22-48). Follow this immediately by embedding the perforated tape over the joint (Fig. 22-49). Fig. 22-50 is a closeup

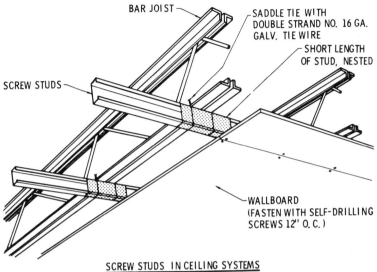

SCREW STUDS IN CEILING SYSTEMS

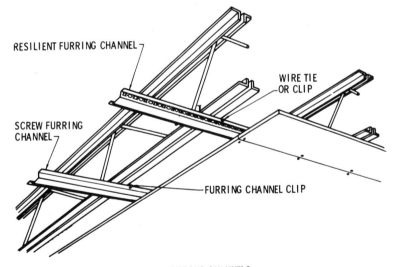

FURRING CHANNELS

Courtesy National Gypsum Co.

Fig. 22-45. Complete details showing two methods of mounting wallboard to suspended ceiling structures.

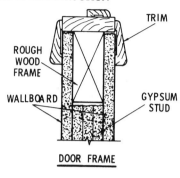

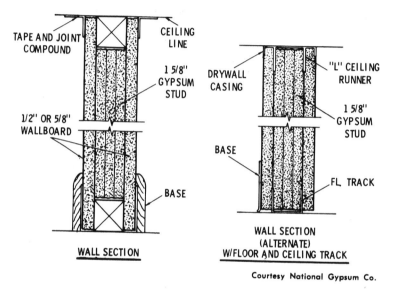

Courtesy National Gypsum Co.

Fig. 22-46. Cross-section view of a semisolid all-plasterboard wall.

view showing the compound squeezed through the perforation of the tape for good keying. Before the compound dries, run the broad knife over the tape to smooth down the compound and to level the surface (Fig. 22-51).

After the first coat has dried, apply a second coat (Fig. 22-52) thinly and feather it out 3 to 4 in. on each side of the joint. Also apply a second coat to the nail spots. When the second coat has

LAMINATING PANEL ASSEMBLIES

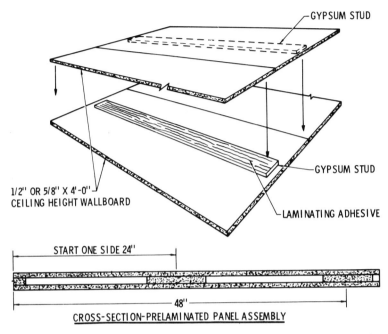

Courtesy National Gypsum Co.

Fig. 22-47. Plasterboards are laminated together with wide pieces of plasterboard acting as wall studs.

dried, apply a third coat also thinly. Feather it out to about 6 or 7 in. from the joint (Fig. 22-53). Final nail spotting is also done at this time.

INSIDE CORNERS

Inside corners are treated in the same way as flat joints, with one exception. The tape must be cut to the proper size and creased

491

down the middle. Apply it to the coated joint (Fig. 22-54) and follow with the treatment mentioned above, but to one side at a time. Let the joint dry before applying the second coat of compound and the same for the third coat.

Courtesy National Gypsum Co.

Fig. 22-48. The first step in finishing joints between plasterboard.

Courtesy National Gypsum Co.

Fig. 22-49. Place perforated tape over the joint and embed it in the jointing compound.

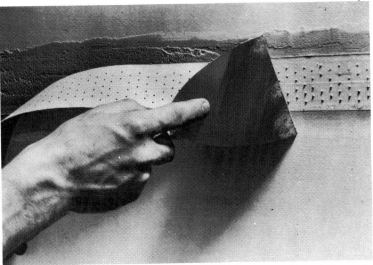

Courtesy National Gypsum Co.

Fig. 22-50. Close-up view showing how a properly embedded tape will show beads of the compound through the perforations.

Courtesy National Gypsum Co.

Fig. 22-51. Using a broad knife to smooth out the compound and feather edges.

493

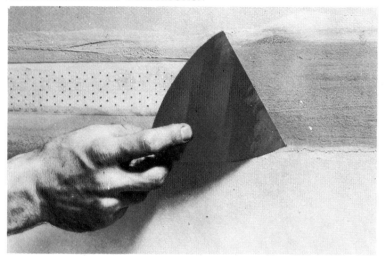

Courtesy National Gypsum Co.

Fig. 22-52. Applying second coat of compound over perforated tape after first coat has dried. This coat must be applied smooth with a good feathered edge.

Courtesy National Gypsum Co.

Fig. 22-53. The final coat is applied with a wide knife, carefully feathering edges. A wet sponge will eliminate need for further smoothing when dry.

Courtesy National Gypsum Co.

Fig. 22-54. Inside corners are handled in the same manner as flat joints. A specially shaped corner knife is also good.

OUTSIDE CORNERS

Outside corners need extra reinforcement because of the harder knocks they may take by the family. Metal corner beads (Fig. 22-55) are used. Nail through the bead into the plasterboard and framing. Apply joint compound over the beading, using a broad knife, as shown in Fig. 22-52. The final treatment is the same as for other joints. The first coat should be about 6 in. wide and the second coat about 9 in. Feather out the edges and work the surface smooth with the broad knife and a wet sponge.

495

Courtesy National Gypsum Co.

Fig. 22-55. Outside corners are generally reinforced with a metal bead. It is nailed in place, going through the plasterboard into the stud.

PREFINISHED WALLBOARD

Prefinished wallboard is surfaced with a decorative vinyl material. Since it is not painted or papered after installation, in order to maintain a smooth and unmarred finish, its treatment is slightly different than that of standard wallboard. In most installations, nails and screws are avoided (except at top and bottom) and no tape and joint compound are used at the joints.

Three basic methods are used to install prefinished wallboard.

1. Nailing to 16 in. on center studs or to furring strips but using colored and matching nails available for the purpose.
2. Cementing to the studs or furring strips with adhesive, nailing only at the top and bottom. These nails may be matching

Courtesy National Gypsum Co.

Fig. 22-56. Applying compound over the corner bead.

  colored nails or plain nails, which are covered with matching cove molding and base trim.
3. Laminating to a base layer on regular wallboard or to old wallboard in the case of existing wall installations.

  Before installing prefinished wallboard, a careful study of the wall arrangement should be made. Joints should be centered on architectural features such as fireplaces and windows. End panels should be of equal width. Avoid narrow strips as much as possible.

  Decorator wallboard is available in lengths to match most wall heights without further cutting. It should be installed vertically, and should be about ⅛ in. shorter than the actual height, so it will not be necessary to force it into place. Prefinished wallboards have

497

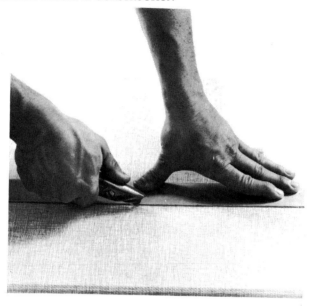

Courtesy National Gypsum Co.

Fig. 22-57. When cutting vinyl-finished wallboard, the cut should be about 1 in. wider than the desired width. This is to allow an inch of extra vinyl for edge treatment.

square edges and are butted together at the joints, with or without the vinyl surface lapped over the edges.

As with standard wallboard, prefinished wallboard is easily cut into a narrow piece by scoring and snapping. Place the board on a flat surface with the vinyl side up. Score the vinyl side with a dimension about 1 in. wider than the width of the panel (Fig. 22-57). Turn vinyl surface face down and score the back edge to the actual dimension desired. In both cases use a good straight board as a straightedge. Place the board over the edge of a long table and snap the piece off (Fig. 22-58). This will leave a piece of vinyl material hanging over the edge. Fold the material back and tack it into place onto the back of the wallboard (Fig. 22-59).

Cutouts are easily made on wallboard with a fine-toothed saw. Where a piece is to be cut out, as for a window, saw along the narrower cuts, then score the longer dimension and snap off the

498

Courtesy National Gypsum Co.

Fig. 22-58. Score the back of the wallboard to the actual width desired, leaving a 1-in. width of vinyl.

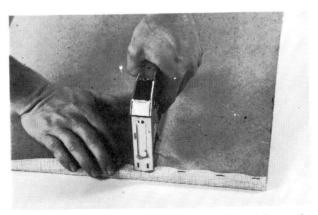

Courtesy National Gypsum Co.

Fig. 22-59. Fold the 1-in. vinyl material over the edge of the wallboard and tack in place. This will provide a finished edge without further joint treatment.

499

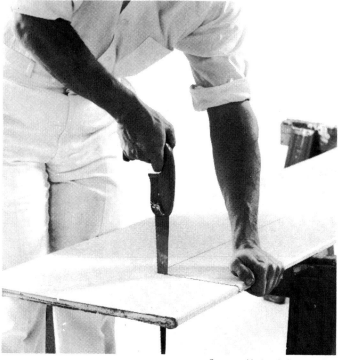

Courtesy National Gypsum Co.

Fig. 22-60. Cutting vinyl-covered wallboard for various openings.

piece (Fig. 22-60). Circles are cut out by first drilling a hole large enough to insert the end of the keyhole saw (Fig. 22-61). Square cutouts for electrical outlet boxes need not be sawed but can be punched out. Score through the vinyl surface, as shown in Fig. 22-62. Give the area a sharp blow and it will break through.

INSTALLING PREFINISHED WALLBOARD

Prefinished wallboard may be nailed to studs or furring strips with decorator type or colored nails. In doing so, however, the job must be done carefully so the nails make a decorative pattern. Space them every 12 in. and not less than ⅜ in. from the panel

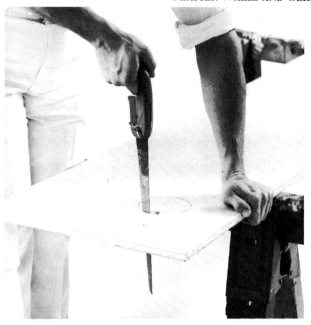

Courtesy National Gypsum Co.

Fig. 22-61. Circle cutouts are made by first drilling a hole large enough to insert a keyhole saw.

edges (Fig. 22-63). When nailing directly to the studs, to be sure the studs are straight and flush. If warped, they may require shaving down at high spots or shimming up at low spots. To avoid extra work, carpenters should be instructed, on new construction, to select the best 2″ × 4″ lumber and do a careful job of placing it. Where studs are already in place, it may be easier to install furring strips horizontally and shim them during installation (Fig. 22-64). Use 1″ × 3″ wood and space them 16 in. apart. Over an existing broken plaster wall, or solid masonry, install furring strips as shown in Fig. 22-65. Use concrete nails to fasten furring strips on solid masonry.

A good way of installing prefinished wallboard is with adhesive, either to vertical studs or horizontal furring strips, or to existing plasterboard or solid walls. In order to get effective pressure over

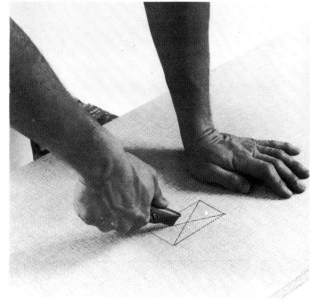

Courtesy National Gypsum Co.

Fig. 22-62. For rectangular cutouts, score deeply on the vinyl side for the outline of the cutout. Tap the section to be removed.

the entire area of the boards for good adhesion, the panels must be slightly bowed. This is done in the manner shown in Fig. 22-66. A stock of wallboard is placed either face down over a center support, or face up across two end supports. The supports can be a 2″ × 4″ lumber, but if they are in contact with the vinyl finish, they must be padded to prevent marring. It may take one day or several days to get a moderate bow, depending on the weather and humidity.

The adhesive may be applied directly to the studs (Fig. 22-67) or to furring strips (Fig. 22-68). Where panel edges join, run two adhesive lines on the stud, one for each edge of a panel. These lines should be as close to the edge of the 2″ × 4″ stud as possible to prevent the adhesive from oozing out between the panel joints. For the same reason, leave a space of 1 in. with no adhesive along the furring strips, where panel edges join.

502

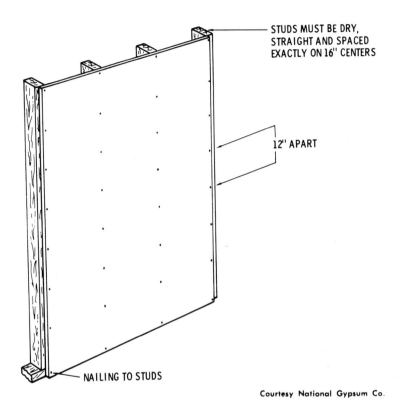

STUDS MUST BE DRY, STRAIGHT AND SPACED EXACTLY ON 16" CENTERS

12" APART

NAILING TO STUDS

Courtesy National Gypsum Co.

Fig. 22-63. Prefinished walls may be nailed directly to studs as shown.

Place the panels in position with a slight sliding motion and nail the top and bottom edges. Be sure long edges of each panel are butted together evenly. Nail to sill and plate at top and bottom only. The bowed panels will apply the right pressure for the rest of the surface.

The top and bottom nails may be matching colored nails or finished with a cove and wall trim. Special push-on trims are available which match the prefinished wall. Inside corners may be left with panels butted but one panel must overlap the other.

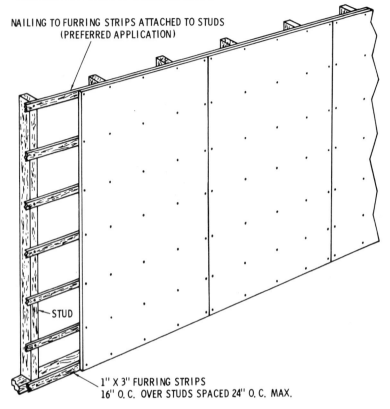

NAILING TO FURRING STRIPS ATTACHED TO STUDS
(PREFERRED APPLICATION)

STUD

1" X 3" FURRING STRIPS
16" O. C. OVER STUDS SPACED 24" O. C. MAX.

Courtesy National Gypsum Co.

Fig. 22-64. Wood furring strips may be installed over wall studs that are not straight.

Outside corners must include solid protection. For both inside corners, if desired, and for outside corners, snap-on trim and bead matching the panels can be obtained (Fig. 22-69). They are applied by installing retainer strips first (vinyl for inside corners and steel for outside corners), which holds the finished trim material in place.

Prefinished panels may be placed over old plaster walls or solid walls by using adhesive. If on old plaster, the surface must be clean and free of dust or loose paint. If wallpapered, the paper must be

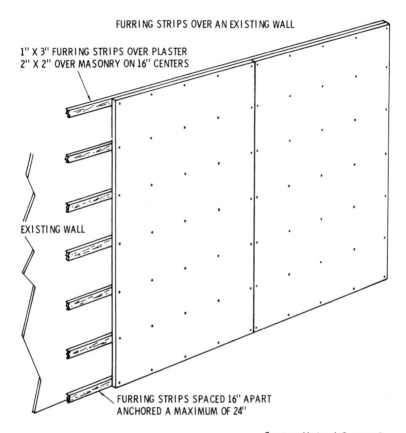

FURRING STRIPS OVER AN EXISTING WALL

1" X 3" FURRING STRIPS OVER PLASTER
2" X 2" OVER MASONRY ON 16" CENTERS

EXISTING WALL

FURRING STRIPS SPACED 16" APART
ANCHORED A MAXIMUM OF 24"

Courtesy National Gypsum Co.

Fig. 22-65. Wood furring strips installed over solid masonry or old existing walls.

removed and walls completely washed. The new panels are bowed, as described before, then lines of adhesive are run down the length of the panels about 16 in. apart (Fig. 22-70). Keep edge lines ¾ to 1 in. from the edges. Apply the boards to the old surface with a slight sliding motion and fasten at top and bottom with 6d

505

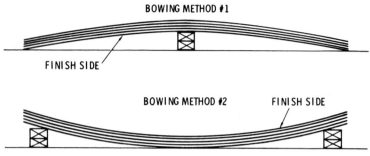

Courtesy National Gypsum Co

Fig. 22-66. Bowing wallboard for the adhesive method of installation.

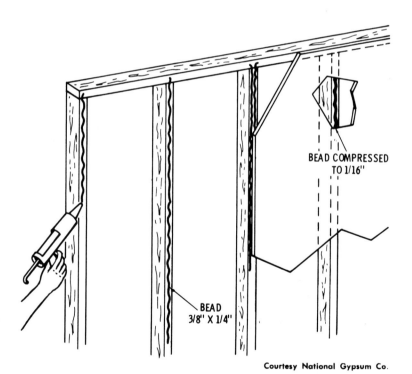

Courtesy National Gypsum Co.

Fig. 22-67. Adhesive applied directly to wall studs. The adhesive is applied in a wavy line the full length of the stud.

LEAVE 1" SPACE AT
JOINT OF PANELS

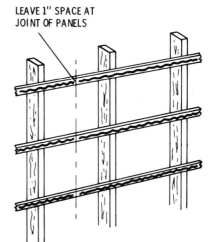

**Fig. 22-68. Adhesive applied
directly to wood
furring strips.**

Courtesy National Gypsum Co.

Courtesy National Gypsum Co.

**Fig. 22-69. Snap-on matching trim is available for prefinished wallboard to
be installed over old plaster walls.**

507

Courtesy National Gypsum Co.

Fig. 22-70. Applying adhesive to prefinished wallboard to be installed over old plaster walls.

nails, or matching colored nails. If boards will not stay in proper alignment, add more bracing against the surface and leave for 24 hours. The top and bottom may be finished with matching trim, as explained before.

CHAPTER 23

Stairs

All craftsmen who have tried to build stairs have found it (like boat building) to be an art in itself. This chapter is not intended to discourage the carpenter, but to impress him with the fact that unless he first masters the principle of stair layout, he will have many difficulties in the construction. Although stair building is a branch of mill work, the craftsman should know the principles of simple stair layout and construction, because he is often called upon to construct porch steps, basement and attic stairs, and sometimes the main stairs. In order to follow the instructions intelligently, the carpenter should be familiar with the terms and names of parts used in stair building.

STAIR CONSTRUCTION

Stairways should be designed, arranged, and installed so as to afford safety, adequate headroom, and space for the passage of furniture. In general, there are two types of stairs in a house—those serving as principal stairs, and those used as service stairs. The principal stairs are designed to provide ease and comfort, and are often made a feature of design, while the service stairs leading to the basement or attic are usually somewhat steeper and constructed of less expensive materials.

Stairs may be built in place, or they may be built as units in the shop and set in place. Both have their advantages and disadvantages, and custom varies with locality. Stairways may have a straight, continuous run, with or without an intermediate platform, or they may consist of two or more runs at angles to each other. In the best and safest practice, a platform is introduced at the angle, but the turn may be made by radiating risers called *winders*. Nonwinder stairways are most frequently encountered in residential planning, because winder stairways represent a condition generally regarded as undesirable. However, use of winders is sometimes necessary because of cramped space. In such instances, winders should be adjusted to replace landings so that the width of the tread 18 inches from the narrow converging end will not be less than the tread width on the straight run.

RATIO OF RISER TO TREAD

There is a definite relation between the height of a riser and the width of a tread, and all stairs should be laid out to conform to the well established rules governing these relations. If the combination of run and rise is too great, the steps are tiring, placing a strain on the leg muscles and on the heart. If the steps are too short, the foot may kick the leg riser at each step and an attempt to shorten the stride may be tiring. Experience has proved that a riser of 7 to 7 ½ in. high, with appropriate tread, combines both comfort and safety, and these limits therefore determine the standard height of risers commonly used for principal stairs. Service stairs may be narrow and steeper than the principal stairs, and are often unduly so, but it is well not to exceed 8 in. for the risers.

As the height of the riser is increased, the width of the tread must be decreased for comfortable results. A very good ratio is pro-

510

vided by either of the following rules, which are exclusive of the nosing:

1. Tread plus twice the riser equals 25.
2. Tread multiplied by the riser equals 75.

A riser of 7 ½ in. would, therefore, require a tread of 10 in., and a riser of 6 ½ in. would require a tread of 12 in. width. Treads are rarely made less than 9 in. or more than 12 in. wide. The treads of main stairs should be made of prefinished hardwood.

DESIGN OF STAIRS

The location and the width of a stairway (together with the platforms) having been determined, the next step is to fix the height of the riser and width of the tread. After a suitable height of riser is chosen, the exact distance between the finish floors of the two stories under consideration is divided by the riser height. If the answer is an *even* number, the number of risers is thereby determined. It very often happens, that the result is *uneven*, in which case the storey height is divided by the whole number next above or below the quotient. The result of this division gives the height of the riser. The tread is then proportioned by one of the rules for ratio of riser to tread.

Assume that the total height from one floor to the top of the next floor is 9′ 6″, or 114 in., and that the riser is to be approximately 7 ½ in. The 114 in. would be divided by 7 ½ in., which would give 15 ⅕ risers. However, the number of risers must be an *equal or whole number*. Since the nearest whole number is 15, it may be assumed that there are to be 15 risers, in which case 114 divided by 15 equals 7.6 inches, or approximately 7 ⁹⁄₁₆ in. for the height of each riser. To determine the width of the tread, multiply the height of the riser by 2 ($2 \times 7 \frac{9}{16} = 15 \frac{1}{8}$), and deduct from 25 ($25 - 15 \frac{1}{8} = 9 \frac{7}{8}$ in.).

The headroom is the vertical distance from the top of the tread to the underside of the flight or ceiling above, as shown in Fig. 23-1. Although it varies with the steepness of the stairs, the minimum allowed would be 6′ 8″.

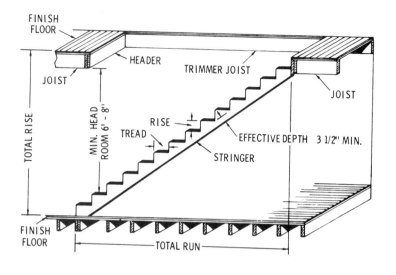

Fig. 23-1. Stairway design.

FRAMING OF STAIR WELL

When large openings are made in the floor, such as for a stair well, one or more joists must be cut. The location in the floor has a direct bearing on the method of framing the joists.

The principles explained in Chapter 14 may be referred to in considering the framing around openings in floors for stairways. The framing members around these openings are generally of the same depth as the joists. Fig. 23-2 shows the typical framing around a stair well and landing.

The headers are the short beams at right angle to the regular joists at the end of the floor opening. They are doubled and support the ends of the joists that have been cut off. Trimmer joists are at the sides of the floor opening, and run parallel to the regular joists. They are also doubled and support the ends of the headers. Tail joists are joists that run from the headers to the bearing partition.

512

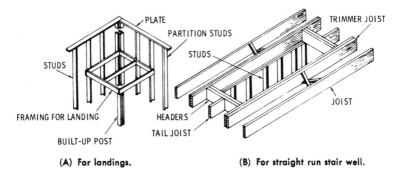

(A) For landings.   (B) For straight run stair well.

Fig. 23-2. Framing of stairways.

STRINGERS OR CARRIAGES

The treads and risers are supported upon stringers or carriages that are solidly fixed in place, and are level and true on the framework of the building. The stringers may be cut or ploughed to fit the outline of the tread and risers. The third stringer should be installed in the middle of the stairs when the treads are less than 1⅛ in. thick and the stairs are more than 2′ 6″ wide. In some cases, rough stringers are used during the construction period. These have rough treads nailed across the stringers for the convenience of workmen until the wall finish is applied. There are several forms of stringers classed according to the method of attaching the risers and treads. These different types are *cleated, cut, built-up,* and *rabbeted.*

When the wall finish is complete, the finish stairs are erected or built in place. This work is generally done by a stair builder, who often operates as a member of separate specialized craft. The wall stringer may be ploughed out, or rabbeted, as shown in Fig. 23-3, to the exact profile of the tread, riser, and nosing, with sufficient space at the back to take the wedges. The top of the riser is tongued into the front of the tread and into the bottom of the next riser. The wall stringer is spiked to the inside of the wall, and the treads and risers are fitted together and forced into the wall stringer nosing, where they are set tight by driving and gluing the wood wedges behind them. The wall stringer shows above the

513

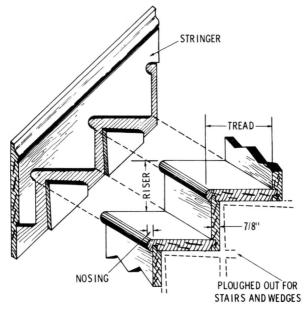

STRINGER

TREAD

RISER

7/8"

NOSING

PLOUGHED OUT FOR
STAIRS AND WEDGES

Fig. 23-3. The housing in the stringer board for the tread and riser.

profiles of the tread and riser as a finish against the wall and is often made continuous with the baseboard of the upper and lower landing. If the outside stringer is an open stringer, it is cut out to fit the risers and treads, and nailed against the outside carriage. The edges of the riser are mitered with the corresponding edges of the stringer, and the nosing of the tread is returned upon its outside edge along the face of the stringer. Another method would be to butt the stringer to the riser and cover the joint with an inexpensive stair bracket.

Fig. 23-4 shows a finish stringer nailed in position on the wall, and the rough carriage nailed in place against the stringer. If there are walls on both sides of the staircase, the other stringer and carriage would be located in the same way. The risers are nailed to the riser cuts of the carriage on each side and butt against each side of the stringer. The treads are nailed to the tread cuts of the carriage and butt against the stringer. This is the least expensive of the types described and perhaps the best construction to use when the treads and risers are to be nailed to the carriages.

514

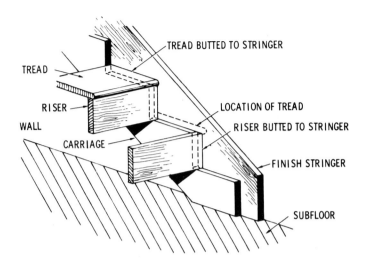

Fig. 23-4. Finished wall stringer and carriage.

Another method of fitting the treads and risers to the wall stringers is shown in Fig. 23-5A. The stringers are laid out with the same rise and run as the stair carriages, but they are cut out in reverse. The risers are butted and nailed to the riser cuts of the wall stringers, and the assembled stringers and risers are laid over the carriage. Sometimes the treads are allowed to run underneath the tread cut of the stringer. This makes it necessary to notch the tread at the nosing to fit around the stringer, as shown in Fig. 23-5B.

Another form of stringer is the cut-and-mitered type. This is a form of open stringer in which the ends of the risers are mitered against the vertical portion of the stringer. This construction is shown in Fig. 23-6, and is used when the outside stringer is to be finished and must blend with the rest of the casing or skirting board. A moulding is installed on the edge of the tread and carried around to the side, making an overlap as shown in Fig. 23-7.

BASEMENT STAIRS

Basement stairs may be built either with or without riser boards. Cutout stringers are probably the most widely used support for the treads, but the tread may be fastened to the stringers by cleats,

515

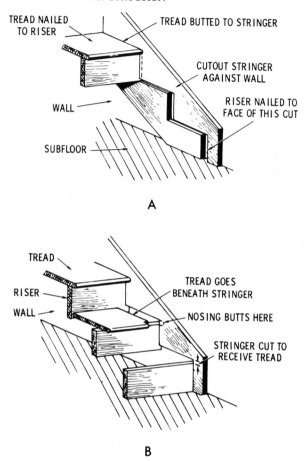

Fig. 23-5. Stringers and treads.

as shown in Fig. 23-8. Fig. 23-9 shows two methods of terminating basement stairs at the floor line.

NEWELS AND HANDRAILS

All stairways should have a handrail from floor to floor. For closed stairways, the rail is attached to the wall with suitable metal brackets. The rails should be set 2′ 8″ above the tread at the riser

516

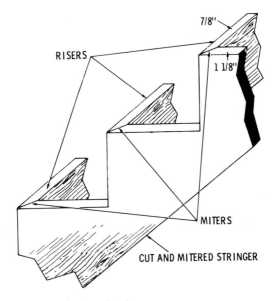

RISERS

7/8"

1 1/8"

MITERS

CUT AND MITERED STRINGER

Fig. 23-6. Cut and mitered stringer.

Fig. 23-7. Use of moulding on the edge of treads.

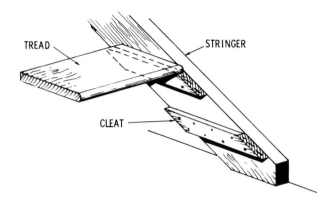

Fig. 23-8. Cleat stringer used in basement stairs.

line. Handrails and balusters are used for open stairs and for open spaces around stairs. The handrail ends against the newel post, as shown in Fig. 23-10.

Stairs should be laid out so that stock parts may be used for newels, rails, balusters, goosenecks, and turnouts. These parts are a matter of design and appearance, so they may be very plain or elaborate, but they should be in keeping with the style of the house. The balusters are doweled or dovetailed into the treads and, in some cases, are covered by a return nosing. Newel posts should be firmly anchored, and where half-newels are attached to a wall, blocking should be provided at the time the wall is framed.

DISAPPEARING STAIRS

Where attics are used primarily for storage, and where space for a fixed stairway is not available, hinged or disappearing stairs are often used. Such stairways may be purchased ready to install. They operate through an opening in the ceiling of a hall and swing up into the attic space, out of the way when not in use. Where such stairs are to be provided, the attic floor should be designed for regular floor loading.

518

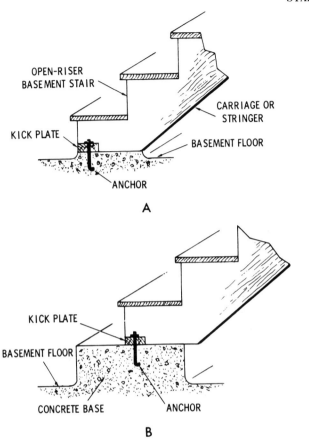

OPEN-RISER
BASEMENT STAIR

CARRIAGE OR
STRINGER

KICK PLATE

BASEMENT FLOOR

ANCHOR

A

KICK PLATE

BASEMENT FLOOR

CONCRETE BASE

ANCHOR

B

Fig. 23-9. Basement stair termination at floor line.

EXTERIOR STAIRS

Proportioning of risers and treads in laying out porch steps or approaches to terraces should be as carefully considered as the design of interior stairways. Similar riser-to-tread ratios can be used, however. The riser used in principal exterior steps should be between 6 and 7 in. The need for a good support or foundation for outside steps is often overlooked. Where wood steps are used, the bottom step should be set in concrete. Where the steps are located

519

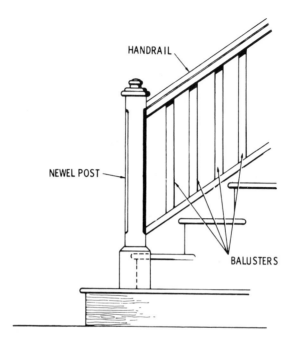

Fig. 23-10. Newel post, balusters, and handrail.

over back fill or disturbed ground, the foundation should be carried down to undisturbed ground. Fig. 23-11 shows the foundation and details of the step treads, handrail, and stringer, and method of installing them. This type of step is most common in field construction and outside porch steps. The materials generally used for this type of stair construction are 2 × 4s and 2 × 6s.

GLOSSARY OF STAIR TERMS

The terms generally used in stair design may be defined as follows.

Balusters—The vertical members supporting the handrail on open stairs (Fig. 23-12).

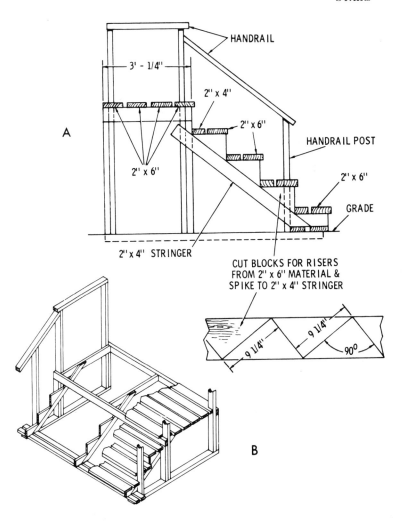

Fig. 23-11. Outside step construction.

Carriage—The rough timber supporting the treads and risers of wood stairs, sometimes referred to as the string or stringer, as shown in Fig. 23-13.

Circular Stairs—A staircase with steps planned in a circle, all the steps being winders (Fig. 23-14).

521

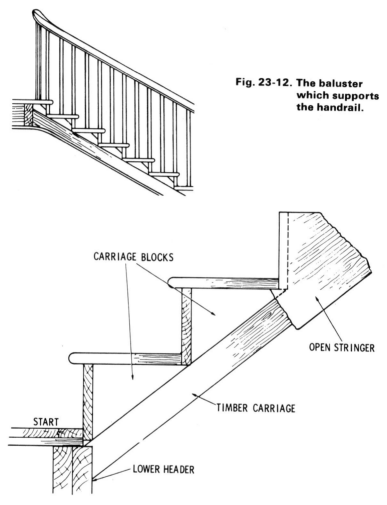

Fig. 23-12. The baluster which supports the handrail.

CARRIAGE BLOCKS

OPEN STRINGER

TIMBER CARRIAGE

START

LOWER HEADER

Fig. 23-13. Carriage blocks connected to a stair stringer.

Flight of Stairs—The series of steps leading from one landing to another.

Front String—The string of that side of the stairs over which the hand rail is placed.

Fillet—A band nailed to the face of a front string below the curve and extending the width of a tread.

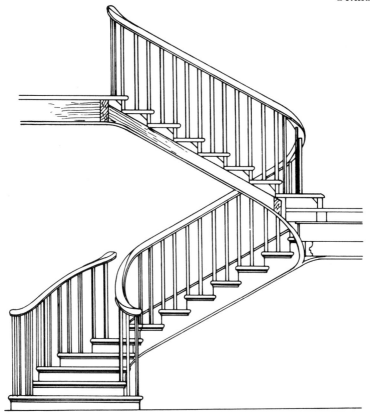

Fig. 23-14. A typical circular staircase.

Flyers—Steps in a flight of stairs parallel to each other.

Half-Space—The interval between two flights of steps in a staircase.

Handrail—The top finishing piece on the railing intended to be grasped by the hand in ascending and descending. For closed stairs where there is no railing, the handrail is attached to the wall with brackets. Various forms of hand rails are shown in Fig. 23-15.

Housing—The notches in the string board of a stair for the reception of steps.

Landing—The floor at the top or bottom of each storey where the flight ends or begins.

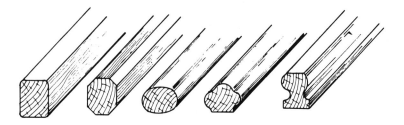

Fig. 23-15. Various forms of handrails.

Newel—The main post of the railing at the start of the stairs and the stiffening posts at the angles and platform.

Nosing—The projection of tread beyond the face of the riser (Fig. 23-16).

Rise—The vertical distance between the treads or for the entire stairs.

Riser—The board forming the vertical portion of the front of the step, as shown in Fig. 23-17.

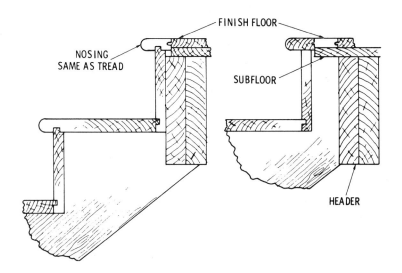

Fig. 23-16. The nosing installed on the tread.

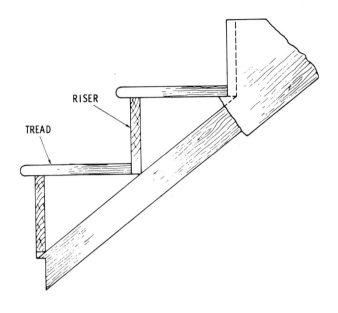

Fig. 23-17. Tread and riser.

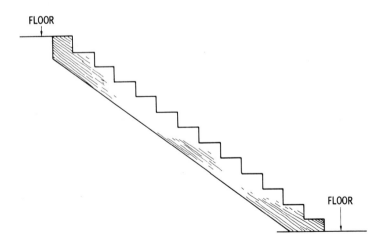

Fig. 23-18. The stair stringer.

525

Run—The total length of stairs including the platform.

Stairs—The steps used to ascend and descend from one storey to another.

Staircase—The whole set of stairs with the side members supporting the steps.

Straight Flight of Stairs—One having the steps parallel and at right angles to the strings.

String or Stringer—One of the inclined sides of a stair supporting the tread and riser. Also, a similar member, whether a support or not, such as finish stock placed exterior to the carriage on open stairs, and next to the walls on closed stairs, to give finish to the staircase. *Open stringers*, both rough and finish stock, are cut to follow the lines of the tread and risers. *Closed stringers* have parallel sides, with the risers and treads being housed into them (Fig. 23-18).

Tread—The horizontal face of a step, as shown in Fig. 23-17.

Winders—The radiating or wedge-shaped treads at the turn of a stairway.

526

CHAPTER 24

Flooring

After the foundation, sills, and floor joists have been constructed, the subfloor is laid diagonally on the joists. The floor
joists form a framework for the subfloor. This floor is called the
rough floor, or subfloor, and may be viewed as a large platform
covering the entire width and length of the building. Two layers or
coverings of flooring material (subflooring and finished flooring)
are placed on the joists. You can use 1 × 4 boards or 1 × 6
tongue-and-groove sheathing or ½-in. plywood for the subfloor.
Plywood is fastest, but usually costs the most. Plywood is also used
in some cases because of its size, weight, and stability, plus the
time and labor saved in application. It will add considerably more
strength to the floor since the weight is distributed over a wider
area. Fig. 24-1 shows the method of laying a board subfloor.

It may be laid before or after the walls are framed, preferably
before, so it can be used as a floor to work on while framing the
walls. The subflooring will also give protection against the
weather for tools and material stored in the basement.

527

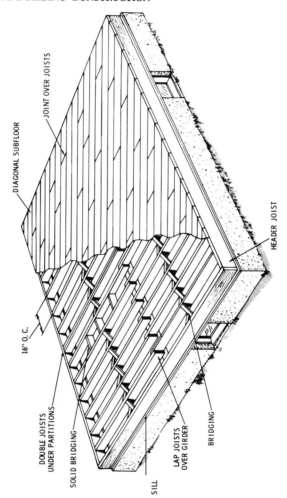

Fig. 24-1. Application of board subflooring.

SUBFLOORING

Tongue-and-groove boards should be installed so that each board will bear at least two joists, and so that there will be no two adjoining boards with end joints occuring between the same pair

528

of joists. Subflooring is nailed to each joist with two eightpenny nails for widths under 8 in. and with three nails for over 8-in. width. The subflooring may be applied either diagonally or at right angles to the joists. When the subfloor is placed at right angles to the joists, the finish floor should be laid at right angles to the subflooring. Diagonal subflooring permits the finish floor to be laid either parallel or perpendicular to the joists.

The joist spacing should not exceed 16 in. on center when finish flooring is laid parallel to the joists or when parquet finish flooring is used. Indeed, it is a good idea to check with local building codes for specs. Where balloon framing is used, blocking should be installed between the ends of the joists at the wall for nailing the ends of diagonal subfloor boards. In areas where rain may occur during construction, square-edge boards should be laid with open joints for drainage. Tongue-and-groove boards should have holes drilled at suitable intervals to allow runoff of rain water.

Table 24-1 shows the thickness of the plywood and joist spacings as suggested by the Federal Housing Administration for plywood subfloor when used as a base for wood finish floors, resilient flooring, or ceramic tile.

When used as a base for parquet wood finish flooring less than $\frac{25}{32}$ in. thick, resilient flooring, or ceramic tile, install solid blocking under all edges at right angles to the floor joists. Nail securely to the joists and blocking with nails 6 in. on center at the edges and 10 in. on center at the intermediate framing members. When used for leveling purposes over other subflooring, the minimum thickness is $\frac{1}{4}$-in. three-ply.

FLOOR COVERINGS

There is a wide variety of finish flooring available, each having properties suited to a particular usage. Of these properties, durability and ease of cleaning are essential in all cases. Specific service requirements may call for special properties, such as resistance to hard wear in storehouses and on loading platforms; comfort to users in offices and shops; and attractive appearance, which is always desirable in residences.

Both hardwoods and softwoods are available as strip flooring in a variety of widths and thicknesses, as well as random-width planks, parquetry, and block flooring. Other materials include those mentioned above. A detailed round-up follows.

Wood Strip Flooring

Softwoods most commonly used for flooring are southern yellow pine, douglas fir, redwood, western larch, and western hemlock. It is customary to divide the softwoods into two classes:

1. Vertical or edge grain.
2. Flat grain.

Each class is separated into select and common grades. The select grades designated as "B and better" grades, and sometimes the "C" grade, are used when the purpose is to stain, varnish, or wax the floor. The "C" grade is well suited for floors to be stained dark or painted, and lower grades are for rough usage or when covered with carpeting. Softwood flooring is manufactured in several widths. In some places, the 2½ in. width is preferred, while in others, the 3½ in. width is more popular. Softwood flooring has tongue-and-groove edges, and may be hollow backed or grooved. Vertical-grain flooring stands up better than flatgrain under hard usage.

Hardwoods most commonly used for flooring are red and white oak, hard maple, beech, and birch. Maple, beech, and birch come in several grades, such as *first, second* and *third.* Other hardwoods that are manufactured into flooring, although not commonly used, are walnut, cherry, ash, hickory, pecan, sweetgum, and sycamore. Hardwood flooring is manufactured in a variety of widths and thicknesses, some of which are referred to as standard patterns, other as special patterns. The widely used standard patterns consist of relatively narrow strips laid lengthwise in a room, as shown in Fig. 24-2. The most widely used standard pattern is $^{25}/_{32}$ in. thick and has a face width of 2¼ in. One edge has a tongue and the other edge has a groove, and the ends are similarly matched. The strips are random lengths, varying from 1 to 16 ft. in length. The number of short pieces will depend on the grade used. Similar patterns of flooring are available in thicknesses of $^{15}/_{32}$ and $^{11}/_{32}$ in. width of 1½ in., and with square edges and a flat back.

The flooring is generally hollow backed. The top face is slightly

Table 1. Plywood Thickness and Joists Spacing

Minimum thickness of five-ply subfloor	Medium thickness of finish flooring	Maximum joist spacing
†½ inch	25/32 inch wood laid at right angles to joists	24 inches
†½ inch	25/32 inch wood laid parallel to joists	20 inches
½ inch	25/32 inch wood laid at right angles to joists	20 inches
½ inch	Less than 25/32 inch wood or other finish	‡16 inches
†⅝ inch	Less than 25/32 inch wood or other finish	‡20 inches
†¾ inch	Less than 25/32 inch wood or other finish	‡24 inches

† Installed with outer plies of subflooring at right angles to joists.
‡ Wood strip flooring, 25/32 inch thick or less, may be applied in either direction.

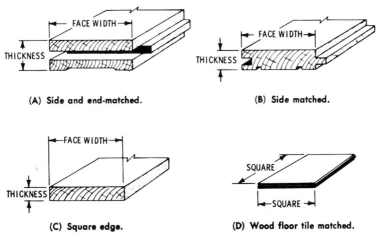

(A) Side and end-matched. (B) Side matched.

(C) Square edge. (D) Wood floor tile matched.
Fig. 24-2. Types of finished hardwood flooring.

wider than the bottom, so that the strips are driven tight together at the top side but the bottom edges are slightly open. The tongue should fit snugly in the groove to eliminate squeaks in the floor.

Another pattern of flooring used to a limited degree is ⅜ in. thick with a face width of 1½ and 2 in., with square edges and a

531

flat back. Fig. 24-2D shows a type of wood floor tile commonly known as parquetry.

INSTALLATION OF WOOD STRIP FLOORING

Flooring should be laid after plastering and other interior wall and ceiling finish is completed, after windows and exterior doors are in place, and after most of the interior trim is installed. When wood floors are used, the subfloor should be clean and level, and should be covered with a deadening felt or heavy building paper, as shown in Fig. 24-3. This felt or building paper will stop a certain amount of dust, and will somewhat deaden the sound. Where a crawl space is used, it will increase the warmth of the floor by preventing air infiltration. The location of the joists should be chalklined on the paper as a guide for nailing.

Strip flooring should be laid crosswise of the floor joist, and it looks best when the floor is laid lengthwise in a rectangular room. Since joists generally span the short way in a living room, that room establishes the direction for flooring in other rooms. Flooring should be delivered only during dry weather, and should be stored in the warmest and driest place available in the house. The

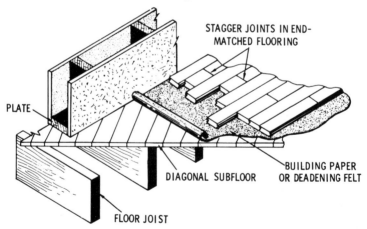

Fig. 24-3. Application of strip flooring showing the use of deadening felt or heavy building paper.

recommended average moisture content for flooring at the time of the installation should be between 6 and 10 percent. Moisture absorbed after delivery to the house site is one of the most common causes of open joints between floor strips. It will show up several months after the floor has been laid.

Floor squeaks are caused by movement of one board against another. Such movement may occur because the floor joists are too light and not held down tightly, tongue fitting too loose in the grooves, or because of poor nailing. Adequate nailing is one of the most important means of minimizing squeaks. When it is possible to nail the finish floor through the subfloor into the joist, a much better job is obtained than if the finish floor is nailed only to the subfloor. Various types of nails are used in nailing various thicknesses of flooring. For $^{25}\!/_{32}$ in. flooring, it is best to use eightpenny steel cut flooring nails; for $\frac{1}{2}$ -in. flooring, sixpenny nails should be used. Other types of nails have been developed in recent years for nailing of flooring, among these being the annularly-grooved and spirally-grooved nails. In using these nails, it is well to check with the floor manufacturer's recommendation as to size and diameter for a specific use. Fig. 24-4 shows the method of nailing the first strip of flooring. The nail is driven straight down through the

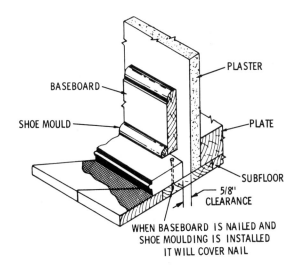

Fig. 24-4. Method of laying the first strips of wood flooring.

board at the groove edge. The nails should be driven into the joist and near enough to the edge so that they will be covered by the base or shoe moulding. The first strip of flooring can be nailed through the tongue.

Fig. 24-5 shows the nail driven in at an angle of between 45° and 50° where the tongue adjoins the shoulder. Do not try to drive the nail home with a hammer, as the wood may be easily struck and damaged. Instead use a nail set to finish off the driving. Fig. 24-6 shows a special nail set that is commonly used for the final driving. In order to avoid splitting the wood, it is sometimes necessary to pre-drill the holes through the tongue. This will also help to drive the nail easily into the joist. For the second course of flooring, select a piece so that the butt joints will be well separated from those in the first course. For floors to be covered with rugs, the

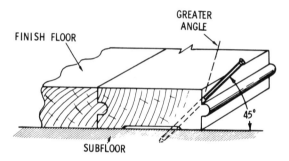

Fig. 24-5. Nailing method for setting nails in flooring.

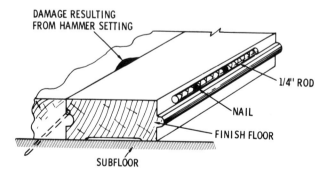

Fig. 24-6. Suggested method for setting nails in flooring.

534

long lengths could be used at the sides of the room and the short lengths in the center where they will be covered.

Each board should be driven up tightly, but do not strike the tongue with the hammer, as this will crush the wood. Use a piece of scrap flooring for a driving block. Crooked pieces may require wedging to force them into alignment. This is necessary in order that the last piece of flooring will be parallel to the baseboard. If the room is not square, it may be necessary to start the alignment at an early stage.

SOUNDPROOF FLOORS

One of the most effective sound resistance floors is called a *floating* floor. The upper or finish floor is constructed on 2 × 2 joists actually floating on glass wool mats, as shown in Fig. 24-7.

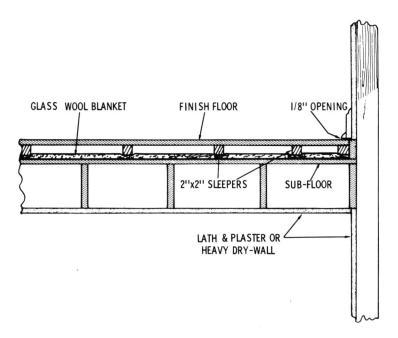

GLASS WOOL BLANKET   FINISH FLOOR   1/8'' OPENING

2''x2'' SLEEPERS   SUB-FLOOR

LATH & PLASTER OR HEAVY DRY-WALL

Fig. 24-7. A sound-resistant floor.

There should be absolutely no mechanical connection through the glass wool mat, not even a nail to either the subfloor or to the wall.

PARQUET FLOORING

Flooring manufacturers have developed a wide variety of special patterns of flooring, including parquet (Fig. 24-8), which is nailed in place or has an adhesive backing. One common type of floor tile is a block 9 in. square and $^{13}/_{16}$ in. thick, which is made up of several individual strips of flooring held together with glue and splines. Two edges have a tongue and the opposite edges are grooved. Numerous other sizes and thicknesses are available. In laying the floor, the direction of the blocks is alternated to create a checkerboard effect. The manufacturer supplies instructions for laying their tile, and it is advisable to follow them carefully. When the tiles are used over a concrete slab, a vapor barrier should be used. The slab should be level and thoroughly aired and dried before the flooring is laid.

CERAMIC TILE

Ceramic tile (Fig. 24-9) is made in different colors and with both glazed and unglazed surfaces. It is used as covering for floors in bathrooms, entryways, kitchens, and fireplace hearths. Ceramic tile presents a hard and impervious surface. In addition to standard sizes and plain colors, many tiles are especially made to carry out architectural effects. When ceramic-tile floors are used with wood-frame construction, a concrete bed of adequate thickness must be installed to receive the finishing layer.

Installation of tile is done with adhesive. See Chapter 30 for specific instructions on tile installations.

OTHER FINISHED FLOORINGS

The carpenter and do-it-yourselfer can select from a wealth of floor coverings. Perhaps the chief development has been resilient

Courtesy United Gilsonite

Fig. 24-8. Parquet flooring.

Fig. 24-9. Ceramic tile is used mainly in bathrooms. The tile comes in many colors and patterns.

flooring, so called because it "gives" when you step on it. There are 12-in. tiles, available for installing with adhesive (Fig. 24-10) as well as adhesive-backed tiles. Today most tile is vinyl and comes in a tremendous variety of styles, colors, and patterns. The newest

538

Fig. 24-10. For the do-it-yourselfer, there are adhesive-backed tiles.

Courtesy Armstrong

Fig. 24-11. Waxless flooring makes life easier for the homemaker.

in flooring is the so-called waxless flooring (Fig. 24-11), which requires renewal with a waxlike material after a certain period of time.

Resilient flooring also comes in sheets or rolls 12 ft. wide. It too, is chiefly vinyl and comes in a great array of styles and colors. It is more difficult to install than tile (Fig. 24-12).

Resilient sheet flooring comes in several qualities, and you should check competing materials before you buy. Vinyl resilient flooring may be installed anywhere in the house, above or below grade, with no worry about moisture problems. An adequate subfloor, usually of particleboard or plywood, is required.

Also available is carpeting. Wall-to-wall carpet installation is a professional job, but carpet tiles can be had. They are available for

Courtesy Congoleum

Fig. 24-12. Sheet flooring is not easy to install correctly.

indoor and outdoor use and are popularly used in kitchens and bathrooms.

Other flooring materials include paint-on coatings and paints. Paint is normally used in areas where economy is most important.

CHAPTER 25

Doors

Doors can be obtained from the mills in stock sizes much cheaper than they can be made by hand. Stock sizes of doors cover a wide range, but those most commonly used are $2'4'' \times 6'8''$, $2'8'' \times 6'8''$, $3'0'' \times 6'8''$, and $3'0'' \times 7'0''$. These sizes are either $1\frac{3}{8}$ (interior) or $1\frac{3}{4}$ in. (exterior) thick.

Paneled Doors

Paneled, or sash, doors are made in a variety of panel arrangements, both horizontal, vertical, and combinations of both. A sash door has for its component parts a top rail, bottom rail, and two stiles, which form the sides of the door. Doors of the horizontal type have intermediate rails forming the panels; and panels of the vertical type have horizontal rails and vertical stiles forming the panels.

The rails and stiles of a door are generally mortised and tenoned, the mortise being cut in the side stiles as shown in Fig. 25-1. Top

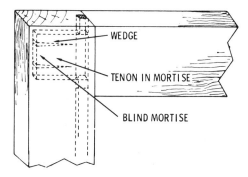

WEDGE

TENON IN MORTISE

BLIND MORTISE

Fig. 25-1. Door construction showing mortise joints.

and bottom rails on paneled doors differ in width, the bottom rail being considerably wider. Intermediate rails are usually the same width as the top rail. Paneling material is usually plywood which is set in grooves or dadoes in the stiles and rails, with the moulding attached on most doors as a finish.

Flush Doors

Flush doors are usually perfectly flat on both sides. Solid planks are rarely used for flush doors. Flush doors are made up with solid or hollow cores with two or more plies of veneer glued to the cores.

Solid-Core Doors

Solid-core doors are made of short pieces of wood glued together with the ends staggered very much like in brick laying. One or two plies of veneer are glued to the core. The first section, about ⅛ in. thick, is applied at right angles to the direction of the core, and the other section, ⅛ in. or less, is glued with the grain vertical. A ¾-in. strip, the thickness of the door is glued to the edges of the door on all four sides. This type of door construction is shown in Fig. 25-2.

Hollow-Core Doors

Hollow-core doors have wooden grids or other honeycomb material for the base, with solid wood edging strips on all four

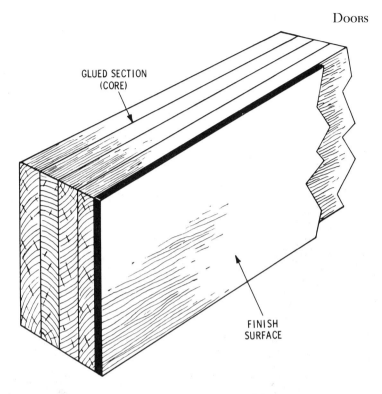

GLUED SECTION
(CORE)

FINISH
SURFACE

Fig. 25-2. Construction of a laminated or veneered door.

sides. The face of this type door is usually 3-ply veneer instead of two single plies. The hollow-core door has a solid block on both sides for installing door knobs and to permit the mortising of locks. The honeycomb-core door is for interior use only.

Louver Doors

This type of door has either stationary or adjustable louvers, and may be used as an interior door, room divider, or a closet door. The louver door comes in many styles, such as shown in Fig. 25-3. An exterior louver door may be used, which is called a *jalousie* door. This door has the adjustable louvers usually made of wood or glass. Although there is little protection against winter winds, a solid-type storm window is made to fit over the louvers to give added protection.

545

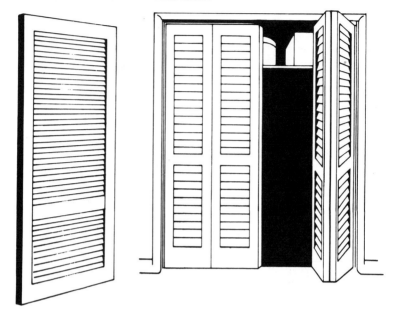

Fig. 25-3. Various styles of louver doors.

INSTALLING DOORS

Before a door can be installed, a frame must be built for it. There are numerous ways to do this. A frame for an exterior door consists of the following essential parts.

1. Sill.
2. Threshold.
3. Side and top jamb.
4. Casing.

These essential parts are shown in Fig. 25-4.

The preparation should be done before the exterior covering is placed on the outside walls. To prepare the openings, square off any uneven pieces of sheathing and wrap heavy building paper around the sides and top. Since the sill must be worked into a portion of the subflooring, no paper is put on the floor. Position the paper from a point even with the inside portion of the stud to a

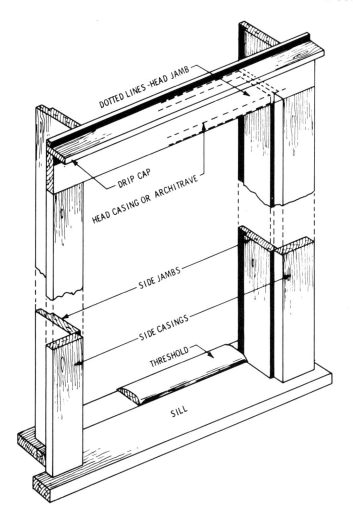

DOTTED LINES - HEAD JAMB

DRIP CAP

HEAD CASING OR ARCHITRAVE

SIDE JAMBS

SIDE CASINGS

THRESHOLD

SILL

Fig. 25-4. View of door frame showing the general construction.

point about 6 in. on the sheathed walls and tack it down with small nails.

In quick construction, there will be no door frame (the studs on each side of the opening act as the frame). The inside door frame is

547

constructed in the same manner as the outside frame, except there is no sill and no threshold.

Door Jambs

Door jambs are the lining of a door opening. Casings and stops are nailed to the jamb, and the door is securely fastened by hinges at one side. The width of the jamb will vary in accordance with the thickness of the walls. The door jambs are made and set in the following manner.

1. Regardless of how carefully the rough openings are made, be sure to plumb the jambs and level the heads when the jambs are set.
2. Rough openings are usually made 2½ in. larger each way than the size of the door to be hung. For example, a 2'8" × 6'8" door would need a rough opening of 2'10½" × 6'10½". This extra space allows for the jamb, the wedging, and the clearance space for the door to swing.
3. Level the floor across the opening to determine any variation in floor heights at the point where the jamb rests on the floor.
4. Cut the head jamb with both ends square, allowing for the width of the door plus the depth of both dadoes and a full ³⁄₁₆ in. for door clearance.
5. From the lower edge of the dado, measure a distance equal to the height of the door plus the clearance wanted at the bottom.
6. Do the same thing on the opposite jamb, only make additions or subtractions for the variation in the floor.
7. Nail the jambs and jamb head together through the dado into the head jamb, as shown in Fig. 25-5.
8. Set the jambs into the opening and place small blocks under each jamb on the subfloor just as thick as the finish floor will be. This will allow the finish floor to go under the door.
9. Plumb the jambs and level the jamb head.
10. Wedge the sides to the plumb line with shingles between the jambs and the studs, and then nail securely in place.
11. Take care not to wedge the jambs unevenly.
12. Use a straightedge 5 to 6 ft. long inside the jambs to help prevent uneven wedging.

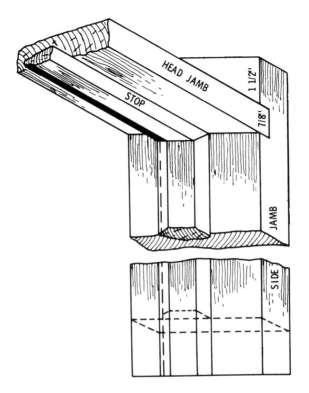

Fig. 25-5. Details showing upper head jamb dadoed into side jamb.

13. Check each jamb and the head carefully. If a jamb is not plumb, it will have a tendency to swing the door open or shut, depending on the direction in which the jamb is out of plumb.

Door Trim

Door trim material is nailed onto the jambs to provide a finish between the jambs and the plastered wall. This is called *casing*. Sizes vary from ½ to ¾ in. in thickness, and from 2½ to 6 in. in width. Most casing material has a concave back, to fit over uneven wall material. In miter work, care must be taken to make all joints clean, square, neat, and well fitted. If the trim is to be mitered at the top corners, a miter box, miter square, hammer, nailset, and

549

block plane will be needed. Door openings are cased up in the following manner.

1. Leave a ¼-in. margin between the edge of the jamb and the casing on all sides.
2. Cut one of the side casings square and even with the bottom of the jamb.
3. Cut the top or mitered end next, allowing ¼ in. extra length for the margin at the top.
4. Nail the casing onto the jamb and set it even with the ¼-in. margin line, starting at the top and working toward the bottom.
5. The nails along the outer edge will need to be long enough to penetrate the casing, plaster, and wall stud.
6. Set all nail heads about ⅛ in. below the surface of the wood.
7. Apply the casing for the other side of the door opening in the same manner, followed by the head (or top) casing.

HANGING MILL-BUILT DOORS

If flush or sash doors are used, install them in the finished door opening as described below.

1. Cut off the stile extension, if any, and place the door in the frame. Plane the edges of the stiles until the door fits tightly against the hinge side and clears the lock side of the jamb about ¹⁄₁₆ in. Be sure that the top of the door fits squarely into the rabbeted recess and that the bottom swings free of the finished floor by about ½ in. The lock stile of the door must be beveled slightly so that the edge of the door will not strike the edge of the door jamb.
2. After the proper clearance of the door has been made, set the door in position and place wedges as shown in Fig. 25-6. Mark the position of the hinges on the stile and on the jamb with a sharp pointed knife. The lower hinge must be placed slightly above the lower rail of the door. The upper hinge of the door must be placed slightly below the top rail in order to avoid cutting out a portion of the tenons of the door rails. There are three measurements to mark—the location of the

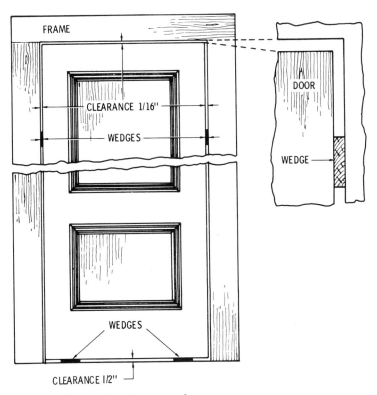

FRAME

CLEARANCE 1/16"

WEDGES

DOOR

WEDGE

WEDGES

CLEARANCE 1/2"

Fig. 25-6. Sizing a door for an opening.

butt hinge on the jamb, the location of the hinge on the door, and the thickness of the hinge on both the jamb and the door.

3. Door butt hinges are designed to be mortised into the door and frame, as shown in Fig. 25-7. Fig. 25-8 shows recommended dimensions and clearances for installation. Three hinges are usually used on full-length doors to prevent warping and sagging.

4. Using the butt as a pattern, mark the dimension of the butts on the door edge and the face of the jamb. The butts must fit snugly and exactly flush with the edge of the door and the face of the jamb. (A device called a butt marker can be helpful here.)

After placing the hinges and hanging the door, mark off the

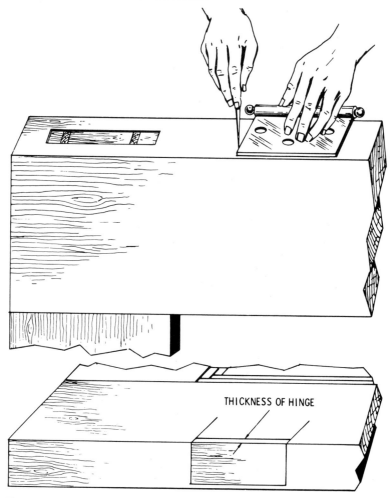

Fig. 25-7. A method of installing hinges.

position for the lock and handle. The lock is generally placed about 36 in. from the floor level. Hold the lock in position on the stile and mark off with a sharp knife the area to be removed from the edge of the stile. Mark off the position of the door-knob hub. Bore out the wood to house the lock and chisel the mortises clean. After the lock assembly has been installed, close the door and mark the jamb for the striker plate.

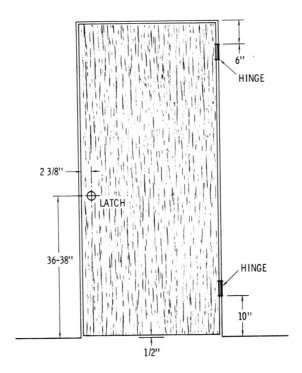

Fig. 25-8. Recommended clearances and dimensions after door is hung.

Installing a door is difficult. A recent trend is for manufacturers to provide not only the door but all surrounding framework (Fig. 25-9). This reduces the possibility of error.

SWINGING DOORS

Frequently, it is desirable to hang a door so that it opens as you pass through from either direction, yet remains closed at all other times. For this purpose, you can use swivel-style spring hinges. This type of hinge attaches to the rail of the door and to the jamb like an ordinary butt hinge. Another type is mortised into the bottom rail of the door and is fastened to the floor with a floor plate. In most cases, the floor-plate hinge, as shown in Fig. 25-10, is

553

Fig. 25-9. Some companies provide the door and surrounding framework to ensure trouble-free installation.

best, because it will not weaken and let the door sag. It is also designed with a stop to hold the door open at right angles, if so desired.

SLIDING DOORS

Sliding doors are usually used for walk-in closets. They take up very little space, and they also allow a wide variation in floor plans. This type of door usually limits the access to a room or closet unless the doors are pushed back into a wall. Very few sliding doors are pushed back into the wall because of the space and expense involved. Fig. 25-11 shows a double and a single sliding door track.

554

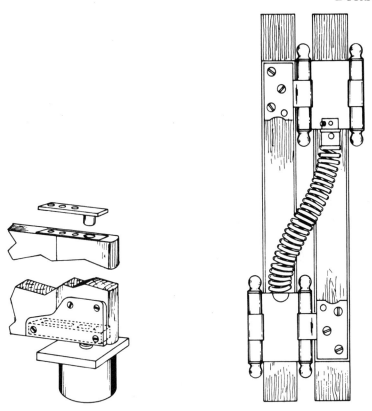

Fig. 25-10. Two kinds of swivel spring hinges.

GARAGE DOORS

Garage doors are made in a variety of sizes and designs. The principal advantage in using the garage door, of course, is that it can be rolled up out of the way. In addition, the door cannot be blown shut due to wind, and is not obstructed by snow and ice.

Although designed primarily for use in residential and commercial garages, doors of this type are also employed in service stations, factory receiving docks, boathouses, and many other buildings. In order to permit overhead-door operation, garage doors of this type are built in suitable hinged sections. Usually 4 to

555

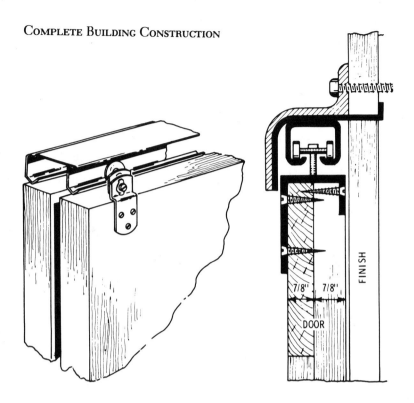

Fig. 25-11. Two kinds of sliding-door tracks.

7 sections are used, depending upon the door height requirements. Standard residential garage doors are usually 9' ×7' for singles and 16' × 7' for a double. Residential-type garage doors are usually manufactured 1³/₄ in. thick unless otherwise requested.

When ordering doors for the garage, the following information should be forwarded to the manufacturer:

1. Width of opening between the finished jambs.
2. Height of the ceiling from the finished floor under the door to the underside of the finished header.
3. Thickness of the door.
4. Design of the door (number of glass windows and sections).
5. Material of jambs (they must be flush).
6. Head room from the underside of the header to the ceiling, or to any pipes, lights, etc.

Courtesy of Overhead Door Corporation

Fig. 25-12. Typical 16-ft. overhead garage door.

7. Distance between the sill and the floor level.
8. Proposed method of anchoring the horizontal track.
9. Depth to the rear from inside of the upper jamb.
10. Inside face width of the jamb buck, angle, or channel.

This information applies for overhead doors only, and does not apply to garage doors of the slide, folding, or hinged type. Doors can be furnished to match any style of architecture and may be provided with suitable size glass windows if desired (Fig. 25-12).

If your garage is attached to your house, your garage door often represents from one-third to one-fourth of the face of your house. Style and material should be considered to accomplish a pleasant effect with masonry or wood architecture. Fig. 25-13 shows three types of overhead garage doors that can be used with virtually any kind of architectural design. Many variations can be created from combinations of raised panels with routed or carved designs as shown in Fig. 25-14. These panels may also be combined with plain raised panels to provide other dramatic patterns and color combinations.

Automatic garage-door openers were once a luxury item, but in

557

(A) Fiberglass.   (B) Steel.

(C) Wood.

Courtesy Overhead Door Corporation

Fig. 25-13. Three kinds of garage doors.

the past few years the price has been reduced and failure mini-
mized to the extent that most new installations include this feature.
Automatic garage-door openers save time, eliminate the need to
stop the car and get out in all kinds of weather, and you also save
the energy and effort required to open and close the door by hand.

The automatic door opener is a radio-activated motor-driven
power unit that mounts on the ceiling of the garage, and attaches to
the inside top of the garage door. Electric impulses from a wall-
mounted push bottom, or radio waves from a portable radio
transmitter in your car, starts the door mechanism. When the door
reaches its limit of travel (up or down), the unit turns itself off and

Fig. 25-14. Variations in carved or routed panel designs.

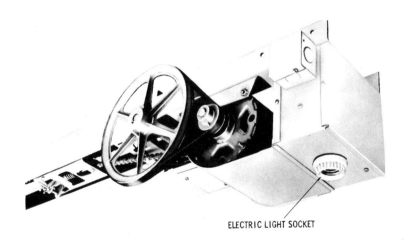

ELECTRIC LIGHT SOCKET

Courtesy of Overhead Door Corporation

Fig. 25-15. Typical automatic garage-door.

559

awaits your next command. Most openers on the market have a safety factor built in. If the door encounters an obstruction in its travel, it will instantly stop, or stop and reverse its travel. The door will not close until the obstruction has been removed. When the door is completely closed, it is automatically locked and cannot be opened from the outside, making it burglar-resistant. Fig. 25-15 shows an automatic garage-door opener which can be quickly disconnected for manual-door operation in case of electrical power failure. Notice the electric light socket on the automatic opener unit, which turns on when the door opens to light up the inside of the garage.

PATIO DOORS

In recent years, a door that has gained increased popularity is the patio door (Fig. 25-16). These doors provide easy access to

Fig. 25-16. Patio door.

easy-living areas. Patio doors are available in metal and wood, with most people favoring wood. They are available with double-insulated glass and inserts that give them a mullion effect. Manufacturers furnish instructions on how to install patio doors.

CHAPTER 26

Miter Work

In treating this important branch of carpentry, it is advisable to mention some of its elements. By definition, a miter is the joint formed by two pieces of moulding, each cut at an angle so as to match when joined angularly; also, to miter means to meet and match together on a line bisecting the angle of junction, especially at a right angle, in other words, to cut and join together the ends of two pieces obliquely at an angle.

MITER TOOLS

To do this with precision, the proper tools are necessary. The first is, of course the saw, which should be a good 20-in. back saw of about eleven or twelve teeth to the inch, filed to a keen edge and rubbed off on the sides with the face of an oil stone. For precision, a manufactured metal miter box should be used. However, a serviceable miter box, such as shown in Fig. 26-1, can be made of suitable hardwood by the carpenter for most of the common miter cuts.

Although several patent miter boxes are now in use, the wooden box will always have its place where the most common miters are

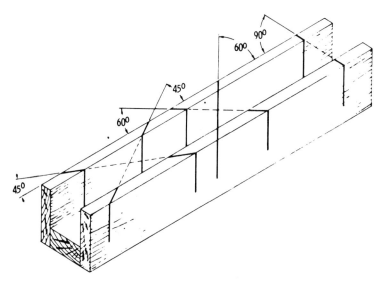

Fig. 26-1. Homemade miter box.

to be cut. The steel miter box is recommended where great precision work is needed (Fig. 26-2). Some of the metal miter boxes have attachments whereby frames may be held firmly in position while the miter is nailed. There are also hand and foot power miter cutters by which both miters are rapidly cut in one operation.

MOULDINGS

In the ornamental side of carpentry construction, various forms of mouldings, of course, are used. Some of these are designed to lay flush or flat against the surfaces to which they are attached, as in Fig. 26-3; others are shaped to lie inclined at an angle to the

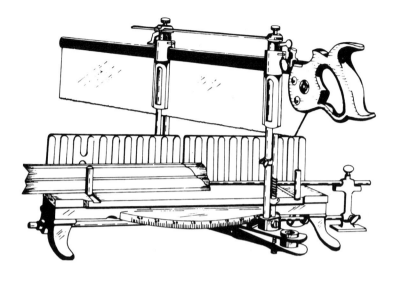

Fig. 26-2. Metal miter box with graduated scales of angle.

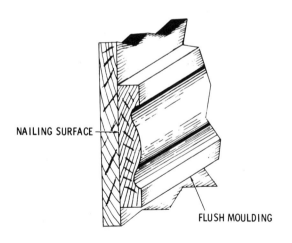

NAILING SURFACE

FLUSH MOULDING

Fig. 26-3. Flush-type moulding.

565

nailing surfaces, as in Fig. 26-4. It is the rake or spring type moulding that is hard to cut.

Mitering Flush Mouldings

Where two pieces of moulding join at right angles, as for instance the sides of a picture frame, the miter angle is 45°. The

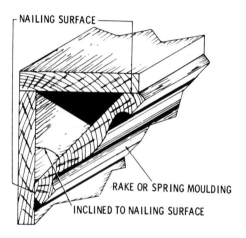

Fig. 26-4. Spring or rake moulding.

term "miter angle" means the angle formed by the miter cut and edge of the moulding, as in Fig. 26-5.

In paneling for a stairway, the mouldings are joined at various angles, as in Fig. 26-6. This is known as varying miters, and a problem arises to find the miter angles. This is easily done by remembering that the miter angle is always half of the joint angle. To find the miter cut, that is the angle at which the miter cut is made, bisect the joint angle. This is done as explained in Fig. 26-7. The triangle ABC corresponds to the triangle A in Fig. 26-6. To find the miter cut at A, describe the arc MS of any radius with A as center. With M and S as centers, describe arcs L and F, intersecting at R. Draw line AR, which is the miter cut required.

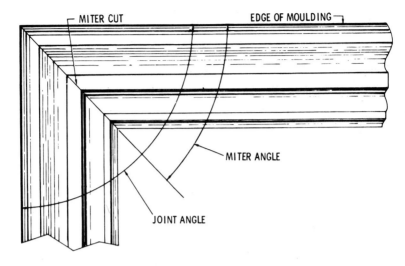

Fig. 26-5. Two pieces of flush moulding joined at 90°.

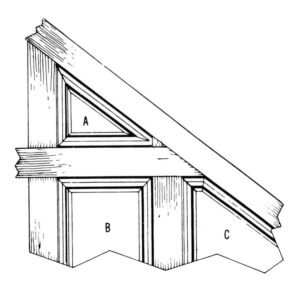

Fig. 26-6. Panel work of a wall showing various miters. In panel (A) each angle is different; (B) both miters are equal; (C) has two different angles.

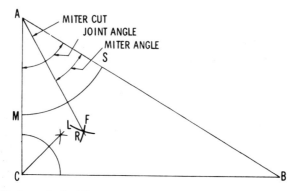

Fig. 26-7. A method of finding various miter cuts for different angles.

Mitering Spring Mouldings

A spring moulding is one that is made of thin material, and is leaned or inclined away from the nailing surface, as explained in Fig. 26-4. These mouldings are difficult to miter, especially when the joint is made with a gable, springs, or raking moulding. The two most unusual forms of miters to cut on spring mouldings are those on the inside and outside angles as shown in Fig. 26-8. The pieces are represented as they would appear from the top sill looking down.

A difficult operation for most carpenters is the cutting of a spring moulding when the horizontal portion has to miter with a gable or raking moulding. The miter-box cuts for such joints are laid out as shown in Fig. 26-9. To lay out these cuts in constructing the miter box, make the "down cuts" BB, the same pitch as the plumb cut on the rake. The "over cuts" OO and $O'O'$, should be obtained as follows. Suppose a roof has a quarter pitch, find the rafter inclination as in Fig. 26-10, by laying off $AB=12$ in. run and $BC=6$ in. rise, giving the roof angle CAB for $1/4$ pitch and rafter length $AC=13.42$ in. per foot run. With the setting 13.42 and 12, lay the steel square on top of the miter box, as shown in Fig. 26-11.

Mitering Panel and Raised Mouldings

The following instructions illustrate how raised and rabbeted mouldings may be cut and inserted in panels. Fig. 26-12 shows a

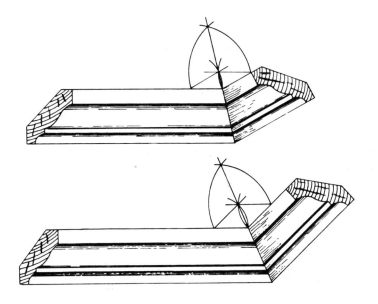

Fig. 26-8. The two most unusual miter forms to cut.

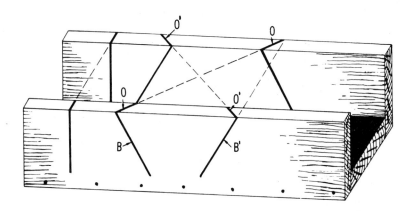

Fig. 26-9. Miter-box layout for cutting a spring moulding when the horizontal portion has to miter with a gable or raking moulding.

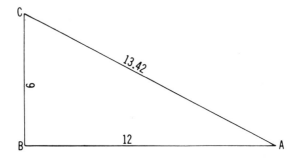

Fig. 26-10. Method of finding the angle for the cuts shown in Fig. 26-9.

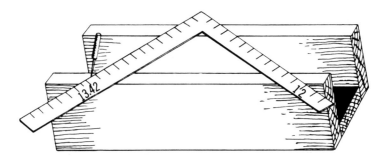

Fig. 26-11. Steel square applied to the miter box with 13, 42 and 12 setting to mark for cuttings.

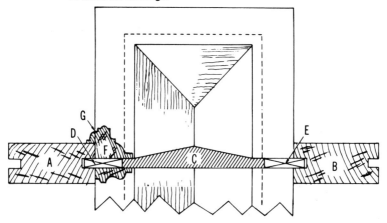

Fig. 26-12. A panel and moulding design.

570

panel and moulding designed for a room or wardrobe door. *AB* denotes the outside frame, and *C*, the raised panel. *D* and *E* are the pine fillets inserted in the plowing, and *F* is the panel moulding which has to be mitered around the inside edges of the frame. Point *G* is the rabbet or lips on the moulding *F*. If the framing *AB* is carefully constructed, and the surfaces are equal, the offset down to the panel will be equal all around, then all that is necessary is to make a hardwood strip or saddle equal in width to the depth of the offset.

The front door shown in Fig. 26-13 has both flush and raised panels. A raised 1-in. moulding is on the outside or street side and an ordinary ogee and chamfer is on the inside. The enlarged section is shown. This door is a good example of mitered mouldings which form an attractive design. The difference between

Fig. 26-13. End and side views of a door with raised moulding.

outside and inside miters must be explained—an *inside* miter is one in which the profile of the moulding is contained, or rather the outside lines and highest parts are contained, within the angle of the framing. An *outside* miter is one which is directly opposite and not contained, but the whole of the moulding is mitered on the panel outside the angle. Both miters are sawed similarly in the box with the exception of the reversing of the intersections.

Cutting Long Miters

In numerous instances, miter cuts must be made that cannot be cut in an ordinary or patent miter box. In such cases the work is facilitated by making a special box if there are several cuts of a kind to be made.

Fig. 26-14 shows a box 13 in. high which has a flare of $3^1/_4$ in. Its construction requires miter cuts which cannot be made on an

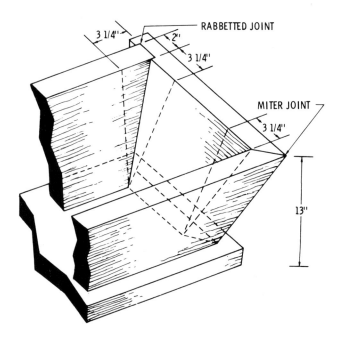

Fig. 26-14. A view of two joints, one showing a rabbet joint and the other a meter joint.

ordinary miter box. One corner is a rabbet joint and the other corner is a miter joint. Each corner can be cut out by the use of an adjustable-table power saw.

Coping

By definition cope means to cover, or match against, a covering. Coping is generally used for mouldings, the square and flat sur-

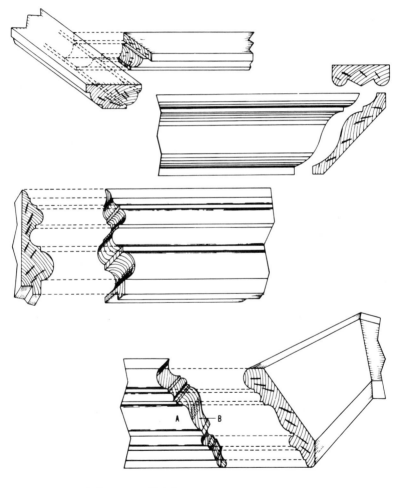

Fig. 26-15. Various coped joints.

573

faces being fitted together, one piece abutting against the other. Against plaster, the inside miter is useless since one piece is almost certain to draw away and open the joint as it is being nailed into the studding. It can be mitered tight enough by cutting the lengths a little full and springing them into place, but it is not advisable except possibly in solid corners. If against plastered walls, plaster may crack. The best way to make this joint is to cope it.

Fig. 26-15 shows various types of coped joints. In order to obtain this joint, the piece of moulding is placed in a miter box and cut to a 45° angle. After this is done, the miter angle is cut by a coping saw along the design of the moulding. If the corners are square, the miter and coped joints will fit perfectly.

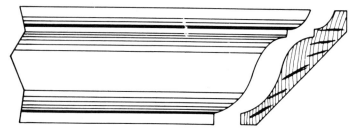

Fig. 26-16. Coped crown or spring moulding.

Fig. 26-16 shows that when a moulding is cut in a miter box for coping, it is always the reverse of the profile, and when cut out to the line thus formed, preferably with a coping saw, it fits to it at every inside corner so as to be invisible. In brief, each curved line and members join and intersect each to each without interruption at any point.

CHAPTER 27

Joints and Joinery

The most challenging job for anyone who works with wood is making joints. The problem is to cut two pieces of wood so that they meet without gaps, or with very little space, so the joint can do the job it is intended to do. It is a job that requires precision and patience, and the ability to make good joints is the hallmark of woodworkers who have mastered their trade.

Before modern glues and clamps, many joints relied on various wedges and keys and stresses to hold the joint together. Today, though, with tenacious glues and proper clamping procedures and equipment—plus the availability of special fasteners, as well as nails and screws and bolts—the job can be done with relative ease. Indeed, most carpenters get by with just a few joints, such as butt, rabbet, dado, and miter.

Following is a presentation of the main joints used today plus a consideration of some joints that the average carpenter or handy-man will rarely see but should know about as part of a well-rounded corpus of knowledge. For craftsmen interested in working on antique furniture or other old items, such as old boats, a knowledge of these joints can be important, even crucial, to the job.

There is no use trying to classify all types of wood joints because their number and descriptions are tremendous, but many of them may be placed under these headings:

1. Straight butt.
2. Dowel.
3. Square, butted, or mitered corners.
4. Dado.
5. Scarf.
6. Mortise and tenon.
7. Dovetail.
8. Wedge.
9. Tongue and groove.
10. Rabbet.

PLAIN EDGE AND BUTTED JOINTS

The plain edge joint is a joint between the edges of boards where the side of one piece is placed against the side of another, whereas the butt joint is a joint in which the square end of one member is placed against the square end of another.

Straight Plain Edge Joint

This type of joint is more or less readily made on a power jointer. The plain edge joint has many uses and is commonly used to build up wide boards for panels, shelves, etc., from narrower pieces. For boat planking, the boards are often curved and slightly beveled so that the joint is left open to be caulked later. Such curved joints must be fitted one edge to the other. In furniture, cabinet, and other fine finish work, the edges are usually glued.

To make a glued edge joint, square and straighten the edges carefully with both fore and jointer planes (Fig. 27-1). Test the edges often with the try square to assure squareness. White glue

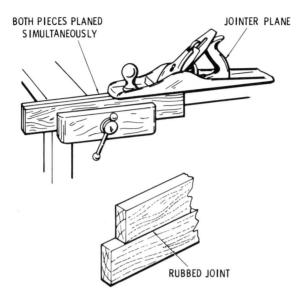

Fig. 27-1. The method of planing both edges together to obtain a straight-butt side joint. This requires a great deal of skill in planing, and it is necessary that the plane be straight on the edge and carefully sharpened and adjusted. After planing, the edges are glued and rubbed together, as shown.

can be used. This glue is easy to use and dries quickly in the air. White glue is also nonstaining, and no special equipment is needed for its application. White glues are now generally used in shops and in furniture factories, though they are not waterproof. If the work is to be exposed to the outside weather, use resorcinol glues; they are entirely waterproof and will air-dry if a catalyst is used, although they are deep red in color and will stain badly. They can be painted over readily. All modern glues will function at room temperatures. If several boards are to be joined edge to edge, as indicated in Fig. 27-2, at least three clamps will usually be necessary—one on one side and two on the other side—to prevent buckling. Take care to make the edges true and even when gluing, or it may unnecessarily require considerable scraping to make the joint flush and smooth.

577

Dowel Joints

It is usually not necessary to dowel a well-fitting glued edge joint, but it is sometimes done to facilitate assembly; the dowels used are usually quite short. For a butt joint into side wood, they are a satisfactory substitute for mortise and tenon joints and are considerably easier to make. When making heavy screen frames, storm sash, etc., dowel joints are satisfactory if they are glued together with an approved waterproof glue. Fig. 27-3 shows the assembly of a typical dowel joint.

The holes must be accurately marked and bored; if these precautions are not taken, the holes will not be in perfect alignment and it will be impossible to assemble the joint, or, when assembled, the pieces will not be in their proper alignment. Jigs designed to hold the bit in alignment are obtainable from several major tool

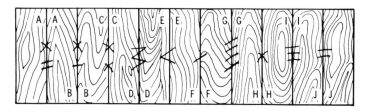

Fig. 27-2. Narrow boards can be jointed and placed together by using a marking system so that the same edges will come together when assembling them.

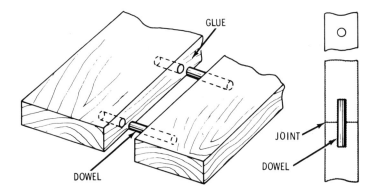

Fig. 27-3. A typical dowel pin joint.

companies, and these devices are a great help when a great amount of doweling is to be done.

The method of making dowel joints without a jig is shown in Fig. 27-4. Dowel rods made of several different types of hardwoods are obtainable; some of them have shallow spiral grooves around them to assist carrying the glue into the hole.

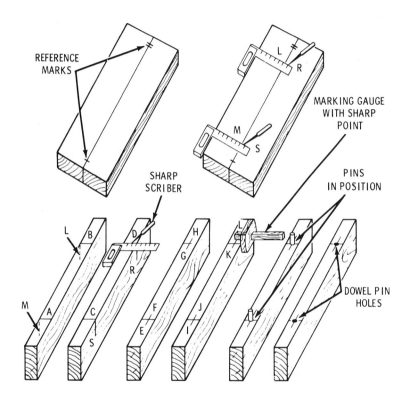

Fig. 27-4. The method of making dowel joints. After making reference marks on the two boards, scribe lines _A_, _B_, _C_, and _D_. Set the marking gauge to half the thickness of the boards, and scribe lines _EF_, _GH_, _IJ_ and _KL_. Bore a hole at the intersection of each of these lines; the holes should be just less than half the thickness of the boards. The dowels should fit tightly in these holes.

579

Square Corner Joint

The two members of a corner joint are joined at right angles, the end of one butting against the side of the other. When making a corner joint, saw to the squared line with a back saw and finish with a block plane to fit. The work should be frequently tested with a try square, both lengthwise and across the joint. The method of marking this type of joint is shown in Fig. 27-5. The joint may be fastened together with nails or screws. When fastening the joint, the pieces should be firmly held in position at a 90° angle by a vise or by some other suitable means.

Mitered Corner Joint

This type of joint is used mostly in making picture frames. To properly make a mitered joint, a picture-frame vise should be used when fastening the pieces together instead of the makeshift method of offset nailing. In fact, a picture-framing shop, to be worthy of the name, should be provided with a picture-frame vise, one type of which is shown in Fig. 27-6.

When cutting the 45° miter, use a miter box. After sawing, dress and fit the ends with a block plane. There are two ways to nail a mitered joint—the correct way with a picture-frame vise and the wrong way with an ordinary vise. Where considerable work is to be done, a combined miter box and vise is desirable; one of these is shown in Fig. 27-7.

The methods of mitering corners shown in Fig. 27-8 are used to a great extent when constructing small drawers for merchandise cabinets, such as those used in drug stores. The joinery shown in Fig. 27-8A is not particularly effective. A much stronger joint may be made by sawing the groove for the feather straight across the corner almost through; then glue in strong hardwood feathers, with their ends cut off and smoothed flush. The method shown in Fig. 27-8B is not too efficient and is difficult to make, since the outside corner of the groove is often chipped in construction. These days, the mitered corner joint has been improved by sawing the grooves on a band saw, or on a jigsaw if there are many to make, and then driving a small patented metal feather with sharp turned edges and a slight taper in from each edge. The feathers draw the joint tight, hold well, and no blocking is necessary.

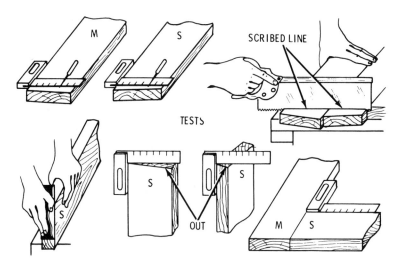

Fig. 27-5. The method of making a corner joint. After squaring and sawing the edges (*M* and *S*), plane the joint surface of one board (*S*), and test the edges with a square until a perfect right-angle fit is obtained.

Fig. 27-6. A picture frame vise. With this tool, any frame can be held in the proper position for nailing.

581

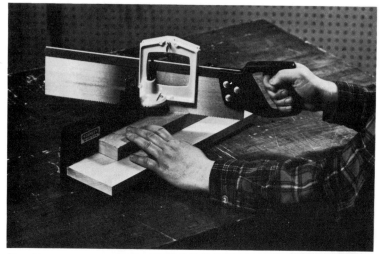

Fig. 27-7. A typical miter machine. With this device, any mitered joint can be cut, glued, and nailed to make tight, close-fitting corners.

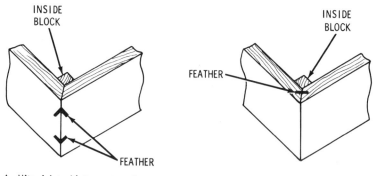

A. Miter joint with two outer-edge feathers.

B. Miter joint with end feathers.

Fig. 27-8. Miter joints reinforced by feathers. These feathers are kept in place by glue; the joint may also be reinforced by an inside block, as shown.

Splined Joint

The form of joint shown in Fig. 27-9 is called a splined joint, or sometimes a slip-tongue joint. In the shop, it is often used for edge-glued joints, since it holds the members in alignment when clamped.

A groove is made in each of the pieces to be joined, and a spline, made as a separate piece, is inserted in both grooves. The main reason for the use of a splined joint is that when two pieces of softwood are joined, a hardwood spline (which should be cut across the grain) will make the joint less likely to snap than if a tongue were cut in the softwood lengthwise with the grain.

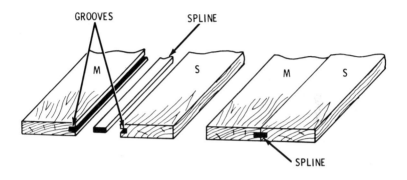

Fig. 27-9. Component parts and assembly of a splined joint. The spline fits into the grooves in *M* and *S*.

Splice Joints

This kind of joint is similar to the familiar double-strap butt joint used on the longitudinal seams of some shell boilers. The two pieces of wood to be joined are placed end to end; they are joined by fish pieces placed on each side and are secured by through-type cross bolts, or nails, as shown in Fig. 27-10. These fish plates may be made either of wood or of iron, and they may have plain or projecting ends.

The plain type, shown in Fig. 27-10A, is normally suitable when the form of stress is compression only; however, if the joint is properly made, it will withstand either tension or compression. If the joint is to be subjected to tension, the fish plates (either wood or iron) should be anchored to the main members by keys or projections, as shown in Fig. 27-10C and D.

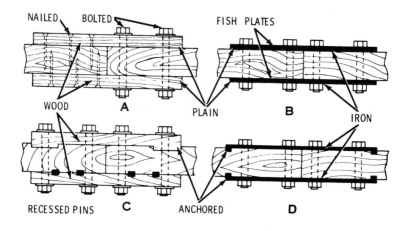

Fig. 27-10. Splice, or fish, joints: (A) plain joint with wooden fish plates; (B) plain joint with iron fish plates; (C) wood plates anchored on the end; (D) iron plates anchored on the end.

LAP JOINTS

In the various joints grouped under this classification, one of the pieces to be joined laps over, or into, the other, hence the name lap joint. Some typical lap joints are shown in Fig. 27-11.

Rabbet Joint

A rabbet joint is cut across the edge or end of a piece of stock. The joint is cut to a depth of about one-half the thickness of the material. The rabbet joint is a common joint in the construction of cabinets and furniture because it allows pieces to be joined so that no seam shows. On a typical cabinet, for example, the recess, or L-shaped section of the joint, can be cut out of the end of each side and the back of the cabinet simply fitted in between the sides and fastened in place. Anyone looking at the cabinet from the sides will not see any seam.

Rabbet joints are also used in the construction of drawers and in the making of boxes of various kinds. A rabbet joint can be made with a wide variety of tools (see Fig. 27-12).

584

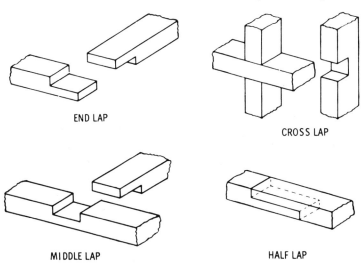

END LAP

CROSS LAP

MIDDLE LAP

HALF LAP

Fig. 27-11. Typical lap joints. The overlapping feature furnishes a greater holding area in the joint and is therefore stronger than any of the butt or plain joints. A half-lap joint, sometimes called a scarf joint, is made by tapering or notching the sides or end of two members so that they overlap to form one continuous piece without an increase in thickness. The joints are usually fastened with plates, screws, or nails and are strengthened with glue.

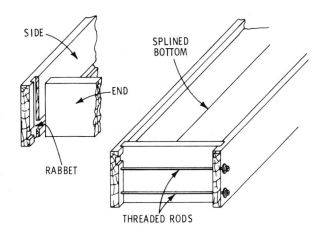

SIDE

SPLINED BOTTOM

END

RABBET

THREADED RODS

Fig. 27-12. The housed, or rabbet, joint.

585

Dado Joint

A dado joint is a groove cut across the grain and will receive the butt end of a piece of stock. The dado is cut to the width of the stock that will fit in it and to a depth of one-half the thickness of the material. The dado joint is a common joint in construction and is used for the installation of shelves, stairs, and kitchen cabinets.

Scarf Joints

By definition, a scarf joint is made by cutting away the ends of two pieces of timber and by chamfering, halving, notching, or sloping, making them fit each other without increasing the thickness at the splice. They may be held in place by gluing, bolting, plating, or strapping.

There are various forms of scarf joints, and they may be classified according to the nature of the stresses which they are designed to resist, as:

1. Compression.
2. Tension.
3. Bending.
4. Compression and tension.
5. Tension and bending.

Compression Scarf Joint—This is the simplest form of scarf joint. As usually made, one-half of the wood is cut away from the end of each piece for a distance equal to the lap, as shown in Fig. 27-13A; this process is called "halving." The length of the lap should be five to six times the thickness of the timber. Mitered ends, as in Fig. 27-13B, are better than square ends, where nails or screws are depended on to fasten the joint. For extraheavy-duty joints, iron fish plates are sometimes provided, thereby greatly strengthening the joint, as shown in Fig. 27-13C; when these are used, mitered ends are not necessary.

Tension Scarf Joint—There are various methods of "locking" joints to resist tension, such as by means of keys, wedges, or so called keys or fish plates with fingers, etc., as shown in Figs. 27-14 and 27-15. The difference between keys and wedges is shown in Fig. 27-16. Today, you will see fish plates used, but keys and wedges are rare in modern joinery.

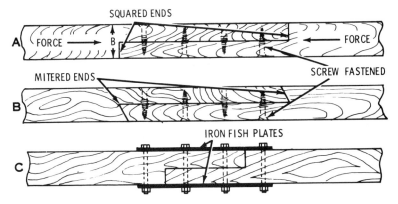

Fig. 27-13. Compression scarf joints: (A) plain, square ends; (B) plain, mitered ends; (C) plain, square ends reinforced with iron fish plates.

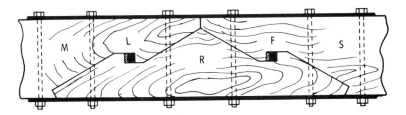

Fig. 27-14. Butt and lap plate scarf joint; this joint is designed to avoid reducing the length of the joined timbers when the timbers are not long enough for a lap joint. Piece *R* is splayed onto timbers *M* and *S*, which are cut as shown. The laps of *M* and *S* on *R* are cut with notches and are provided with wedges *L* and *F* to handle any tension stress. The joint is bolted and often reinforced with fish plates.

Bending Scarf Joint—When a beam is acted on by a transverse, or bending stress, the side on which the bending force is applied is subjected to a compression stress, and the opposite side is subjected to a tension stress. Thus, in Fig. 27-17A, the upper side is in compression, and the lower side is in tension. At L, the end of the joint may be square, but at F, it should be mitered. If this end were square (as at F', Fig. 27-17B), the portion of the lap M between the bolt and F' would be rendered useless to resist the bending force.

587

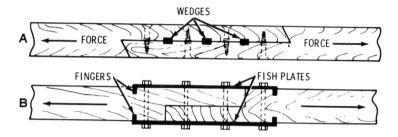

Fig. 27-15. Tension scarf joints: (A) mitered ends fastened with screws; tension stress resisted by wedges; (B) square ends bolted and reinforced with iron fish plates; tension stress resisted by fingers on the fish plates.

Fig. 27-16. Key and wedges.

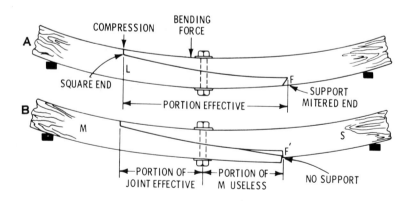

Fig. 27-17. Bending scarf joints. One end of the joint should be mitered to provide adequate support for the stresses applied to the joint.

When designing a bending scarf joint, it is important that the thickness at the mitered end be ample, otherwise the strain applied at that point might split the support. Gluing normally helps prevent such stresses from developing.

Mortise and Tenon Joints

A mortise is defined as a space hollowed out in a member to receive a tenon, and a tenon is defined as a projection, usually with a rectangular cross section, at the end of a member which is to be inserted into a socket, or mortise, in another timber to make a joint.

Mortise and tenon joints are frequently called simply tenon joints. The operation of making mortise and tenon joints is also termed tenoning, which also implies mortising.

There are many different mortise and tenon joints, as illustrated in Figs. 27-18 through 27-24, and they may be classified with respect to:

1. Shape of the mortise
2. Position of the tenon
3. Degree in which tenon projects into mortised member
4. Degree of mortise housing
5. Number of tenons
6. Shape of tenon shoulders
7. Method of fastening the tenon

The mortise and tenon must exactly correspond in size; that is, the tenon must accurately fit into the mortise. The position of the tenon is usually at the center of the timber, but sometimes it is located at the side, depending, except in special cases, on the degree of housing. The tenon may project partly into, or through, the mortised timber. When the tenon and mortise do not extend through the mortised timber, the joint is called a stub tenon. This form of tenon is used for jointing the framework of partitions and is also employed in work where the joint will not be subjected to any tension.

The term "degree of housing" signifies the degree in which the tenon is covered by the mortise, that is, the number of sides of the mortise. The number of tenons depends on the shape of the timbers, whether they are square or rectangular, with considera-

589

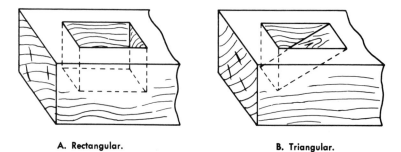

A. Rectangular.   B. Triangular.

Fig. 27-18. Mortise and tenon joints—shape of the mortise.

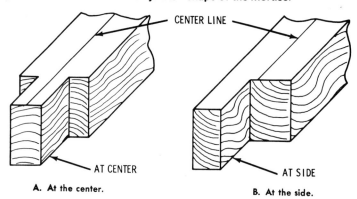

CENTER LINE

AT CENTER   AT SIDE

A. At the center.   B. At the side.

Fig. 27-19. Mortise and tenon joints—position of the tenon.

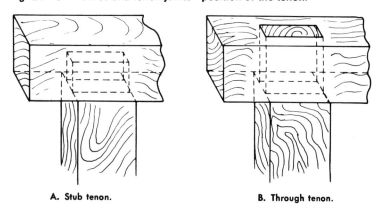

A. Stub tenon.   B. Through tenon.

Fig. 27-20. Mortise and tenon joints—degree to which the tenon projects
        into mortised timber.

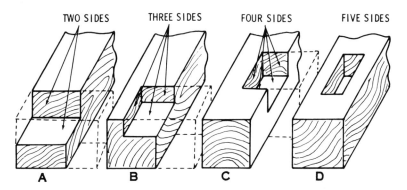

Fig. 27-21. Mortise and tenon joints—degree of mortise housing: (A) two sides; (B) three sides; (C) four sides; (D) five sides.

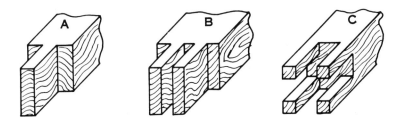

Fig. 27-22. Mortise and tenon joints—number of tenons: (A) single tenon; (B) double tenon; (C) multitenon.

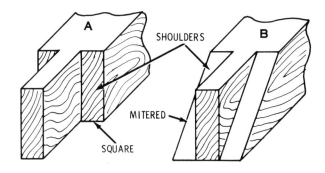

Fig. 27-23. Mortise and tenon joints—shape of the tenon shoulders: (A) square; (B) mitered.

591

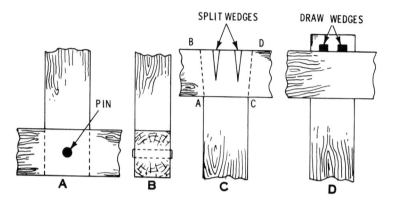

Fig. 27-24. Mortise and tenon joints—methods of fastening the tenons: (A) side view, tenon secured by a pin; (B) front view, tenon secured by a pin; (C) tenon secured by internal, or split, wedges; sides *AB* and *CD* are tapered, thus securely wedging the tenon into the mortise; (D) tenon secured by external, or draw, wedges, which are driven into rectangular holes beyond the mortise.

ble width and little thickness, etc. The tenon shoulders are usually at right angles with the tenon as they are when the two timbers are joined at right angles, but they may be mitered to some smaller angle, such as 60° or 45°, as in the case of a brace.

There are several ways of fastening mortise and tenon joints, such as with pins or wedges. When making a mortise and tenon joint, the work is first laid out to given dimensions, as shown in Fig. 27-25.

Cutting the Mortise—Select a chisel that is as near to the width of the mortise as possible. This chisel, especially for large work, should be a framing or mortise chisel. Bore a hole the same size as the width of the mortise at the middle point. If the mortise is for a through tenon, bore halfway through from each side. In the case of a large mortise, most of the wood may be removed by boring several holes, as shown in Fig. 27-26. When cutting out a small mortise with a narrow chisel, work from the hole in the center to each end of the mortise, holding the chisel firmly at right angles with the grain of the wood. At the ends of the mortise, the chisel must be held in a vertical position, as shown in Fig. 27-27, with the flat side facing the end of the mortise.

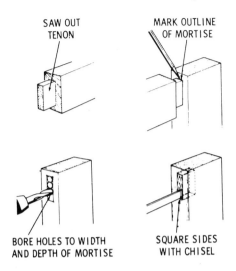

Fig. 27-25. The method of laying out and making a small mortise and tenon joint.

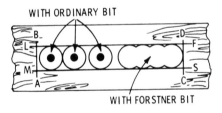

Fig. 27-26. The method of boring holes when making a large mortise.

Always loosen the chisel by a backward movement of the handle; a movement in the opposite direction would injure the ends of the mortise. Never make a chisel cut parallel with the grain, since the wood at the side of the mortise may split. When cutting a through mortise, cut only halfway through on one side, and finish the cut from the other side. After cutting, test the sides of the mortise by using a try square, as shown in Fig. 27-28; this procedure will check the accuracy with which the work was laid out.

Cutting the Tenon—A back saw is used for cutting out the wood on each side of the tenon, and, if necessary, a finishing cut may be taken with a chisel. After the wood has been cut away, the tenon should be pointed by chiseling all four sides.

Fig. 27-29 shows the appearance of the tenon before and after the pointing operation; if this operation were omitted, a tight-fitting tenon would be difficult to start into the mortise and could splinter the sides of the mortise when driven through. Do not cut off the point until the tenon is finally in place and the pin is driven home.

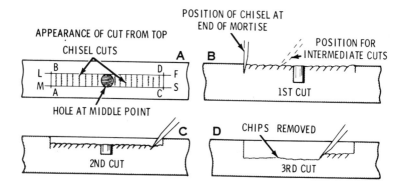

Fig. 27-27. The method of cutting a small mortise. After laying out the mortise, bore a hole at the center (A) and work toward each end with a chisel. Chisel cuts should always be made against the grain.

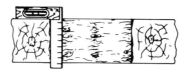

Fig. 27-28. Test the end with a square after cutting the mortise.

Draw Boring—The term "draw boring" signifies the method of locating holes in the mortise and tenon that are eccentric with each other so that when the pin is driven in, it will "draw" the tenon into the mortise, thereby forcing the tenon shoulders tightly against the mortised timber. The holes may be located either by accurately laying out the center, as shown in Fig. 27-30C, or by boring the mortise and finding the center for the tenon hole, as in Fig. 27-30D.

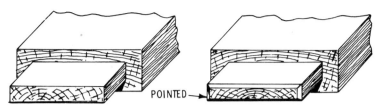

Fig. 27-29. Appearance of the tenon before and after pointing.

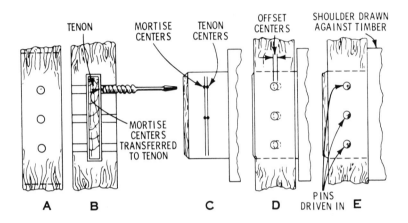

Fig. 27-30. The method of transferring pin centers from the mortise holes to the tenon by draw boring. When laying out the tenon-hole centers, make the offset toward the tenon shoulder.

Considerable experience is necessary to properly locate the tenon hole. If too much offset is given, an undue strain will be brought to bear on the joint; this strain is frequently sufficient to split the joint. It is much better to accurately lay out the work and make a tightly fitting pin than to depend on draw boring.

Dovetail Joints

A dovetail joint may be defined as a partially housed tapered mortise and tenon joint, the tapered form of mortise and tenon forming a lock which securely holds the parts together. The word "dovetail" is used figuratively; the tenon expands in width toward the tip and resembles the fan-like form of the tail of a dove. The

595

various forms of dovetail joints, some of which are shown in Fig. 27-31, may be classed as:

1. Common
2. Compound
3. Lap, or half-blind
4. Mortise, or blind

Common Dovetail Joint—This is a plain, or single "pin," joint. In dovetail joints, the tapered tenon is called the *pin*, and the mortised part that receives this joint is call the *socket*. Where strength rather than appearance is important, the common dovetail joint is used. The straight form of this joint is shown in Fig. 27-32, and the corner form is shown in Fig. 27-33; the proportions of the joint are shown in Fig. 27-33A and B.

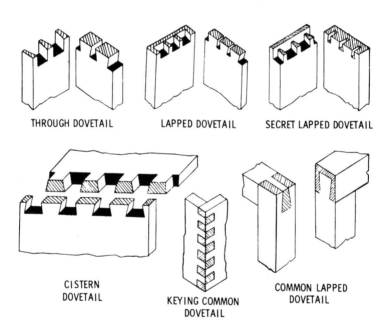

THROUGH DOVETAIL   LAPPED DOVETAIL   SECRET LAPPED DOVETAIL

CISTERN DOVETAIL

KEYING COMMON DOVETAIL

COMMON LAPPED DOVETAIL

Fig. 27-31. Dovetail joints are used principally in cabinetmaking, drawer fronts, and fine furniture work. They are a partly housed and tapered form of tenon joint in which the taper forms a lock to hold the parts together securely.

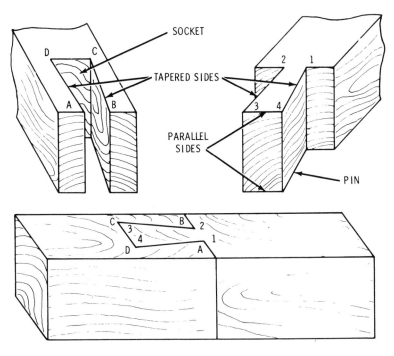

Fig. 27-32. The straight form of the common, or plain, dovetail joint. By noting the positions of the letters and numbers, you can see how the socket and pin are assembled.

Compound Dovetail Joint—This is the same as the common form but has more than one pin, thereby adapting the joint for use with wide boards. When making this joint, both edges are made true and square; a gauge line is run around one board at a distance from the end equal to the thickness of the other board, and the other board is treated similarly. Two methods are commonly followed. Some mark and cut the pins first; others mark and cut the sockets first.

In the first method, the pins are carefully spaced, and the angles of the tapered sides are marked with the bevel. Saw down to the gauge line, and work the spaces in between with a chisel and a mallet. Then, put B on top of A (in Fig. 27-34), and scribe the mortise. Square over, cut down to the gauge line, clean out, and fit together.

597

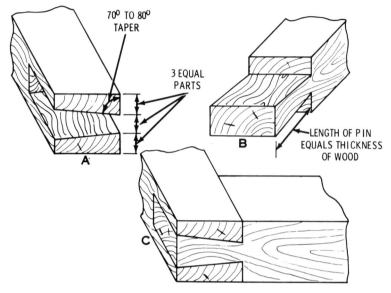

70° TO 80° TAPER

3 EQUAL PARTS

A

B

LENGTH OF PIN EQUALS THICKNESS OF WOOD

C

Fig. 27-33. The corner form of the common, or plain, dovetail joint, with the proper proportions for the socket and pin.

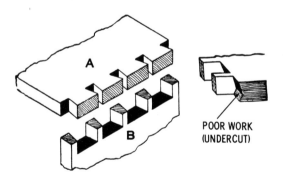

A

B

POOR WORK (UNDERCUT)

Fig. 27-34. A multiple dovetail joint, with a poorly cleaned joint shown in detail.

The second method is to first mark the socket on A (sometimes on common work, the marking is dispensed with, and the worker uses his eyes as a guide); then, run the saw down to the gauge line, put A on B, and mark the pins with the front tooth of the saw. Cut

598

the pins, keeping outside of the saw mark sufficiently to allow the pins to fit tightly; both pieces may then be cleaned out and tried together.

When cleaning out the mortises and the spaces between the pins, the woodworker must cut halfway through, then turn the board over and finish from the other side, taking care to hold the chisel upright so as not to undercut, as shown in Fig. 27-34, which is sometimes done to ensure the joint fitting on the outside.

Lap or Half-Blind Dovetail Joint—This joint is used in the construction of drawers on the best grades of work. The joint is visible on one side but not on the other, as shown in Fig 27-35, hence the name "half-blind." Since this form of dovetail joint is used so extensively in the manufacture of furniture, machines have been devised for making the joint, thus saving time and labor.

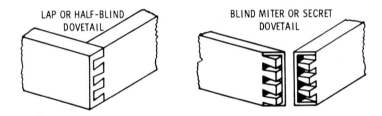

Fig. 27-35. Half-blind and blind dovetail joints. These joints are used in the best grades of drawer and cabinetwork, since the joint is visible on only one side. They should be exceptionally well fitted because of the frequent pull on the front piece.

Blind Dovetail Joint—This is a double lap joint; that is, the joint is covered on both sides, as shown in Fig 27-35, and is sometimes called a secret dovetail joint. The laps may be either square, as in Fig. 27-36, or mitered, as in Fig. 27-37. Because of the skill and time required to make these joints, they are used only on the finest work. The mitered form is the more difficult of the two to assemble.

Spacing—The maximum strength would be gained by having the pins and sockets equal; however, this is rarely done in practice, since the mortise is made so that the saw will just clear at the narrow side with the space from eight to ten times the width of the

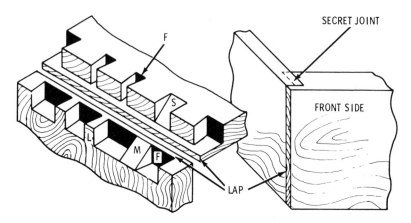

Fig. 27-36. The blind, square-lap dovetail joint. Two forms of pins and sockets are used: mitered (*MS*) and square (*LF*).

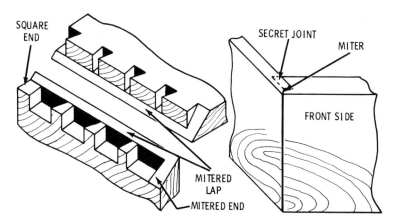

Fig. 27-37. The blind, mitered-lap dovetail joint.

widest side. Small pins are used for the sake of appearance, but fairly large ones are preferable. The outside pin should be larger than the others and should not be too tight or there will be the danger of splitting, as shown in Fig. 27-38 at point A. The angle of taper should be slight (70° to 80°) and not acute as shown, otherwise there is the danger of pieces L and F in Fig. 27-38 being split off in assembling.

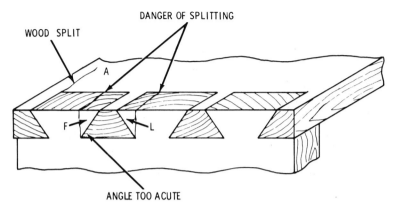

Fig. 27-38. A badly proportioned common dovetail joint can result in splitting.

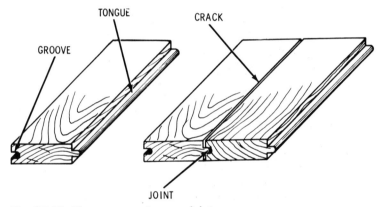

Fig. 27-39. The tongue-and-groove joint.

Position of Pins—When boxes are made, the pins are generally cut on the ends with the sockets on the sides. Drawers have the pins on the front and back. The general rule is to locate the tapered sides so that they are in opposition to the greatest stress that may be applied on the piece of work to which the joint is connected.

Tongue-and-Groove Joint—In this type of joint, the tongue is formed on the edge of one of the pieces to be joined, and the groove is formed in the other, as shown in Fig. 27-39.

601

CHAPTER 28

Cabinetmaking Joints

The most challenging joints the carpenter and do-it-yourselfer can make are for cabinets. They are, as it were, unforgiving. If you make an error, it can show up badly, not only marring the structural integrity of a piece but making for a poor appearance.

Yet cabinet joints can be done perfectly, even by the beginner. All it takes is a lot of patience—and the right tools.

THE TOOLS

A full set of good carpenters' tools is necessary in the cabinet-making shop, including a set of firmer chisels, a set of iron bench planes, a set of auger bits with slow-feed screws that range in size from $\frac{1}{4}$ to 1 in., and a set of numbered bits for the electric drill. In addition, the following tools will be found useful almost continually.

1. Router plane
2. Plow plane
3. A substantial cast-iron miter box
4. Several bar clamps with varying lengths of bars
5. Hand clamps (those with wood jaws are most useful in the shop, but malleable C-clamps are often used)
6. A high-speed ¼-in. electric drill
7. A chute board of wood or iron.

In addition to the regular vise, a workbench with an end vise, for holding material between stops on top of the bench will be found quite convenient.

JOINTS

Many of the joints seen earlier in this book are used in cabinetry. There is a great variety used, and they are usually classified according to their general characteristics, as glued, halved and bridle, mortise-and-tenon, dovetail, mitered, framing, hinging and shutting. Under each of these classifications is grouped a variety of joints that will be considered separately and will also be briefly explained.

Glued Joints

In cabinetwork and furniture making practically all joints are, or should be, glued. Several glues are suitable for use in the cabinet shop:

White glue—This is one of the most popular, perhaps *the* most popular, glue in the woodworker's shop. It is strong and dries clear in a short amount of time. It is also relatively inexpensive.

Yellow glue—This is a woodworker's or carpenter's glue and is relatively new. It is strong and is good when you want the glue to dry more quickly than white glue.

Resorcinol—This is the glue to use outdoors. It is strong and weatherproof, but you must use it carefully because it is brown and can stain wood.

Contact cement—This is for use in applying plastic laminate and the like. It bonds things instantly.

Hot-melt—This is extruded from a glue gun and bonds within a minute. It does not require clamping, as other types (except contact cement) do.

Some of the joints used are as follows:

Beveled Joints—In this type of joint, the sides of the pieces fit together to form angles, or corners, as shown in Fig. 28-1. An infinite amount of planing and dressing can be saved by first ripping the edges roughly, by hand if necessary, on a tilting-arbor table saw if one is available. If ripped on a power saw, the angle can be adjusted precisely, and only a small amount of hand dressing will be necessary. Try the bevel continually with a T-bevel while dressing to assure that the joints fit; they must fit properly if the joint is to be glued. The joints must be clamped, and without special clamps this is troublesome. The woodworker's ingenuity will usually suggest a method. Short pieces of chain with bolts through the end links are useful, if the chains can be passed around the work. Beveled joints do not usually require exceedingly high pressure.

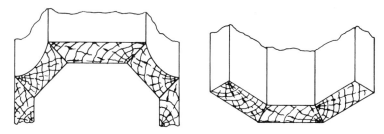

Fig. 28-1. Typical beveled joints.

Plowed-and-Splined Joints—This method of jointing is commonly used in cabinetwork and is similar to the spline joint in the preceding chapter, except that in cabinetwork, the splines are cut *across* the grain. When the thickness of the material will permit, two splines are used instead of one, as in Fig. 28-2B, because of the additional gluing surface afforded and the increased strength to the joint. Splines are cut lengthwise with the grain.

605

To make this joint successfully, the pieces should be properly faced, and the edges should be squared and straightened with a jointer so that they fit perfectly. Put reference marks on the face side so that the same edges will come together when assembled. Set the plow plane with the iron projecting approximately $\frac{1}{32}$ in. below the bottom plate. Set the depth gauge to one-half the width of the spline, and adjust the fence so that the cutter will be the required distances from the edge between the two sides. Fasten the piece securely in the bench vise so that the groove can be plowed from the face side. Begin plowing at the front (Fig. 28-3A) and work backward; finish by going right through from back to front, as in Fig. 28-3B. Hold the plow plane steady, otherwise an irregular groove will result.

For the cross-grain spline, cut off the end of a thin board of hardwood; mark it, and carefully saw off a strip across its width that is the required width of the spline, approximately $\frac{3}{4}$ in. wide. Plane the spline to the desired thickness in a tonguing board. Then, assemble the parts and glue them up, as shown in Fig. 28-4.

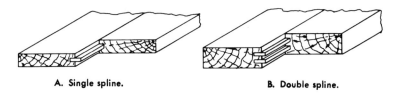

A. Single spline.   B. Double spline.

Fig. 28-2. Single- and double-splined joints.

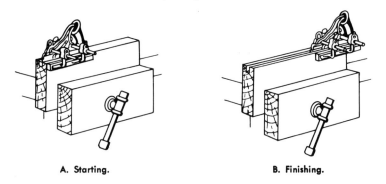

A. Starting.   B. Finishing.

Fig. 28-3. The method of plowing a single spline groove.

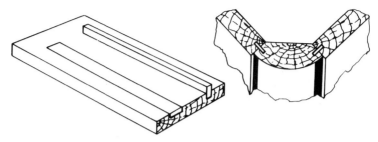

A. A tonguing board.　　　　　**B. Plowed-and-splined joints.**

Fig. 28-4. The tonguing board is a simple and handy device when used to
overcome the difficulty of holding a narrow piece of thin material
steady while planing. To make the board, use a piece of 7/8-in.
faced material, 8 to 10 in. wide and longer than the tongues to be
planed. Cut the grooves as indicated with a tenon saw; clean out
the grooves with a chisel and a router. The wider groove should
be slightly shallower than the thickness of the finished tongue to
allow for planing both sides of strips placed in it.

Hidden Slot Screwed Joints—This joint is an effective way of
fastening brackets and shelves to finished work where the
fastening must be concealed. The joint consists of a screw which is
driven part way into one piece and a hole and slot cut into the
opposite piece. The joint is effected by fitting them together with
the head of the screw in the hole and then forcing the screw back
into the slot. Fig. 28-5A shows the slot and screw in relation to each
other, and Fig. 28-5B gives a cross-sectional view of the completed
joint. This joint is also used in interior work for fastening pilasters
and fireplaces to walls, for paneling, and for almost every kind of
work requiring secure and concealed fastening.

To make a joint as shown in Fig. 28-5B, gauge a center line on
each of the pieces. Determine the position of the screws and insert
them; they should project approximately ⅜ in. above the surface.
Hold the two pieces evenly together, and, with a try square, draw
a line from the back of the screw shank across the center line of the
opposite piece. From this line, measure ⅞ in. forward on the
center line; with this point as the center, bore a hole to fit the screw
head that is just slightly deeper than the amount that the screw
projects above the surface. Cut a slot from the hole back to the line
from the screw shank; this slot should be as wide as the diameter of

the shank and as deep as the hole. As a general rule, the total length of the slot and hole should be slightly more than twice the diameter of the head of the screw.

The process of fastening pilasters to fireplaces by this method is as follows: First, mark the position of the piece and the place on the wall for the screws. In brick and cement walls, holes are drilled and wooden plugs are driven in flush with the surface to hold the screws. The plugs are shaped as illustrated in Fig. 28-6. Plugs cut as shown seldom work loose. Turn the screws into the plugs, and allow them to project approximately ⅜ inch from the surface; the screw heads should be smeared with moist lampblack. Put the piece in position and press it against the screw heads; this pressure will leave black impressions. Bore holes to fit the screw heads approximately ⅝ in. below the impressions, and cut the slot to receive the shank of the screw. Replace the piece, with the heads in the holes, and force it down.

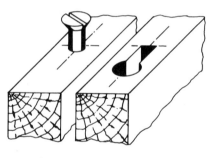

A. The joint before assembly.

B. Cross-sectional view of assembled joint.

Fig. 28-5. The hidden slot screwed joint.

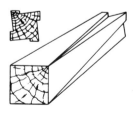

Fig. 28-6. A plug is used to hold the screw when hidden slot screwed joints are utilized on mortar and brick walls.

Dowel Joints—There are many variations of dowel jointing commonly used in cabinetwork. The basic principle and method of making a dowel joint is explained and illustrated in the preceding chapter.

To accurately fix the position of dowels in a butt joint, such as the one shown in Fig. 28-7, make all measurements and gauge lines from the edge of the faced sides. For example, with material that is 4 in. square, mark diagonal lines from the corners with a scratch awl, intersecting at the center. Then, from the edge of the faced sides, mark off 1 in. and 3 in. as shown in Fig. 28-8A. From the same sides, gauge the lines as in Fig. 28-8B. The intersection of the lines is the center for the holes, as shown in Fig. 28-8C.

Where the ordinary means of aligning dowel holes cannot be used, a dowel template or pattern is used. The template is usually made of a strip of zinc or plywood, with a small block of wood fastened to one end to act as a shoulder; the position of the dowel points is then pierced through the zinc or plywood pattern with a fine awl. Various types of templates are made and used as the occasion requires. Fig. 28-9A shows a template that is used for making doweled rails in furniture; it is made to fit the section of the rail in Fig. 28-9B. While held in position, a line is gauged down through the middle; the position of the dowels is indicated on the line. The template is laid flat on a board, and the dowel points are pierced through the surface with a fine awl. When in use, the template is placed in position on the piece, and the dowel positions are marked with an awl through the holes, as shown in Fig. 28-9C. A "bit gauge" should be used to regulate the depth of the bore when doweling. If a great amount of doweling is to be done, a doweling jig, which insures accurate boring of holes from ¼ to ¾ in., will be found useful.

Dowels are glued into one piece, cut to length, and sharpened with a dowel sharpener. As a precaution against splitting the joint, cut a V-shaped groove down the side of the dowel with a chisel; this groove permits the glue and air to escape.

Coopered Joints—These are so named because of the resemblance to the joints used in barrels made by coopers, and are used for practically all forms of curved work; they are usually splined before gluing, although dowels are used occasionally. Fig. 28-10

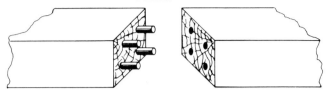

Fig. 28-7. The use of dowels in a butt joint adds strength to the joint. This construction is frequently found in cabinetwork to lengthen large mouldings and where the cross grain prevents tenoning.

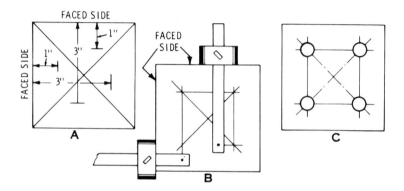

Fig. 28-8. The method used for marking the position of the dowels in a butt joint.

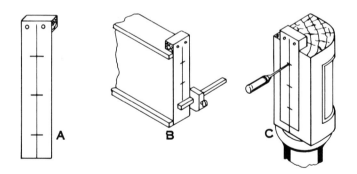

Fig. 28-9. The use of a template, or pattern, for marking dowel-pin locations: (A) the template; (B) the template is made to fit the section of rail; (C) marking dowel positions on leg.

610

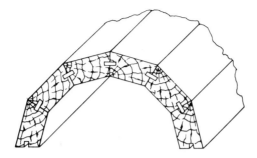

Fig. 28-10. Coopered joints are employed to form curvatures in cabinet-work.

shows the coopered joint in semicircular form, with the segments beveled at an angle of 15 degrees. They are clamped after gluing and planed to shape.

Halved and Bridle Joints

These are lap joints with each of the pieces halved and shoul-dered on opposite sides, so that they fit into each other. They are the simplest joints used in cabinetwork. Fig. 28-11A shows the common halved angle, which is the one most frequently used. Fig. 28-11C illustrates the oblique halved joint, which is used for oblique connections. Fig. 28-11D represents the mitered halved joint, which is useful when the face or frame piece is moulded. Figs. 28-11E, F, and G shows the joints that are used for cross connections having an outside strain. Fig. 28-11H illustrates the blind dovetail halved joint, which is used in places where the frame edge is exposed. Bridle or open-tenon joints are used to connect parts of flat and moulded frames. The joint in Fig. 28-12B is used where a strong framed groundwork, which is to be faced up, is required. The joint in Fig. 28-12D is used as an inside frame connection.

Mortise and Tenon Joints

Many variations of the mortise and tenon joint are used in cabinetwork. They differ in size and shape according to the requirements of the location and the purpose for which the joint is

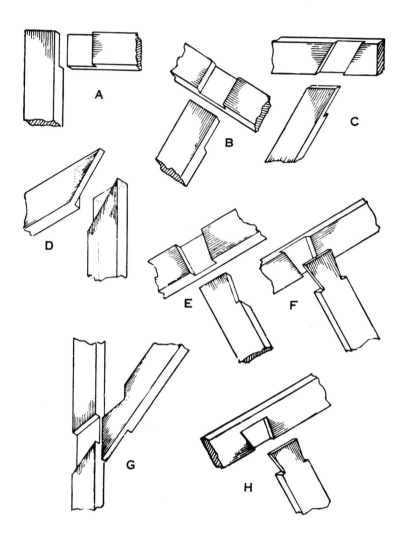

Fig. 28-11. Halved joints: (A) halved angle joint; (B) halved tee joint; (C) oblique halved joint; (D) mitered halved joint; (E) dovetail halved joint; (F) dovetail halved joint; (G) oblique dovetail halved joint; (H) blind dovetail halved joint.

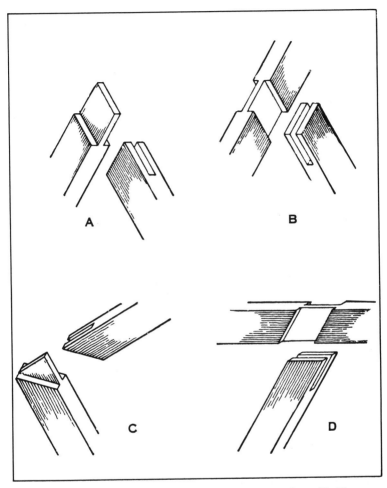

Fig. 28-12. Halved joints generally used to make flat and moulded frames: (A) angle bridle joint; (B) tee bridle joint; (C) mitered bridle joint; (D) oblique bridle joint.

used. The most frequently used mortise and tenon joint is the stub tenon, so called because it is short and penetrates only part way through the wood; Fig. 28-13A illustrates the type most generally used in doors and furniture framing. Fig. 28-13B shows the stub tenon with a mitered end; this type of construction is often neces-

613

sary when fitting rails into a corner post. Fig. 28-13C shows the rabbeted or "haunched" tenon, which is considered a stronger joint because of the small additional tenon formed by the rabbet; it is often mitered, as shown in Fig. 28-13D, to conceal the joint when used on outside frames. The joint in Fig. 28-14A is the same as that shown in Fig. 28-13C, but it is shouldered on one side only; it is sometimes called a "barefaced" tenon and is used when the connecting rail is thinner than the stile into which it is joined. Fig. 28-14B shows the long and short shoulder tenon; this joint is used when connecting a rail into a rabbeted frame, since it has one shoulder cut back so as to fit into the rabbet. Fig. 28-14C illustrates the double tenon joint, which increases the lateral strength of the stile into which it is jointed. It is simply a stub tenon that is rabbeted and notched to form two tenons, and, when glued, it makes an exceptionally strong joint. Fig. 28-14D represents a type of through mortise and tenon that is sometimes used for mortising partitions into the top or bottom of wardrobes, cabinets, etc.; the partitions are wedged across the tenon and glued.

Laying Out the Mortise and Tenon—The general practice when laying out a mortise and tenon is to square the mortise lines across the edge of the stile in pencil and then scribe two lines for the sides of the mortise with a mortise or slide gauge between the pencil lines. If the tenon is to be less than the full width of the rail, square the rail lines across the edge, in addition to the mortise lines, as shown in Fig. 28-15A. This procedure insures greater accuracy when designating the position of the mortise. When two or more stiles are to be mortised, they are clamped together, and the lines are squared across all the edges simultaneously.

For a through mortise, continue the pencil lines across the face side and onto the back edge; gauge the mortise lines from the faced side. With the gauge set for the mortise, scribe the lines for the tenon on both edges and the end of the rail, and with the aid of a try square, mark the shoulder lines with a knife or chisel on all four sides.

The proportions of stub and through mortises and tenons are usually considered as about one-third the thickness of the wood, and they should be cut with a mortise chisel of the required size. If the chisel is not exactly one-third the thickness of the material, it is better to make the mortise more than one-third rather than less. Set the mortise gauge so that the chisel fits exactly between the points, as shown in Fig. 28-15B. Make a chisel mark in the center of the edge to be mortised, and adjust the head of the gauge so that the points coincide with this mark.

Mortise cutting in cabinetwork is usually done entirely with a mortise chisel, beginning at the center and working toward the near end with the flat side of the chisel toward the end. Remove the core as you proceed; then reverse the chisel, and cut to the far end, being careful to keep the chisel in a perpendicular position when cutting the ends. Through mortises are cut halfway through

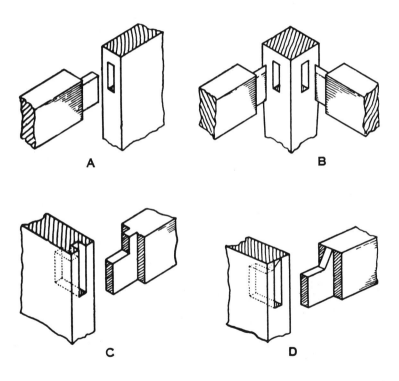

Fig. 28-13. Variations of mortise and tenon joints are frequently used in doors and in the framing of furniture: (A) stub mortise and tenon; (B) mitered stub tenon; (C) haunched, or rabbeted, mortise and tenon; (D) haunched and mitered mortise and tenon.

from one side, and the material is then removed and cut through from the opposite side.

A depth gauge for stub mortises is made by gluing a piece of paper or tape on the side of the chisel, as shown in Fig. 28-16. If the method of boring a hole in the center from which to begin the cutting of the mortise is used for stub mortises, it is advisable to use a bit gauge to regulate the depth of the bore. A small firmer chisel is used to clean out stub mortises.

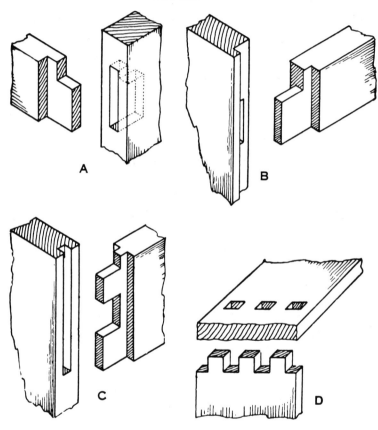

Fig. 28-14. Other mortise and tenon joints used in furniture construction: (A) barefaced mortise and tenon; (B) long and short shoulder mortise and tenon; (C) double mortise and tenon; (D) pinned mortise and tenon.

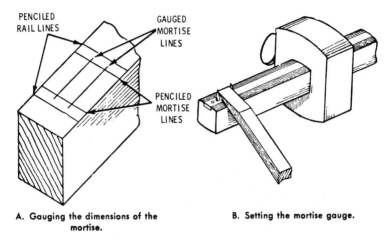

A. Gauging the dimensions of the mortise.

B. Setting the mortise gauge.

Fig. 28-15. Laying out the mortise.

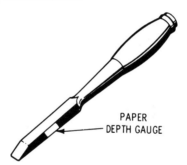

PAPER DEPTH GAUGE

Fig. 28-16. The depth of the mortise joint may be controlled by fastening a piece of paper or tape on the side of the chisel.

Cutting the Tenon—Fasten the piece firmly in the bench vise. Start the cut on the end grain, and saw diagonally toward the shoulder line, as in Fig. 28-17A. Finish by removing the material in the vise and cutting downward flush with the edge, as in Fig. 28-17B. The diagonal saw cut acts as a guide for the finishing cut and provides greater accuracy. Small tenons are usually cut with a dovetail saw.

Cutting the Shoulder—After making the tenon cuts, and to overcome any difficulty in cutting the shoulders, place the piece

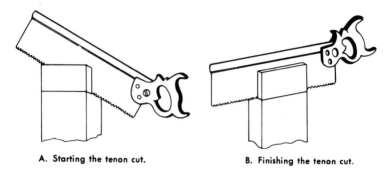

A. Starting the tenon cut.   B. Finishing the tenon cut.

Fig. 28-17. Making a tenon cut.

on the shoulder board or bench hook, and carefully chisel a V-shaped cut against the shoulder line, as shown in Fig. 28-18A. Hold the work firmly against the stop on the board; place the saw in the chiseled channel, and begin cutting by drawing the saw backward and then pushing it forward with a light stroke. Hold the thumb and forefinger against the saw, as in Fig. 28-18B, and keep the saw in an upright position. A straightedge can be placed against the shoulder line to act as a guide when cutting wide shoulders. In the case of extremely wide tenons and shoulders, a rabbet plane and shoulder plane are used; the straightedge is used as a guide for the rabbet plane.

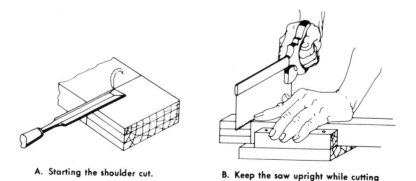

A. Starting the shoulder cut.   B. Keep the saw upright while cutting the shoulder.

Fig. 28-18. Making the shoulder cut.

Dovetail Joints

The method of making dovetail joints is described in the previous chapter. In common dovetailing, it is a matter of convenience whether to cut the pins or the dovetails first. However, where a number of pieces are to be dovetailed, time can be saved by clamping them together in the vise and cutting the dovetails first.

Dovetail Angles—For particular work where the joint is exposed, the dovetails should be cut at an angle of 1 in 8, and for heavier work, 1 in 6. To find the dovetail angle, draw a line square with the edge of a board, and divide it into 6 or 8 equal parts as desired; from the end of the line and square with it, mark off a space equal to one of the divisions, and set the bevel as shown in Fig. 28-19.

A dovetail template, as shown in Fig. 28-20A, will be found quite handy if there is a great deal of dovetailing to be done. To make the template, take a rectangular piece of ¾-in. material of any desired size, and square the edges; with the mortise gauge set for a ¼-in. mortise at ¼ in. from the edge, scribe both edges and one end. With the bevel set as shown (Fig. 28-19), mark the shoulder lines across both sides of the lower portion, and cut it with a tenon saw. Make one cut for each of the two angles. The template may also be made by gluing a straightedge, at the required angle, across both sides at one end of a straight piece of thin material. The use of a dovetail template saves time and ensures uniformity. Place the shoulder of the template against the edge, as shown in Fig. 28-20B,

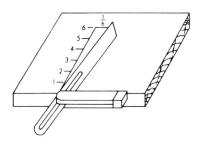

Fig. 28-19. Finding the angle of the dovetail.

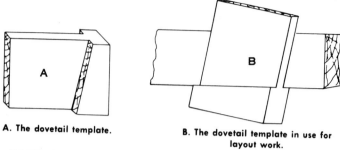

A. The dovetail template.

B. The dovetail template in use for layout work.

Fig. 28-20. A template is invaluable for dovetail work.

and mark one side of the dovetail along its edge. Reverse the template, place the other shoulder at the same edge, and mark the other side of the dovetail.

Beveled Dovetailing—The joint shown in Fig. 28-21 is sometimes required in cabinetwork, and a template is a great help for marking it. To use the template for marking beveled dovetails, cut a wedge-shaped piece of material, as shown in Fig. 28-22A, that is beveled at the same angle as the bevel of the material to be dovetailed. Insert this wedge between the edge of the material and the template, with the square edge of the wedge against the shoulder of the template, as in Fig. 28-22B. Mark the dovetail as described, but do not reverse the wedge-shaped piece.

The common, or through dovetail, shown in Fig. 28-23A, is primarily used for dovetailing brackets and frames which are subject to a heavy downward pressure. Fig. 28-23B illustrates the common lapped, or half-blind, dovetail as it is applied to a curved door frame; it is used in all locations of this type where mortise and tenon joints would not be effective. The common lapped dovetail joint may also be used for purposes similar to those described for the common dovetail joint.

Fig. 28-24A illustrates the common housed "bareface" dovetail; it is shouldered on one side only. The joint in Fig. 28-24B is shouldered and dovetailed on both sides and is another of the same type with the dovetailing parallel along its entire length. These are the simplest forms of housed dovetailing. Their application to the framing of furniture is shown in Fig. 28-29, under the section on framing joints discussed later in this chapter.

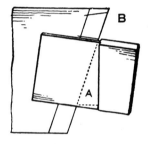

Fig. 28-21. The beveled dovetail joint.

A

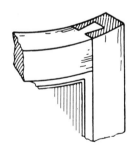

B

A

A. A wedge is used with the dovetail template to mark the desired bevel.

B. Laying out beveled dovetails with the template and wedge.

Fig. 28-22. The method of laying out work with the aid of a dovetail template.

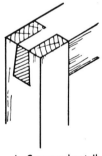

A. Common dovetail.

B. Common lapped dovetail.

Fig. 28-23. Two typical dovetail joints.

Fig. 28-24C illustrates a shouldered dovetail housing joint with the dovetail tapering along its length; as with the two preceding joints, this joint can be shouldered on one side or both sides. The tapered dovetail makes this joint particularly adaptable for con-

621

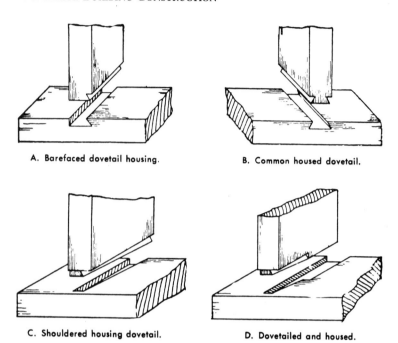

A. Barefaced dovetail housing.

B. Common housed dovetail.

C. Shouldered housing dovetail.

D. Dovetailed and housed.

Fig. 28-24. Housed dovetails of the single- and double-shouldered types.

necting fixed shelves to partitions, because the dovetails prevent the partitions from bending. A dovetailed and housed joint, frequently called a "diminished" dovetail, is shown in Fig. 28-24D; it is principally used on comparatively small work, such as small fixed shelves and drawer rails.

Making a Diminished Dovetail—Square division lines across the ends into which the shelf is to be housed and dovetailed as far apart as the thickness of the shelf, and gauge the depth of the housing on the back edge. Gauge lines ⅜ and 4½ in. from the front edge between the division lines; the space between these gauge lines is the length of the actual dovetail, as shown in Fig. 28-25A.

Cut out the section indicated at A with a chisel, and undercut side B to form a dovetail; insert a tenon saw, and cut the sides across to the edge. Remove the core with a firmer chisel, and finish to depth with a router. Gauge lines on both ends of the shelf on the

622

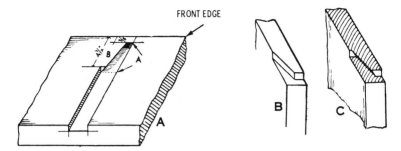

Fig. 28-25. The method of making a diminished dovetail joint.

side and end for the depth of the dovetail, and square across the
end the distances from the front edge as given; cut away the
surplus wood with a tenon saw, as shown in Fig. 28-25B, and finish
the cut with a chisel, carefully testing until it fits hand tight. Fig.
28-25C shows the completed end. The average length of the actual
dovetail of this type is slightly less than one-quarter of the total
length of dovetail and housing.

Mitered Joints

Mitering is an important part of cabinetwork, in the framing of
furniture, and in paneling, where many difficult mouldings must
be mitered into place. Fig. 28-26A illustrates a plain miter with a
cross tongue (or spline) inserted at right angles to the miter. This
joint is principally used for mitering end grain and is additionally
strengthened by gluing a block to the internal angle, as shown in
Fig. 28-29.

Tonguing a Miter—One practical way of tonguing this joint is to
fasten two miters together in a vise so as to form a right angle, as in
Fig. 28-26B, thus providing an edge from which to gauge the
position of the tongues and plow the grooves. If the pieces are not
over 6 inches in width, the grooves are cut with a dovetail saw and
chiselled to depth.

Fig. 28-27A is a variation of plain mitering and, like the preceding joint, is most generally used for end-grain jointing. For this
joint, the tongue should be approximately one-third the thickness
of the material, and it may extend all the way through or only part

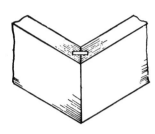

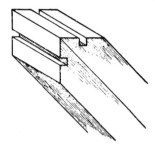

A. A tongued miter.

B. Method of tonguing a miter.

Fig. 28-26. Mitered joints.

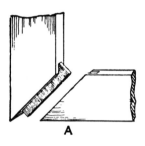

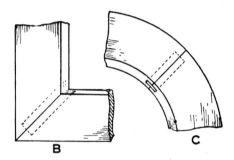

A

B

C

Fig. 28-27. The tongued miter joint in various stages and types of construction: (A) before completion; (B) completed joint; (C) for connecting segments in curved work.

way, as shown in Fig. 28-27B. This joint is especially useful in cabinetwork for connecting and mitering various types of large mouldings around the tops of pieces, for mitering material for tops and panels, and for connecting sections in curved work, as shown in Fig. 28-27C.

Screwed Miter Joint—Fig. 28-28A illustrates a plain miter with a screw driven at right angles to the miter across the joint through a notch cut in the outside of the frame. This type of joint is used principally in light moulded frames.

A common method of clamping a tongued miter is to glue blocks to the piece and hand-screw the joint together, as shown in Fig. 28-28B. The blocks are glued on and allowed to dry before

624

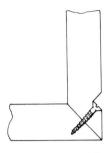

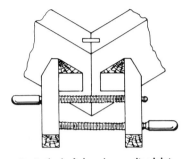

A. Cross-sectional view of the completed joint.

B. Method of clamping a miter joint.

Fig. 28-28. The screwed miter joint.

gluing up the joint; when the joint is dry, the blocks are knocked off, and their marks are erased.

Fig. 28-29 is a part plan for the base of a breakfront cabinet; it shows the application of mitered and housed dovetail joints to furniture construction.

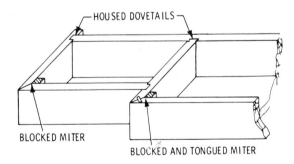

Fig. 28-29. Application of various forms of joints in cabinet construction.

Framing Joints

The term "framed," or "framing," as used in cabinetwork, indicates work that is framed together, as in Fig. 28-29. It also refers to the "grounds" for securing panelling to walls.

Fig. 28-30 represents various joints that are used to connect

625

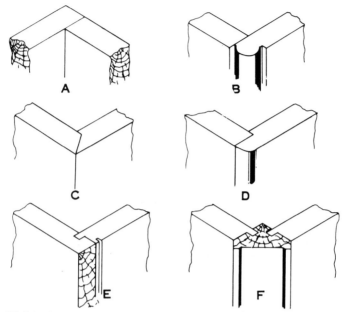

Fig. 28-30. Commonly used framing joints: (A) butt, or square, joint; (B) return bead and butt joint; (C) rabbet and miter joint; (D) rabbet and round joint; (E) barefaced tongue joint; (F) splayed corner joint.

angles for paneling. The joints shown in Fig. 28-30A and B are identical except for the return bead, which is worked on one of the pieces. These two joints are usually glued and nailed, but, when used as an external angle, they are secret-screwed before painting; that is, the screws are sunk below the surface, and plugs or pellets of wood are glued in the holes and beveled off. The joint represented in Fig. 28-30C may be used to connect framing at any angle; the rabbet prevents slipping while being nailed or screwed. Fig. 28-30D shows an ordinary rabbeted joint with the corner rounded off and the pieces glued together; because of its rounded corner, it is often used in furniture for children's nurseries. Fig. 28-30E illustrates a joint that is shouldered on one side only. A bead is worked on the tongue piece to hide the joint. It is used for both internal and external angles, with or without the bead. The joint in Fig. 28-30F is the splayed corner tongued joint, which is used for joining sides into a pilaster corner.

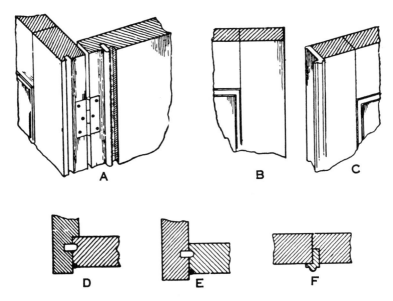

Fig. 28-31. Hinging and shutting joints commonly used in cabinet construction: (A) rabbeted dustproof joint; (B) astragal shutting joint; (C) beaded shutting joint; (D) plain hinged joint; (E) plain butted hinge dustproof joint; (F) beaded shutting joint.

Fig. 28-32. A single-door dustproof joint. Both ends are dustproof when the door is closed.

Hinging and Shutting Joints

The dustproof joint and its applications are considered a necessary part of cabinetmaking. The rabbeted dustproof joint, shown in Fig. 28-31A, is applied to a butt-hinged double door closet. A bead is glued into the rabbeted end behind the hinge, which is sunk flush, and a corresponding groove is cut in the door stile to fit

627

over the bead when the door is closed. The beads are sometimes covered with felt or rubber, thereby making the joint absolutely dustproof.

Fig. 28-31B and C illustrate the shutting joint with a beaded strip glued to the stile; the joint is rabbeted to project over and thus conceal any shrinkage in the adjoining stile. The joint at the hinged ends (when closed) is shown in Fig. 28-31D. Fig. 28-31E represents a plain butted hinged dustproof joint, and Fig. 28-31F shows the beaded shutting joint when closed but with the astragal rabbeted into the stile. Fig. 28-32 shows a single door dustproof at both ends.

CHAPTER 29

Paint and Wallpaper

For many craftsmen, the most satisfying part of building anything is the final touches. And when it comes to home building
construction, this is often the application of paint or the installation
of wallcovering.

What follows is a round-up of tips and techniques for painting
interior and exterior surfaces, including paints to use and how to
select and hang wallcovering.

Exterior Latex Paints—These are by far the easiest, fastest, most
convenient, and most popular paints to use. They are tough, long-lasting,
quick-drying, mildew-resistant, and allow easy tool and splatter
clean-up with soap and water. When dry, the paint repels rain,
hose water, dew, and other water, but allows water vapor from
inside the building to pass through to the outside air without
peeling.

Technological development in these paints has been rapid, and
the most widely available now is the acrylic latex. This paint is

available under a number of brand names and formulations, all of which have the characteristics mentioned in varying degrees, and a flat or slightly glossy, nonchalking finish. Some need no primer over new wood, but most do require one for best adhesion and prevention of bleed-through.

Two finish coats are usually ample, but they must be applied as full coats and not brushed or rolled out thinly (they go on with so little drag on the brush that the tendency is to brush them out too much). Allow an hour between coats or a little more if the humidity is high, even though the paint will dry in 20 to 30 minutes. Acrylic latex can be applied over damp surfaces with no problem, but not after or during a heavy rain. To avoid damage to the paint film, stop painting in time to let the paint dry before it is struck by rain or heavy dew. Bonding and curing of the paint is adversely affected by relatively low temperatures, and almost all manufacturers specify that the paint not be applied if atmospheric or surface temperatures are 50° or lower. Apply the paint as it comes from the can, even if it seems too heavy; some brands are almost jelly-like, and all are thicker than oil-base paint. Thin it only for spray application, and then according to instructions on the can. Generally, spray-thinning should amount to no more than ½ -pint of water per gallon of paint.

Acrylic latex comes in a number of stock colors which vary from brand to brand, and most paint and hardware stores will custom-mix special colors for a small charge. Glossy trim paints (which are also acrylic latex) are available in matching or contrasting colors.

For best results, and to save time, work, and money, plan on applying latex only over latex or oil only over oil unless previous coats are stripped to bare wood.

Interior Latex Paint—Many of the qualities mentioned for exterior latex also apply to its interior counterpart: it is quick-drying, easy to apply, shows no lap marks, is available in literally hundreds of shades and tones of color, is durable and washable, and has an odor not usually regarded as objectionable (but it should be applied and allowed to dry with plenty of ventilation). It is by far the most popular interior paint sold, and is available in flat and semigloss.

Like exterior latex, the interior variety should be used as it comes from the can, without thinning; some brands are called "dripless" latex and are heavy-bodied to prevent brush drips and roller splatter. One manufacturer's advertising has shown his gel-paint to be so thick that holes could be scooped into a can of it—yet it was just right for application.

New work requires at least two coats of paint. After that, depending on the color, you may be able to use only one. As a general rule, yellows have the poorest hiding power and blues the best. If you decide you want white ceilings, use the latex "ceiling" paint. It has extra pigment for good hiding power, and can cover a previously painted ceiling in one coat. If it is unavailable, have your paint dealer add just a small amount of his darkest blue tinting color to his densest tinting base—the result will be indistinguishable from pure white, but will cover very well.

COLOR SELECTION

The diagram shown in Fig. 29-1 shows the three primary colors, their secondaries, and what may be called the tertiary colors.

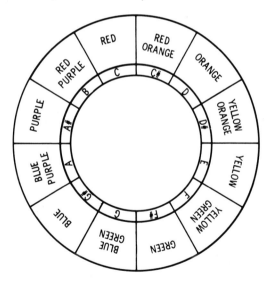

Fig. 29-1. The three primary colors with secondary and tertiary colors shown.

Opposite each of these there has been placed one of the notes of the chromatic music scale forming a perfect octave. It is interesting to note that the claim has been made, and with much insistance, that any scheme of color that may be selected and which may be struck as a chord will, if the chord is harmonious, become a harmonious scheme of color. If this chord produces a discord of music, there will be a discord of color.

By definition, primary colors are those that cannot be made by mixing two or more colors together. The three primary colors are *red*, *blue*, and *yellow*. Fig. 29-2 shows a diagram of the primary and secondary colors. The colors obtained from mixing any two of the primary colors together are called secondary colors. There are three secondary colors: *purple, green,* and *orange.*

Red and blue gives *purple.*
Blue and yellow gives *green.*
Red and yellow gives *orange.*

By mixing any two of the secondary colors together you get what are called tertiary colors, which are: *citrine, olive,* and *russet.*

Orange and green gives *citrine.*
Green and purple gives *olive.*
Orange and purple gives *russet.*

Black and white are not regarded as colors. A good black can be produced by mixing the three primary colors together in proper proportions. By adding white to any color you produce a tint of that color. By adding black to any color you get a shade of that color. That is the difference between *tint* and *shade.* The use of

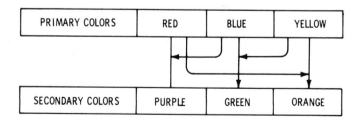

Fig. 29-2. Primary and secondary colors.

black subdues or lowers the tone of any color to which it is added. To preserve the richness of colors when you wish to darken them, use the primary colors instead of black. To make a yellow darker, use red or blue, and to darken blue, add red and yellow, and so on.

Every shade or tint of color required by the painter can be made from red, blue, and yellow with black and white. To make any of the umbers or siennas lighter in color and to preserve the clear richness of tone, always use lemon chrome instead of white. If you wish a subdued or muddy umber or sienna color, then use white.

The most useful primary colors are:

Yellows—*lemon chrome, deep ocher*
Reds—*vermilion, Venetian red, crimson lake*
Blues—*Prussian blue, ultramarine*

Gold or silver leaf harmonizes with all colors and, with black and white in small quantities, can be used to bring into harmony the most glaring colors. The old time heraldic painters knew the value of outlining their strong primary colors with gold, silver, black, or white, to bring them into harmony with each other. The Egyptians and other ancient people made use of the same knowledge in their decorative schemes. Yellow ocher is the most useful color the painter possesses. In its pure state it is admirable for large wall spaces, and if you are in doubt as to what color to use to complete a color scheme, you will find ocher or one of its shades or tints will supply the missing link.

Red on walls makes a room look smaller and will absorb light. Yellow gives light and airiness to any room and it will also reflect light. Useful colors in large quantities for churches, public halls, etc., are:

Primrose red
Terra cotta (white, burnt sienna, lemon chrome)
All tints of ocher
Flesh colors (white and burnt sienna)
Pea green, apple green
Grey green (white, paris green and a touch of black)
Ivory shades (white, lemon chrome, or ocher)
Old rose (white, ocher, venetian red, or pure indian red and black)

Nile blue and nile green (white, prussian blue, lemon chrome)
Light citrine, light olive, light russet

For ceilings the best tints are the creams and ivory tints, and grey. Creams and ivory tints are made from white tint with one or more of these colors.

Lemon chrome
Orange chrome
Ocher
Raw sienna

To produce a warm tone, add a small quantity of burnt sienna, vermilion, or venetian red. To produce a colder tone, use a little green, black, raw umber, or blue. Greys are made from white tint with either black, black and green, blue and umber, black and red, red and blue, or burnt sienna and blue. Light colors are used for ceilings in preference to dark colors. Contrasting colors are better for ceilings than a lighter tint of the wall color.

FINISHING WOOD FLOORS

Wood floors exposed to the weather, such as porch floors, are commonly painted with two coats of porch and deck paint made for the purpose. Generally two coats are needed. Natural finishes are rarely considered durable enough for floors exposed to weather. Concrete floors may be painted, if properly treated.

Interior floors made of hardwoods most often are given a clear finish, though new wood can be stained and coated. Shellac used to be the most common finish, and when it is used, three coats are applied and the surface is gone over lightly with steel wood after the first and again after the second coat has been applied. Shellacked floors may be waxed if desired.

The most popular clear finish today is polyurethane. It dries clear and hard and does not require waxing (Figs. 29-3 and 29-4).

Some woods need to be filled before being finished. Oak flooring should be filled with wood filler, either natural or colored, before a clear coating is applied. When floor-seal finish is used, application of filler is optional. Some forms of wood, laminated

Courtesy United Gilsonite

Fig. 29-3. Polyurethane finish was used on this floor. It stands up very well.

wood, or plywood-finish flooring can be purchased with the finish applied.

PAINTING PLYWOOD AND WALLBOARD

The painting and finishing characteristics of plywood are essentially those of the kind of wood with which the plywood is faced. In large sheets of plywood, any wood checking is more objectionable in appearance than it is on boards of lumber. Plywood with hardwood faces usually is not much more prone to checking than lumber of the same species, but plywood faced with most softwoods is likely to check to some extent even when

635

Fig. 29-4. Clear coatings are most popular for floors today.

used indoors with protective finishes. Such checking is more conspicuous with some finishes of light color, but somewhat less as the gloss diminishes and also as the color is darkened greatly.

With opaque finishes, the checking is least objectionable when flat finishes with rough surfaces are chosen, such as stippled, textured, or sanded finishes. Checking is less readily observable with natural finishes than with opaque finishes, because the grain pattern of the wood distracts attention from the checking. Softwood plywood may be used without danger that checking will mar the finish if the exposed face of the plywood is covered with paper, cloth, or resin-impregnated paper firmly glued in place. For exterior exposure, of course, the glue must be thoroughly weather resistant. Plasterboard is paintable with the same material and methods that are suitable for plaster.

CHARACTERISTICS OF WOODS FOR PAINTING

Some raw woods allow a wide choice in paints and painting procedures, whereas others are more exacting in their requirements if fully satisfactory experience is to be obtained. Woods differ also in the appearance of natural finish obtainable with them in the ways in which they behave if exposed to the weather without protective coating or treatment. Woods that offer the greatest freedom of choice for painting are the woods that are low in density, slowly grown, cut to expose edge grain rather than flat grain on the principal surface, are either free from pores or have pores smaller than those in birch, and are free from defects such as knots and pitch pockets. Flat-grain boards hold paint better on the back side than the pitch side. All of the pines, spruces, and douglas firs are improved for painting by kiln-drying to fix the resin in the wood so it will not exude through the coating.

Eastern red cedar and spanish cedar contain slowly volatile oils that prevent proper drying of most paints and varnishes unless most of the oils are removed by exceptionally thorough kiln-drying. Incense cedar, port oxford cedar, and cypress may sometimes give similar troubles if they contain too much of their characteristic oils. The method of seasoning on most other woods has little or no effect on paintability. Water-soluble extractives in redwood, western red cedar, walnut, chestnut, and oak may retard drying of coatings applied while the wood is damp or wet and may discolor paint if the wood becomes thoroughly wet after painting. Otherwise, such extractives contribute to better durability of paint.

On exterior surfaces, softwoods that expose wide bands of summer wood require extra care in painting because it is over such bands of summer wood that the paint coating, when embrittled by the action of the sun and rain, begins to wear away by crumbling or flaking. Repainting is usually considered before too much summer wood becomes exposed. Coatings on interior surfaces seldom wear away by crumbling or flaking. But greater smoothness of painted surfaces may be demanded on interior surfaces. The presence of wide bands of summer wood makes it necessary to take extra care in smoothing the surface before painting and in applying undercoats to keep the wood from showing ridges in the coating, commonly called raised grain.

Sometimes the easiest woods to paint smoothly may be considered too soft and too easily dented to be a good choice for interior woodwork, in which case a compromise must be reached between the properties that favor smooth painting and those that give enough hardness for the intended use. For good service of paints or other finishes, wood must be kept reasonably dry, in which case it is not subject to decay. But there are some places, such as outside steps, railings, fence post, porch columns, and exterior paneled doors, where joints in the structure may admit enough water to permit stain or decay to develop in nondurable kinds of wood. In such cases the performance of both wood and paint may often be improved by treating the wood with preservatives or water-repellent preservatives before painting. Treatment of the wood with water-repellent preservatives at least several days before erection has been found effective in minimizing such penetration of water.

PAINTING TOOLS

A satisfactory paint job requires cleaning, sanding, and scraping. To prepare a surface for painting, the proper tools should be on hand and in good condition and ready for use.

Brushes

Classic large paint brushes were called "pound brushes" by the professional, and were termed 4/0, 6/0, and 8/0 in size. They were made both round and flat, and the hog-hair bristles were bound with string or copper wire. The next smaller brushes were available in a dozen different sizes, and were called "tools" or "sash tools." Some of these were bound with string, and others had their bristles fixed in metal. The smallest hog-hair brushes were called "fitches" and were used where tools would not fit. Smallest brushes of all were the camel-hair brushes, made with long or short hair, according to the work to be done.

Modern brushes available to the consumer are made with flagged nylon or polyester bristles for use with latex paints and stains or with natural bristle for use with oil-base paints and stains, varnish, enamel, or synthetic finishes. Nylon can also be used with oil-base paint. They range in size from ½ in. to 6 in. wide, have their bristles set in adhesive or rubber, and may have wooden or plastic handles. Their quality, aside from bristle composition, is determined largely by the length of their bristles (the longer the better), the number of bristles per square inch, endways (the denser the better), and the springiness or life of the bristles (not limp, but not so stiff as to leave tracks in the paint).

Brushes in use should be dipped no more than two-thirds of their bristle length into the paint, and should be tapped on the inside of the paint bucket to remove excess paint instead of being stroked across the rim. Brushes in use for several hours should be washed out periodically during the day to get rid of paint buildup below the ferrule and restore their springiness. If brush use is interrupted for short periods (overnight, too) the brush should be wiped off and the bristle end carefully wrapped in airtight plastic film to preserve the lay of the bristles and to prevent the paint on the bristles from drying until the brush is put back into use.

After the day's painting is through, synthetic brushes used with latex should be gently but thoroughly washed in cool or lukewarm soapy water. Use a paintbrush comb to make sure all bristles are separated and all paint is removed. Rinse the brush thoroughly to remove all soap, shake the water out of it, and shape the damp bristles into a wedge-shape. Hang the brush up to dry. Brushes used in oil-base paint are treated the same way, but are thoroughly cleaned in mineral spirits to remove the paint before washing and drying. If the brush is not to be used for a long time, place it in its original wrapping, if available, and store in a cool, dry place (preferably hung up). Unwrapped brushes should be hung in a dust-free cabinet.

Brushes allowed to become stiff with hardened paint can be cleaned, but never seem to have the shape or feel they had before they hardened up. Oil-base paint can be softened by soaking the brush in linseed oil for 24 hours (or until it begins to soften) and then transferring it to a can of mineral spirits and working the paint loose the rest of the way by hand. If this conservative method doesn't work, try suspending the brush in a closed can, on the bottom of which is a shallow container of lacquer thinner. Fumes from the thinner will usually liquefy the hardened paint in the bristles, but will also strip the paint off the handle of the brush. The fume method for cleaning paint brushes is shown in Fig. 29-5. Failing all else, use a good liquid brush cleaner according to the directions on the can. This is a strong liquid with powerful fumes and potent action. Use it with care, and in a well-ventilated place. It will remove hardened latex as well as oil paint.

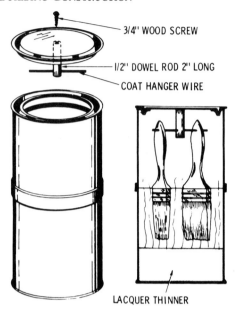

3/4'' WOOD SCREW

1/2'' DOWEL ROD 2'' LONG

COAT HANGER WIRE

LACQUER THINNER

Fig. 29-5. Suspending paint brushes above lacquer thinner in a closed can.

Rollers

The paint roller is an efficient, rapid means for applying paint; nearly anyone can get professional-quality results with practice, and perfectly acceptable results with care, using either latex or oil-base paint. Rollers can be used to coat almost any material inside or outside the house.

Most popular consumer roller-size is the 9-in. for walls, ceilings, and similar large surfaces, although a longer roller is available in some paint stores. The 7-in. is more economical of paint and easier to handle on beveled siding, fences, and smaller or cramped flat surfaces. Besides paint, rollers can also apply glue, insecticide, contact cement, tile mastic, roof coatings, and blacktop sealer, among other things.

The first roller covers were made from lambswool, which is still used by some professional painters to apply oil-base paint, but most are now made with man-made fiber with naps of assorted lengths. In general, the rougher the surface to be painted, the longer the nap of the roller should be in order to hold enough paint and to be able to reach the bottom of voids in rough surfaces. The roller packages will be marked as to the surface it should be used on.

Roller frames, or the handle assembly on which the roller sleeve mounts, should be constructed of heavy wire for maximum strength at minimum weight and should have a plastic handle molded in place (wood will loosen up and is too heavy). The cage over which the roller sleeve fits can be made of wire or flat steel and should allow the sleeve to be slipped on and off easily. The end of the plastic handle should form a threaded socket into which an extension handle can be turned for ceiling or floor work.

A roller tray is necessary. It should be sturdily made, and have a deep paint well. Liners are available so the same tray can be used for different color paints without cleaning it.

One roller-load of paint will cover an area about three roller-widths wide and four feet high. Roll it on easily in the pattern of a large "W," crisscross the strokes to cover the unpainted segments, and repeat on the adjacent wall areas until an area about 4 ft. wide and the full height of the wall is covered. Then, go over this area lightly, using no more pressure than the weight of the roller, in one direction only. Resume painting the next area before the "wet edge" of the preceding area has dried, in order to prevent streaks and lap marks. Do ceilings and upper areas first; finish-stroke ceilings from the source of light rearward, or parallel with the short dimension of the room. Lift the roller gradually from the wet surface as it nears the end of the stroke; a roller which is stopped and then lifted from the surface can leave a mark.

Sash brushes, edging pads, trim rollers, shields and similar devices are available to work the paint around trim, windows, corners, and ceiling lines when using a paint roller. They can be used before or after the main area is painted (sometimes both), but are more successful if used in advance. Figs. 29-6 to 29-10 give some painting tips for brush and roller.

SPRAY PAINTING

With the construction of modern equipment, spraying has become an increasingly popular method of applying paint to all

641

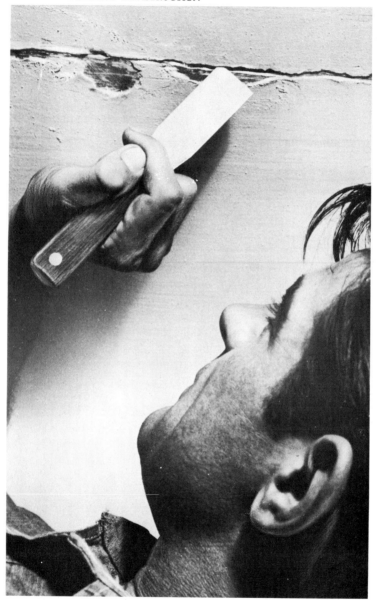

Fig. 29-6. Patch and smooth before painting.

642

Courtesy Benjamin Moore

Fig. 29-7. Cut in ceiling . . .

types of interior and exterior surfaces. It is much faster than brushing, both in application and in the use of faster drying finishes. Small spraying units of the better types are very efficient, performing perfect work with paint, lacquer, varnish, and synthetics.

Airless Sprayers

For large-surface work both inside and out, the traditional compressed-air spray gun has been almost entirely replaced by the high-pressure airless sprayer which is capable of applying enormous quantities of paint with virtually no spray drift. So efficient is

643

Fig. 29-8. . . . then roll.

this machine that one man can use it to paint the entire exterior of an unpainted brick ranch style home with two coats of paint in one weekend, including clean-up time.

An airless sprayer consists of a motor which drives a pump which, in turn, sucks paint out of a reservoir and forces it through the nozzle on a spray gun. The motor may be electric, hydraulic, pneumatic, or internal combustion; the pump is a special recipro-cating paint pump; the spray gun is equipped with a nozzle and valves made from tungsten carbide or similar material in order to withstand paint friction at pressures of 3,000 psi or higher.

Nothing comes out of the spray gun but paint, which is driven to the surface to be painted in a precise, fan-shaped spray pattern. So

Courtesy Benjamin Moore

Fig. 29-9. Long stick with roller can make painting simpler. Do walls after ceiling.

precisely formed is this pattern that at distances of less than 3 or 4 in. from the nozzle of a commercial unit, the spray will cut through bare skin like a knife; for this reason, the guns have safety tip guards or similar devices to help prevent serious accidents. Paint applied by airless spray will penetrate cracks and crevices on the surface to be painted which no other method of painting could reach, making it ideal for rough masonry or wood as well as smooth surfaces. The paint will not bounce off the surface, and is not wasted through drifting, as is the case with compressed-air sprayers.

Fig. 29-10. Painting window sash with brush at angle shown makes job easier.

Most airless sprayers are rental items because their high price makes purchase practical only for painting contractors or building maintenance people; small, relatively inexpensive units are available to the individual, but these should be investigated with care before purchase.

Airless units do not have a paint cup on the spray gun, as in compressed-air sprayers. A single hose up to 50 ft. long (or longer) transports paint under pressure from the pump to the spray gun; the pump unit, which is easily portable, can mount directly on any 5-gal. paint bucket, sit on the ground, or be designed with wheels, like a hand truck. In the latter instances, the pump draws paint

646

from a separate container through a special suction hose or rigid pipe. Clean-up is simple and economical, since paint in the hose and most of the cleaning solvent (if used) can be recovered for later use. Almost any kind of paint can be applied to almost any surface by airless spray, but the use of lacquer is not recommended.

Compressed-Air Sprayers

These are traditional spraying units which consist of an air compressor, spray gun, and connecting hose.

Air Compressor—There are two kinds of air compressors—the *piston* type and the *diaphragm* type. The piston type consists essentially of a metal piston working inside a cylinder, very much like the piston in an automobile engine. On the down-stroke of the piston, air is drawn into the cylinder and on the up-stroke, this air is compressed. A piston-type stationary paint sprayer is shown in Fig. 29-11, and a portable type shown in Fig. 29-12.

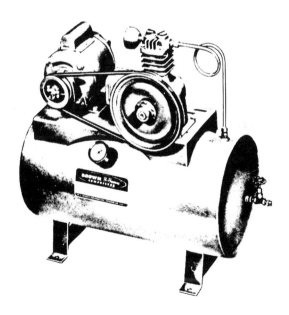

Fig. 29-11. Typical 1/3 horsepower stationary paint sprayer.

Fig. 29-12. A portable 1/2 horsepower paint sprayer.

The diaphragm-type compressor differs from the piston type mainly in that it is equipped with a rubber diaphragm taking the place of the piston. Because of the fact that the diaphragm type of compressor is rather inexpensive and also requires less maintenance, it is usually preferred in the spraying applications that are performed around the home where the air requirements are small.

The motor supplying the power for the compressor is furnished in fractional horsepower, and it also determines the size of the compressor. For home spraying, ¼ to ⅓ horsepower is the smallest motor practical to use for satisfactory spraying. Most compressors should be able to deliver a constant-volume supply of air to the gun at a pressure of about 40 psi.

Spray Guns—In operation, the spray gun atomizes the material which is being sprayed and the operator controls the flow of the material with the trigger attachment on the gun, as shown in Fig. 29-13. Spray guns are supplied with various sizes of nozzles and air gaps to permit the handling of different types of material, as

648

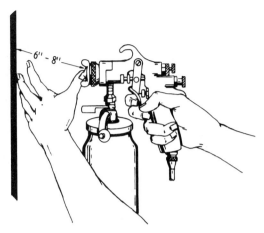

Fig. 29-13. A typical paint spray gun.

shown in Fig. 29-14. An air adjustment built into the gun makes possible a change in the shape and size of the spray pattern, ranging from a wide fan-shaped spray to a small round spray as the occasion requires.

Spray guns are either *external mix* or *internal mix*. In the external-mix type of gun, the material leaves the gun through a hole in the center of the cap, while air from an annular ring surrounding the fluid stream (as well as from two or more holes or orifices set directly opposite each other) engages the material at an angle. Spray guns in which the air and material are mixed before leaving the orifice of the gun are of the internal-mix type. The suction-feed gun, as the name implies, syphons the material from a cup attached to the gun and is generally used for radiators, grills, and other small projects which require light bodies and only a small amount of material.

The pressure-feed gun forces the material by air pressure from the container. Air directed into the container puts the material under pressure and forces it to the nozzle. If compressed-air spraying is employed, the pressure-type gun is undoubtedly the best all-around gun for use with small compressors and with such materials as house paint, floor enamels, wall paints, and the like. Any spray gun is either a bleeder or non-bleeder type. A bleeder type is one which is designed for passage of air at all times,

649

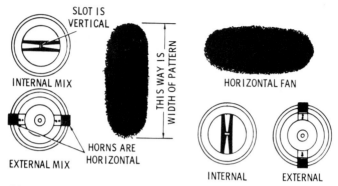

Fig. 29-14. Various nozzles used for different patterns of spray.

The pressure-feed gun forces the material by air pressure from the container. Air directed into the container puts the material under pressure and forces it to the nozzle. If compressed-air spraying is employed, the pressure-type gun is undoubtedly the best all-around gun for use with small compressors and with such materials as house paint, floor enamels, wall paints, and the like. Any spray gun is either a bleeder or non-bleeder type. A bleeder type is one which is designed for passage of air at all times, whereas a nonbleeder gun cannot pass air until the trigger is pulled. This nonbleeder type of gun is used only when air is supplied from a tank or from a compressor equipped with pressure control. Bleeder guns are always used when working directly from small compressors, such as those used in small workshops or for spraying jobs around the home.

Air Compressor Accessories—One of the most useful accessories for home use spraying is the angle head. An angle head fitted to the pressure gun prevents excessive tilting of the fluid cup when spraying floors, ceilings, and similar surfaces. A respirator is seldom required for work of short duration, but it is a very useful item when spraying for long periods in confined quarters.

Another useful accessory is the condenser which filters water, oils, and various foreign particles out of the air supply. If the condenser is fitted with an air regulator, the combination is termed a transformer. With a condenser-regulator combination, it is possible to have the air filtered, while the regulator allows setting the air pressure at a predetermined rate as required for the particular job. Additional useful accessories consist of an air-duster gun and extra lengths of air hose with couplings to facilitate an increase in the working radius.

SPRAY BOOTHS

Where spray equipment is installed in a permanent location and the work to be sprayed is of such a size that it can be conveniently done in one place, exhaust equipped spray booths are usually provided for the removal of vaporized paint and odors, Fig. 29-15.

The ordinary floor-type booth may accumulate a certain amount of waste material after the spraying has been completed. These should be removed immediately, since cleanliness is important to good spray results. When spraying is confined to an area, it is advisable to have a portable fan secured in a frame which can be placed in a window to provide proper ventilation.

Turntables of simple construction are used in spray booths to facilitate rotation of the work while spraying. A suitable turntable eliminates a lot of body movement as well as handling of the work during the spraying process.

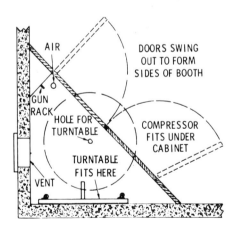

Fig. 29-15. A typical corner spray booth suitable for the average home workshop.

651

Before constructing such a unit, you should check with your local building code about fire restrictions.

PLACING THE WORK

Prior to the actual spraying, it is necessary to arrange the work properly, and one of the fundamental rules relating to placement is to get the work off the floor. There are several methods to accomplish this, the most common of which is to provide a couple of sawhorses bridged with suitable boards or planking. Another method to get the work off the floor and in a comfortable spraying position is to suspend the work with a rope or hooks from a clothesline or other convenient overhead support.

Small wooden articles can be readily sprayed if speared on an icepick, or other special holding device, such as illustrated in Fig. 29-16.

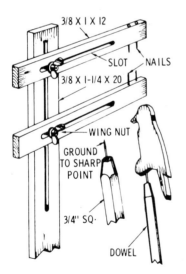

3/8 X 1 X 12

SLOT   NAILS

3/8 X 1-1/4 X 20

WING NUT

GROUND
TO SHARP
POINT

3/4" SQ.

DOWEL

Fig. 29-16. A holding device for spraying articles of wood or other light material.

For general indoor spraying, an adjustable backstop stand, such as shown in Fig. 29-17, will prove very satisfactory. This stand uses newspapers or paper rolls and provides a suitable surface for testing the spray gun as well as a backstop when spraying the

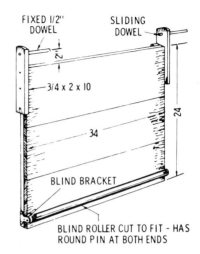

FIXED 1/2" DOWEL

SLIDING DOWEL

2"

3/4 x 2 x 10

24

34

BLIND BRACKET

BLIND ROLLER CUT TO FIT - HAS ROUND PIN AT BOTH ENDS

Fig. 29-17. A typical backstop stand suitable for spray pattern.

work. Pedestal turntables with adjustable heights are good for placement of small furniture and similar articles to be sprayed. Various base styles may be used to suit individual preferences, as shown in Fig. 29-18. Other practical supports and work-holding devices may be improvised to suit almost any condition and space requirement. Large articles of wood or metal are sometimes suspended by a block-and-tackle arrangement, permitting the work to be hoisted out of the way for drying after spraying.

COMPRESSED-AIR SPRAYING TECHNIQUE

Prior to commencing the actual spraying job, it is good practice to test spray a few panels in order to obtain the feel of the gun and to make any necessary adjustments. Such practice spraying can be done on old cartons or on sheets of newspaper tacked to a carton or box.

With the spray cup filled with water-mixed paint or other inexpensive material, practice spraying can be done directly on any scrap material, as previously noted. It will be useful to experiment with the full range of fluid adjustment, starting with the fluid needle screw backed off from the closed position just far enough to obtain a small pattern an inch or so wide when the trigger is

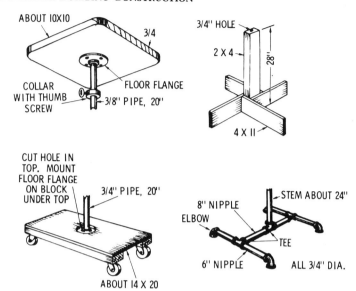

Fig. 29-18. A pedestal turntable for placing furniture or similar articles for spraying.

pulled all the way back. The spray-gun stroke is made by moving the gun parallel to the work and at right angles to the surface. The speed of stroking should be about the same as brushing. the distance from gun to work should be between 6 and 8 in., as shown in Fig. 29-19. Practice should be done with straight uniform strokes moving back and forth across the surface in such a way that the pattern laps about 50 percent on each pass. An appreciable variation of distance between gun and work or angle of the gun will result in uneven coatings. Therefore, always hold the gun without tilting and move it as described. Spraying is merely the action of the wrist and forearm and the proper technique is readily and easily acquired. Fairly large pieces of scrap wood remaining from the project are excellent for practicing spray patterns, because this gives the operator an idea of any problems he might expect in actually spraying the final project itself.

Release the trigger at the end of each stroke to avoid piling up material. Avoid pivoting or circular movements of the wrist or forearm or any oblique spraying which will cause material to rebound from the surface. It causes excessive mists and waste

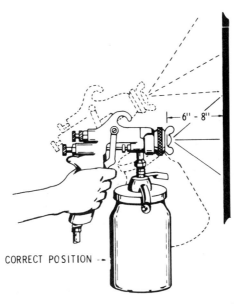

CORRECT POSITION

Fig. 29-19. The correct position of spray gun when spraying.

material as well as a patchy, unsatisfactory job.

Cleaning the Gun—To obtain continued and satisfactory service, a spray gun (like any other tool) requires constant care. It should not be left in a thinner overnight, because such practice removes the lubricant from the packing through which the needle moves, thereby causing the needle to stick and the gun to "spit." A dirty air gap causes a defective spray. Thus, it is good practice to clean it promptly after the spraying job is finished. At the end of the day, remove the gun from the cup and hose, and thoroughly clean it. Clean the entire gun, the outside as well as the inside. The cup or tank should likewise be cleaned.

WALLPAPER

They don't call it only wallpaper anymore, even though paper-based wallcoverings are still available (Fig. 29-20). "Wallcovering" has become the generally accepted industry terminology as paper has given way to a good degree to vinyl, vinyl-coated fabric,

Fig. 29-20. For sheer good looks it is difficult to top wallcovering.

burlap, grass, metal foil, and many other substances which could be successfully pasted and stroked onto an interior wall.

One popular wallcovering, as of now, is vinyl plastic in combination with paper or either woven or unwoven cloth. It is all sturdier, more colorful, and easier to apply and remove than the old-fashioned wallpaper; the better grades are tough enough to withstand any abuse and scrubbing normally found in an average, active household. Vinyl wallcovering takes no longer to apply than two coats of paint, yet has none of paint's objectionable fumes or odor, and it can be used successfully in any room, including bathrooms. The job can be interrupted and restarted days later with no problem.

Getting Ready to Hang It

New plaster should age 30 days and then be coated with a good primer-sealer. Plasterboard must likewise be coated with a primer-sealer so that the wall covering may be removed later without damage to the paper surface of the plasterboard. All surfaces, primed or not, must then be coated with glue sizing before hanging the wall covering.

Most vinyl wallcovering not intended for professional application—and some that is—is supplied pretrimmed and prepasted. If the paper selected is not pretrimmed (that is, if the selvage has not been cut off at the factory), arrangements should be made at the time the covering is ordered to have it trimmed before delivery in order to save some work during hanging. It can be trimmed during hanging, but is a nuisance that should be avoided if possible. Prepasting is a mixed blessing; it can save time, work, and mess, but not all manufacturers put on enough paste to let the paper be slipped into position and worked easily. For this reason, many experienced hangers go right ahead and coat the paper with paste, as if it had not been prepasted, in order to assure a good job. A paste specifically intended for vinyl wallcovering, or any similar, nonporous wallcovering, must be used.

Hanging the First Strip

The first strip of wallcovering should be applied to the right of a corner, a door frame, or a window casing. This allows the joint between the first and last strips to be as unnoticeable as possible, in case of a minor pattern mismatch at the end of the job.

To locate this strip, measure to the right of the corner, door, or window trim a distance equal to the width of the wallcovering, minus one inch, and mark the distance with a small dot on the wall. Drop a plumb line from the top of the wall, through the dot, to the

657

baseboard and mark the wall at the top and bottom of the plumb line. Replace the plumb line with a chalk line and snap a vertical reference mark from floor to ceiling. Since the chalk mark can be easily wiped away by accident, go over it lightly with a pencil and straightedge. That way, if the chalk is wiped during or before hanging the first strip, there will still be a vertical reference line.

Unroll a length of the wallcovering on the floor or pasting table and study the pattern to determine which end is the top and to decide if there is some feature in it that you would like to have at a particular height in this first strip. If the latter is the case, measure from the center of the feature toward the top end of the strip a distance equal to the distance from the ceiling line to the center of where you would like the feature to be, add 3 in. and mark the covering. Measure the distance from the center of the feature to the top of the baseboard (you are going in the opposite direction now), add 3 in., mark the covering, and cut the strip square across at each mark. If the pattern-feature positions are unimportant (as most are), cut a strip of covering as long as the distance from the ceiling line to the baseboard, plus 6 in. Lay the strip on the floor, pattern side up.

Cutting the Strips to Be Hung

Unroll a second strip to the right of the first, match the pattern and cut it off so that there is at least a 3-in. surplus at top and bottom. The pattern drop will probably be such that the third strip will match the first; if so, measure the perimeter of the room in inches and divide this figure by the width of one strip to get the total number of strips needed. Cut half this number of strips identical to the first strip and the remainder identical to the second. If the strips do not begin to duplicate until the fourth strip, cut one-third of the strips identical to the first, one-third to the second, and so forth. If there is no pattern drop, as with straight-pattern or random pattern coverings, cut all strips alike.

Hanging Prepasted Covering

Strips of prepasted coverings to be hung without additional pasting are loosely rolled pattern-side in and top edge out, arranged in the order they were cut (the first one cut is the first one hung). Soak the first strip for one minute, or according to directions packed with each roll of wallcovering, in a waterbox filled to one inch from the top with warm water. It is best to put the rolled strip in the water so that when the top edge is lifted, the paste side will be toward the wall. Place the waterbox against the baseboard, immediately under the wall area where the strip is to go, and place a stepladder so that its front legs touch the room side of the box.

After one minute, grasp the top edge of the strip firmly (it will be slick with paste), walk up the ladder with it, and lay it on the wall so that 3 in. of it laps onto the ceiling and the right-hand edge of it lays exactly along the vertical chalk or pencil mark on the wall. Smooth it to the wall with a smoothing brush or damp sponge, tap it gently into the corner with the ends of the bristles or a wet finger, and temporarily stick the 3-in. overlap to the ceiling.

With a single-edge razor blade, make a cut in the overlap at the ceiling corner so that the flaps are separated to each wall. Lay a broadknife on the covering so that its edge rests in the junction between the ceiling and the wall; trim the covering with a razor blade run along the edge of the broadknife. Strip the trimmed waste off the ceiling and wipe off the paste with a damp sponge. Trim the covering at the baseboard, using the broadknife and razor blade in the same manner as at the ceiling, and run a seam roller (easier to clean than wood) down the full length of each vertical edge. Wipe the surface of the strip with the damp sponge to remove all paste smears before they dry—they are nearly impossible to remove later on some coverings. Rinse the sponge frequently in clean, warm water, and don't be miserly with razor blades. As soon as one does not cut cleanly, throw it away and use a new one because cuts must sometimes be made under difficult circumstances and sawing away at wet wall covering anytime is bad business. Blades sell for about two cents apiece in boxes of 100 at your dealer.

If you feel that you were able to slide the strip where you wanted it once it was on the wall, that it laid smoothly to the wall and corners without stretching and that it had a limber, "cushiony" feel to it, proceed to hang the remaining strips to the right, in the same manner. If the covering seemed to be stiff, like it was not wet enough, or was hard to manage, soak the next strip as before, but instead of putting it on the wall, lay it paste-side up on the table or floor. Fold one end back to the middle, paste-to-paste, do the same with the other end, roll the whole thing up loosely, and leave it alone for 5 minutes. Unroll it, unfold the top end, and hang it to the right of the first strip. Unfold the remaining end, slip the strip left until its left edge is neatly butted against the right edge of the first strip, and smooth it to the wall. If that strip handled better than the first strip, use the soak-and-fold procedure when hanging the remaining strips. Several can be soaked and folded at a time, but if their edges start to dry out, you are waiting too long between soaking and hanging; work faster, or don't moisten so many in advance. If the second strip was just as hard to manage as the first, treat the remainder as unpasted paper.

Pasting Wallcovering

Pasting wallcovering is not difficult, but be sure that the right paste for the covering is used so that you don't come into the room later and find your wall covering all over the floor or have the rich aroma of mildew all over the place. In general, old-fashioned wheat paste will work with anything porous, and vinyl adhesive will work with anything nonporous. However, certain coverings take special adhesives; to be absolutely certain, ask the dealer when you buy the paper, or follow the instructions that come with the rolls.

Unless otherwise instructed, add the paste powder to the water, stir thoroughly, use an old eggbeater to get rid of the lumps, mix to the consistency of a fairly thick milkshake, and let it sit for awhile before use. Stir it again before spreading.

Spread the paste over two-thirds the length of a strip of covering, using a wide paste brush, and fold the pasted end back to the middle, paste-to-paste, as previously described. Paste the remainder of the strip and fold the other end back. If you have untrimmed wall covering, now is the time to trim it. Fold again, pattern-to-pattern this time, make sure the edges are even, lay a metal straightedge along the pattern at the inside edge of the selvage, and trim it off with a razor blade or rotary trimmer. Paste several strips, but not so many that the edges dry out before the strips are hung, and roll each one loosely. Apply to the wall, smooth, and trim the same way as prepasted paper (obviously, without the waterbox).

Fitting at Inside Corners

At inside corners, measure from the last strip to the corner and from the corner onto the second wall for a total distance equal to the width of a strip of covering, less one inch. Mark the wall with a dot and establish a vertical reference mark from floor to ceiling, as was done in the beginning. Hang a strip around the corner; measure one inch from the corner onto the second wall and slit the wall covering here from floor to ceiling. Strip the right-hand piece from the wall and reposition it so that its right-hand edge lies on the vertical reference mark and its left-hand edge is in the corner, covering the one-inch overlap. Make another floor-to-ceiling cut, freehand this time, through both layers at once, about ½ in. from the corner. Strip and discard both pieces of waste, butt the cut edges, and roll lightly with a seam roller for an invisible joint. This strip provides a vertical edge for hanging wallcovering on the second wall and compensates for a corner out-of-plumb.

Fitting at Outside Corners

Outside corners can be tough, but are usually more plumb than inside corners. If the corner is accurate, just paste the covering around it and go right on. If it is not, and a break in the pattern would not be too obvious or objectionable, take the covering around the corner for an inch and cut if off vertically. Reposition the cut piece so that it overlaps the one-inch overlap by a bare minimum on the left and is perfectly vertical on the right (use another vertical reference mark if necessary). With a new razor blade held perpendicular to the wall, make a freehand cut through both left-hand layers at once, remove both strips of waste, and roll the resulting butt joint with a seam roller. If such a resulting mismatch would be objectionable, since it is on a flat surface, take the covering around the corner ½-in. and cut it off as previously described. Reposition the cut piece so that its left edge comes right up to the corner (and hangs off if necessary) and the right-hand edge is vertical. Trim the left edge even with the corner and cement it to the covering underneath with a special cement which will bond vinyl to vinyl.

Fitting at Window Openings

At window and door openings, hang the covering to overlap the opening. Cut away the majority of the waste, bisect each corner angle with a cut, and lay the covering tight to the wall. Trim around the casing, using a broadknife and razor blade as previously described.

Hints and Kinks

Remember that all joints must butt and not overlap, since vinyl will not stick to vinyl unless specially cemented. Also, small rooms, such as bathrooms, go much more slowly and are more difficult to do than large rooms, so you may want to keep this in mind when estimating how long a job will take.

Remove all switch and outlet plates, wall register grilles, and wall-mounted fixtures before hanging wall covering. Slit the covering to allow switches and outlets to protrude and cut the larger opening for them later with a razor blade. Don't cut the opening too large, or the plate won't cover it, and be careful that the blade does not short against live electric terminals or slice the insulation on a wire. Openings for plumbing pipes should be rough-cut and the covering slit to the nearest edge in order to accommodate them. Bevels are installed over the pipes for a neat finish. It is usually not necessary to remove a toilet tank because there is enough space behind it to install the wall covering. If the bathroom is a small second bath, however, you may want to remove the whole toilet to gain more working room. It can be reinstalled later with little trouble.

CHAPTER 30

Ceramic Wall and Floor Tile

Ceramic tile is not unlike brick in its origins and general method of production. The ingredients are essentially earth clays that are baked hard, like brick. In particular, however, tile is pressed into much thinner bisques and smaller surface sizes and then baked. The baking and face glazing produce a variety of finishes that make tile suitable for decorative purposes. Unlike brick which is used for structural purposes, tile is used to surface either walls or floors. It has a hard glossy surface that is impervious to moisture and resists soiling. It is easy to keep clean.

Tile generally is used for floors, walls, roof coverings and drain pipe. Some roof coverings are made of baked clays, the same as the drain tile used underground. This chapter discusses ceramic tile used for wall and floor coverings and its installation.

The term ceramic tile distinguishes the earth clay tiles from the metal and plastic tiles. The use of ceramic tile dates back four to six thousand years ago. It was used in the early Egyptian, Roman, and

Greek cultures and, because of its durability, is sometimes the only part of structures left to study from the diggings into the ruins of those ancient days (Fig. 30-1).

Fig. 30-1. Reconstructed Egyptian hall showing tile floor.

The introduction of ceramic tile in the United States began about the time of the Philadelphia Centennial Exposition in 1876. Up to then, tile was made and used in Europe. Some English manufacturers showed tile products at the exposition, which intrigued American building material producers. Subsequently a factory was started in Ohio, then in Indiana.

TILE APPLICATIONS

The uses of tile range from the obvious to the exotic. The obvious begins with its use in a bathroom (Fig. 30-2), because water does not affect it and glazed tile harmonizes with tubs and

Fig. 30-2. A bathroom is a perfect place for ceramic tile.

other fixtures. Other areas of the home where moisture may affect conventional wall finishes are likely areas for tile. Kitchen-sink tops and back-splashes (Fig. 30-3) are a natural for tile. Fig. 30-4 shows an entire wall in tile, in a breakfast nook. For the same reason, tile is ideally suited to swimming pools. Walls and floors, as well as the pool itself, are covered with ceramic tile in the public

665

Fig. 30-3. Tile is also used for countertops.

pool. You'll even find tile on a swimming pool on a deluxe ocean cruiser.

TYPES OF TILE

Not so long ago, tile was available in a relatively limited number of colors and shapes and finishes; a few smooth, pastel-colored

Fig. 30-4. Tile is available in block style for floors and walls.

tiles was about it. But no more. Today, there is a variety available. For walls, there is the standard $4\frac{1}{4}'' \times 4\frac{1}{4}''$ tile, but you can also get tile in a variety of shapes—diamond, curved, and so on—and in a great variety of colors and textures, from sand to smooth.

For the floor there is block tile, a tile that comes in shapes, and mosaic tile with tiles in individual 1-in. squares secured to a mesh background.

Formerly, tile was installed with portland cement—a so-called mud job—but today the job is usually done, on floors and walls, with adhesive. There is a wide variety of adhesives available, and you should check with a tile dealer to ensure that you select the proper one.

Fig. 30-5. Guidelines are essential for installing tile on walls and floors.

INSTALLING TILE

Dealers also carry detailed instructions—in some cases, color films—that show how to install tile. What follows, however, are the basic steps involved for tiling walls and floors:

For walls, the job usually involves standard tub surround—three walls. Start by drawing horizontal and vertical lines in the center of each wall. On one wall, lay a line of tile along the top of the tub. Look at the tile in relation to the lines. If you will have to install tile that is less than one tile wide, move the lines so that when you install, you won't have to. Tile pieces that are less than one-half a tile don't look good. In some cases, of course, you will not be able to install tiles of the proper size, but you should try.

Fig. 30-6. Tile being applied to adhesive that has been put on with a toothed (notched) trowel.

Apply Adhesive

Use a notched trowel to apply adhesive; your dealer will tell you the type to use. Hold the trowel at a 45° angle and apply only a limited amount of adhesive at a time so that it doesn't dry out before the tile can be applied.

Place each tile with a twisting motion to embed it well in the cement. Press the tile down, tapping it firmly in place. Keep the joints straight.

Standard tile comes with nibs or spacers on the edges so that the

669

Fig. 30-7. Tile nippers are good for making irregular cuts in the tile.

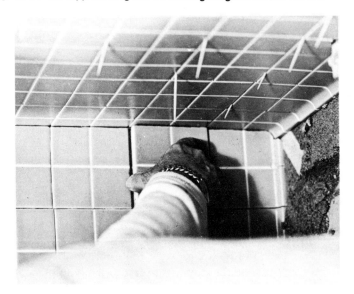

Fig. 30-8. If tile does not have spacers, you must provide them.

670

Fig. 30-9. Tape tile carefully to be sure it is seated properly.

Fig. 30-10. A rented tile cutter works well for making square cuts.

671

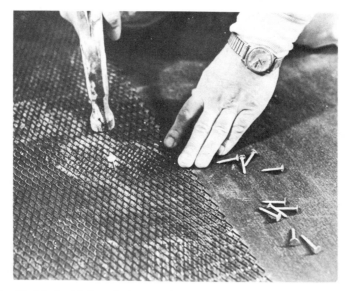

Fig. 30-11. In this installation, mesh is secured to floor.

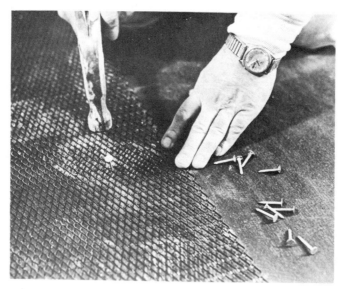

Fig. 30-12. Cement is to be used. It is an old-fashioned "mud job."

Fig. 30-13. A squeegee or rubber trowel is good for smoothing grout.

proper joint distance will be maintained. If this is not the case, you will have to be very careful. Use small sticks of the same thickness to maintain joint distance. However you do it, make sure that the tiles are straight.

Cutting Tile

At some point in the installation it is highly likely that you will have to cut the tile. For this you can use a rented tile cutter—it works like a paper cutter. For tiles that must have small pieces nibbled from them to fit around pipes and the like, you can use nippers—tile dealers usually have these.

When cutting tile, follow these procedures:

1. Cut and place sections one at a time. Cutting all the tile sections you need can lead to errors, because few corners are plumb.
2. Nip off very small pieces of the tile when cutting. Attempting to take a big chunk at one time can crack the tile.

673

Applying Grout

Grout is the material that fills the joints. It once was available only in limited colors—indeed, just white—but today you can get grout in many different colors, and it makes a nice accent material to the tile.

The package containing the grout—it comes as a powder you mix with water—will give instructions on its use. It should be applied at least 24 hours after the tile. A squeegee makes a good applicator: lay the grout on and squeegee off the excess. To shape the joint lines, you can use the end of a toothbrush. Excess grout can be cleaned off with a damp sponge. When the joints' grout has dried, spray it with silicone and polish with a cloth.

Doing a Floor

Installing floor tile follows essentially the same procedure: draw lines, move them as needed, apply adhesive, set tiles. If you are tiling a countertop and have both a horizontal and vertical surface, it is easier to install tile on the horizontal area first.

CHAPTER 31

Paneling

A major development in wallcovering over the last twenty-five years is paneling. It can take walls in any condition—from ones that are slightly damaged to ones that are virtually decimated—and give them a sparkling new face. Paneling is a key improvement material for the carpenter and do-it-yourselfer.

Paneling is available in an almost limitless variety of styles, colors, and textures (Fig. 31-1). Prefinished paneling may be plywood or processed wood fiber products. Paneling faces may be real hardwood or softwood veneers finished to enhance their natural texture, grain, and color. Other faces may be printed, paper overlaid, or treated to simulate wood grain. Finishing techniques on both real wood and simulated wood-grain surfaces provide a durable and easily maintained wall. Cleaning usually consists of wiping with a damp cloth.

Courtesy Georgia Pacific

Fig. 31-1. This paneling looks like it came off a barn. Paneling is available in an array of colors and styles.

HARDWOOD AND SOFTWOOD PLYWOOD

Plywood panelings are manufactured with a face, core, and back veneer of softwood, hardwood or both (Fig. 31-2). The face and back veneer wood grains run vertically, the core horizontally. This lends strength and stability to the plywood. Hardwood and softwood plywood wall paneling is normally ¼ in. thick with some paneling ranging up to ⁷⁄₁₆ in. thick.

The most elegant and expensive plywood paneling has real hardwood or softwood face veneers—walnut, birch, elm, oak, cherry, cedar, pine, fir. Other plywood paneling may have a veneer of tropical hardwood. Finishing techniques include embossing, antiquing, or color toning to achieve a wood grain or decorative look. Panels may also have a paper overlay with wood grain or patterned paper laminated to the face. The panel is then grooved and finished. Most of these panelings are 5/32 in. thick.

Processed wood fiber—particleboard or hardboard—wall paneling is also available with grain-printed paper overlays or printed surfaces. These prefinished panels are economical yet attractive.

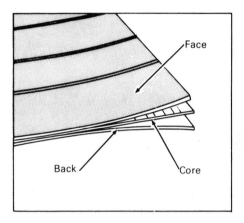

Fig. 31-2. Typical panel with core, back, and face.

Thicknesses available are ⁵⁄₃₂, ³⁄₁₆, and ¼ in. Wood fiber paneling requires special installation techniques. Look for the manufacturer's installation instructions printed on the back of each panel.

Other paneling includes hardboard and hardboard with a tough plastic coating that makes the material suitable for use in high-moisture areas, such as the bathroom.

GROOVE TREATMENT

Most vertical wall paneling is "random grooved" with grooves falling on 16-in. centers so that nailing over studs will be consistent. A typical random-groove pattern may look like the one illustrated in Fig. 31-3.

Other groove treatments include uniform spacing (4, 8, 12, or 16 in.) and cross-scored grooves randomly spaced to give a "planked" effect. Grooves are generally striped darker than the panel

677

Courtesy Georgia Pacific

Fig. 31-3. Random groove paneling.

surface. They are cut or embossed into the panel in V-grooves or channel grooves. Less expensive paneling sometimes has a groove "striped" on the surface.

BUYING TIPS

Like most products for the home, paneling is available in a wide range of prices, from as little as $5 per panel to over $30 per panel retail. Contractors and carpenters get a discount (usually 10 percent). Generally, paneling with a simulated wood-grain finish is less expensive than real wood-surfaced panels. Printed or plywood overlaid paneling is available for about $9 to $15 per panel. Paneling with wood fiber substrate costs $5 to $8 per sheet.

How to Figure the Amount of Material Needed

There are several ways to estimate the amount of paneling you will need. One way is to make a plan of the room. Start your plan on graph paper that has ¼-in. squares. Measure the room width, making note of window and door openings. Translate the measurements to the graph paper.

Example: Let us assume that you have a room 14′ × 16′ in size. If you let each ¼-in. square represent 6 in., the scale is 1 in. equals 2 ft., so you end up with a 7″ × 8″ rectangle. Indicate window positions, doors (and the direction in which they open), a fireplace, and other structural elements. Now begin figuring your paneling requirements. Total all four wall measurements: 14′ + 16′ + 14′ + 16′ = 60′. Divide the total perimeter footage by the 4-ft. panel width: 60′ divided by 4′ equals 15 panels. To allow for door, window, and fireplace cutouts, use the deductions as follows (approximate): door, ½ panel (A); window, ¼ panel (B); fireplace, ½ panel (C) (Fig. 31-4).

To estimate the paneling needed, deduct the cutout panels from your original figure: 15 panels − 2 panels = 13 panels. If the perimeter of the room falls between the figures in Table 31-1, use the next higher number to determine panels required. These figures are for rooms with 8-ft. wall heights or less. For higher walls, add in the additional materials needed above the 8-ft. level.

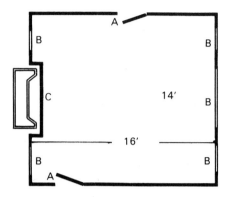

Fig. 31-4. See the text for details on calculating panel requirements as depicted in this drawing.

New paneling should be stored in a dry location. In new construction, freshly plastered walls must be allowed to dry thoroughly before panel installation. Prefinished paneling is moisture-resistant, but like all wood products, it is not waterproof and should not be stored or installed in areas subject to excessive moisture. Ideally the paneling should be stacked flat on the floor with sticks

Table 31-1

Perimeter	No. of 4' × 8' Panels Needed (Without Deductions)
36'	9
40'	10
44'	11
48'	12
52'	13
56'	14
60'	15
64'	16
68'	17
72'	18
92'	23

between sheets (Fig. 31-5) to allow air circulation, or it should be propped on the 8-ft. edge. Panels should remain in the room 48 hours prior to installation to permit them to acclimatize to temperature and humidity.

INSTALLATION

Installing paneling over existing straight walls above grade requires no preliminary preparation. First, locate studs. Studs are usually spaced 16 in. on center, but variations of 24-in. centers and other spacing may be found. If you plan to replace the present molding, remove it carefully—you may reveal nailheads used to secure plaster lath or drywall to the studs. These nails mark stud locations. If you cannot locate the studs, start probing with a nail or small drill into the wall surface to be paneled, until you hit solid wood. Start 16 in. from one corner of the room with your first try, making test holes at ¾-in. intervals on each side of the initial hole until you locate the stud (Figs. 31-6 and 31-7).

Studs are not always straight, and so it is a good idea to probe at several heights. Once a stud is located, make a light pencil mark at floor and ceiling to position all studs, then snap a chalk line at 4-ft. intervals (the standard panel width).

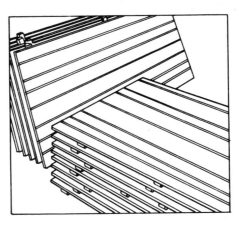

Fig. 31-5. If possible, store panels for a day or so with sticks between panels so that they can become accustomed to room conditions.

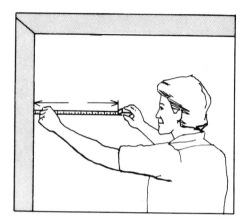

Fig. 31-6. Measure 16 in. from corner of room . . .

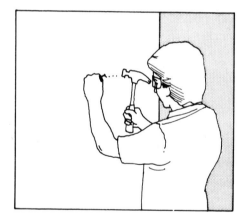

Fig. 31-7. . . . then make test holes to locate stud.

Measuring and Cutting

Start in one corner of the room and measure the floor-to-ceiling height for the first panel. Subtract ½ in. to allow for clearance top and bottom. Transfer the measurements to the first panel and mark the dimensions in pencil, using a straightedge for a clean line. Use a sharp crosscut handsaw with ten or more teeth to the inch, or a plywood blade in a table saw, for all cutting. Cut with the panel

682

face-up. If you are using a portable circular saw or a saber saw, mark and cut panels from the back.

Cutouts for door and window sections, electrical switches, and outlets or heat registers require careful measurements. Take your dimensions from the edge of the last applied panel for width, and from the floor and ceiling line for height. Transfer the measurements to the panel, checking as you go. Unless you plan to add moulding, door and window cutouts should fit against the surrounding casing. If possible, cutout panels should meet close to the middle over doors and windows.

An even easier way to make cutouts is with a router. You can tack a panel over an opening—a window, say—after removing the trim. Then use the router to trim out the waste paneling in the opening, using the edge of the opening as a guide for the router bit to ride on. Then you can tack the moulding or trim back in place.

For electrical boxes, shut off the power and unscrew the protective plate to expose the box. Then paint or run chalk around the box edges and carefully position the panel. Press the paneling firmly over the box area, transferring the outline to the back of the panel (Fig. 31-8). To replace switchplate, a $\frac{1}{4}$-in. spacer or washer may be needed between the box screw hole and the switch or receptacle.

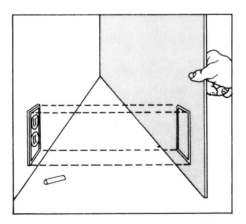

Fig. 31-8. A good way to get the outline is to transfer the chalk mark from plate to back of panel.

683

Drive small nails in each corner through the panel until they protrude through the face. Turn the panel over, drill two ¾-in. holes just inside the corners, and use a keyhole or saber saw to make the cutout (Fig. 31-9). The hole can be up to ¼ in. oversize and still be covered when the protective switchplate is replaced.

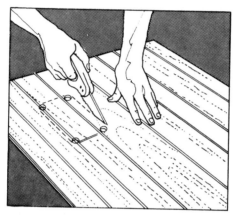

Fig. 31-9. Drill starter holes, then cut out plate section with a keyhole or saber saw.

Securing Panels

Put the first panel in place and butt to an adjacent wall in the corner. Make sure it is plumb and that both left and right panel edges fall on solid stud backing. Most corners are not perfectly true, however, so you will probably need to trim the panel to fit into the corner properly. Fit the panel into the corner, checking with a level to be sure the panel is plumb vertically. Draw a mark along the panel edge, parallel to the corner. On rough walls like masonry, or on walls adjoining a fireplace, scribe or mark the panel with a compass on the inner panel edge, then cut on the scribe line to fit (Fig. 31-10). Scribing and cutting the inner panel edge may also be necessary if the outer edge of the panel does not fit directly on a stud. The outer edge must fall on the center of a stud to allow room for nailing your next panel. Before installing the paneling, paint a stripe of color to match the paneling groove color on the wall location where panels meet. The gap between panels will not show.

Most grooved panels are random grooved to create a solid lumber effect, but there is usually a groove located every 16 in. This allows most nails to be placed in the grooves, falling directly on the 16-in. stud spacing. Regular small-headed finish nails or colored paneling nails can be used. For paneling directly onto studs 3d (1¼ in.) nails are recommended; but if you must penetrate backer board, plaster, or drywall, 6d (2-in.) nails are needed to get a solid bite in the stud. Space nails 6 in. apart on the panel edges and 12 in. apart in the panel field (Fig. 31-11). Nails should be countersunk slightly below the panel surface with a nailset, then hidden with a matching putty stick. Colored nails eliminate the need to do this. Use 1-in. colored nails to apply paneling to studs, 1⅝-in. nails to apply paneling through plasterboard or plaster.

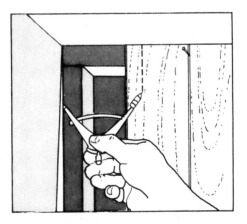

Fig. 31-10. On rough walls, use scriber as shown to get wall outline on paneling. Trim as required.

Adhesive Installation

Using adhesive to install paneling eliminates countersinking and hiding nailheads. Adhesive may be used to apply paneling directly to studs or over existing walls as long as the surface is sound and clean.

Paneling must be cut and fitted prior to installation. Make sure

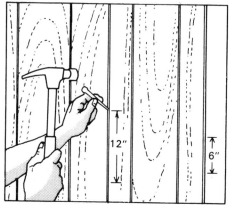

Fig. 31-11. Nails should be 12 in. apart vertically.

the panels and walls are clean—free from dirt and particles—before you start. Once applied, the adhesive makes adjustments difficult.

A caulking gun with adhesive tube is the simplest method of application. Trim the tube end so that a ⅛ -in.-wide adhesive bead can be squeezed out. Once the paneling is fitted, apply beads of adhesive in a continuous strip along the top, bottom, and both ends of the panel. On intermediate studs, apply beads of adhesive 3 in. long and 6 in. apart (Fig. 31-12).

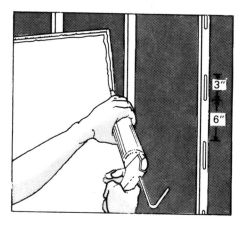

Fig. 31-12. Apply adhesive as shown.

686

With scrap plywood or shingles used as a spacer at the floor level, set the panel in place and press firmly along the stud lines spreading the adhesive on the wall. Using a hammer with a padded wooden block or rubber mallet, tap over the glue lines to assure a sound bond between panel and backing.

Some adhesives require panels to be placed against the adhesive, then gently pulled away and allowed to stand for a few minutes for the solvent to "flash off" and the adhesive to set up. The panel is then repositioned and tapped home.

Uneven Surfaces

Most paneling installations require no preliminary preparations. When you start with a sound level surface, paneling is quick and easy. Not every wall is a perfect wall, however. Walls can have chipped, broken, and crumbly plaster, peeling wallpaper, or gypsum board punctured by a swinging door knob or a runaway tricycle on a rainy afternoon. Or walls can be of rough poured concrete or cinderblock. Walls must be fixed before you attempt to cover them with paneling.

Most problem walls fall into one of two categories. Either you are dealing with plaster or gypsum board applied to a conventional wood stud wall, or you are facing an uneven masonry wall—brick, stone, cement, or cinderblock, for example. The solution to both problems is the same, but getting there calls for slightly different approaches.

On conventional walls, clean off any obviously damaged areas. Remove torn wallpaper, scrape off flaking plaster and any broken gypsum board sections. On masonry walls, chip off any protruding mortar. Don't bother making repairs; there is an easier solution.

The paneling will be attached to furring strips. Furring strips are either 1″ × 2″ lumber or ⅜- or ½-inch plywood strips cut 1½ in. wide. The furring strips are applied horizontally 16 in. apart on center on the wall (based on 8-ft. ceiling) with vertical members at 48 in. centers where the panels butt together.

Begin by locating the high spot on the wall. to determine this, drop a plumb line (Fig. 31-13). Fasten your first furring strip, making sure that the thickness of the furring strips compensates for the protrusion of the wall surface. Check with a level to make

Fig. 31-13. Wall must be plumbed before you apply furring strips.

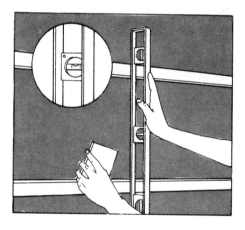

Fig. 31-14. Use level to determine levelness of wall. Use shim as shown.

sure that each furring strip is flush with the first strip (Fig. 31-14). Use wood shingles or wedges between the wall and strips to assure a uniformly flat surface (Fig. 31-15). The furred wall should have a ½-in. space at the floor and ceiling with the horizontal strips 16 in. apart on center and the vertical strips 48 in. apart on center (Fig. 31-16). Remember to fur around doors, windows, and other openings. (Fig. 31-17).

688

Fig. 31-15. An overall view of shimmed furring strips.

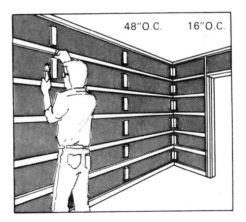

48"O.C.   16"O.C.

Fig. 31-16. Secure vertical nailers as shown.

On stud walls, the furring strips can be nailed directly through the shims and gypsum board or plaster into the studs. Depending on the thickness of the furring and wallcovering, you'll need 6d (2-in.) or 8d (2½-in.) common nails.

Masonry walls are a little tougher to handle. Specially hardened masonry nails can be used, or you can drill a hole with a carbide tipped bit, insert wood plugs or expansion shields, and nail or bolt

689

the shimmed frame into place (Fig. 31-18). If the masonry wall is badly damaged, construct a 2" × 3" stud wall to install paneling. Panels may be directly installed on this wall (Fig. 31-19).

DAMP WALLS

Masonry walls, besides being uneven, often present a more difficult problem: dampness. Usually, damp walls are found partially or fully below ground level. Damp basement walls may result from two conditions: (1) seepage of water from outside walls; or (2) condensation, moist warm air within the home that condenses or beads up when it comes in contact with cooler outside walls. Whatever the cause, the moisture must be eliminated before you consider paneling.

Seepage may be caused by leaky gutters, improper grade, or cracks in the foundation. Any holes or cracks should be repaired with concrete patching compound or with special hydraulic cement designed to plug active leaks.

Weeping or porous walls can be corrected with an application of masonry waterproofing paints. Formulas for wet and dry walls are available, but the paint must be scrubbed onto the surface to penetrate the pores, hairline cracks, and crevices.

Fig. 31-17. Don't forget to install furring around windows.

Fig. 31-18. If wall is very rough, a 2 × 3 wall can be built, plumbed up, and paneling attached to it.

Fig. 31-19. Before installing panels, line them up for best color matching.

Condensation problems require a different attack. Here, the answer is to dry out the basement air. The basement should be provided with heat in the winter and cross-ventilation in the summer. Wrap cold-water pipes with insulation to reduce sweating. Vent basement washer, dryer, and shower directly to the outside. It may pay big dividends to consider installing a dehumidifying system to control condensation.

691

When you have the dampness problem under control, install plumb level furring strips on the wall. Apply insulation if desired. Then line the walls with a polyvinyl vapor moisture barrier film installed over the furring strips. Rolls of this inexpensive material can be obtained and installed, but be sure to provide at least a 6-in. lap where sections meet.

Apply paneling in the conventional method. If the dampness is so serious that it cannot be corrected effectively, then other steps must be taken. Prefinished hardwood paneling is manufactured with interior glue and must be installed in a dry setting for satisfactory performance.

PROBLEM CONSTRUCTION

Sometimes the problem is out-of-square walls, uneven floors, or studs placed out of sequence. Or trimming around a stone or brick fireplace may be required, or making cutouts for wall pipes, or handling beams and columns in a basement.

Any wall to be paneled should be checked for trueness. If the wall is badly out-of-plumb, it must be corrected before you install paneling. Furring strips should be applied so that they run at right angles to the direction of the panel application. Furring strips of $\frac{3}{8}'' \times 1\frac{1}{2}''$ plywood strips or $1'' \times 2''$ lumber should be used. Solid backing is required along all four panel edges on each panel. Add strips wherever needed to ensure this support. It is a good idea to place the bottom furring strip $\frac{1}{2}$ in. from the floor. Leave a $\frac{1}{4}$-in. space between the horizontal strips and the vertical strips to allow for ventilation.

If the plaster, masonry, or other type of wall is so uneven that it cannot be trued by using furring strips and shimming them out where the wall bellows, $2'' \times 3''$ studs may be necessary. You can use studs flat against the wall to conserve space. Use studs for top and bottom plates and space vertical studs 16 in. on center. Apply paneling direct to studs or over gypsum or plywood backboard.

Out-of-sequence studs (i.e., studs not on regular centers) require a little planning, but usually they are not a serious problem. Probe to find the exact stud locations as described earlier, snap chalk lines, and examine the situation. Usually you will find a normal stud sequence starting at one corner; then perhaps where the carpenter framed for a mid-wall doorway, the spacing abruptly changes to a new pattern. Start paneling in the normal manner, using panel adhesive and nails. When you reach the changeover point, cut a filler strip of paneling to bridge the odd stud spacing then pick up the new pattern with full-size panels.

Occasionally you will find a stud or two applied slanted. The combination of panel adhesive plus the holding power of the ceiling and baseboard mouldings generally solve this problem.

Uneven floors can usually be handled with shoe moulding. Shoe, $\frac{1}{2}'' \times \frac{3}{4}''$, is flexible enough to conform to moderate deviations. If the gap is greater than $\frac{3}{4}$ in., the base moulding or panel should be scribed with a compass to conform to the floor line. Spread the points slightly greater than the gap, hold the compass vertically, and draw the point along the floor, scribing a pencil line on the base moulding. Remove the moulding trim to the new line with a coping or saber saw and replace. If you have a real washboard of a floor, then you have a floor problem, not a wall problem. Either renail the floor flat or cover it with a plywood or particleboard underlayer before paneling.

The compass trick is also used to scribe a line where paneling butts into a stone or brick fireplace. Tack the panel in place temporarily, scribe a line parallel to the fireplace edge, and trim. Check to be sure that the opposite panel edge falls on a stud before applying.

CHAPTER 32

Solar Energy

One area that is increasingly likely to require the services of carpenters and builders, and an area in which more and more do-it-yourselfers are getting involved, is solar energy. Thus it is important for the craftsman to have a fundamental understanding of the systems involved.

The basic heating plant in any solar system is, of course, the sun. But one form of the system is active—hardware is required, as we will detail later—while the other is passive, and no moving hardware is needed.

PASSIVE SYSTEMS

Passive-system solar buildings absorb and store solar heat in masonry floors, walls, or room dividers or in water-filled tubes, drums, or other containers. Enough heat can be stored by a well-designed passive building to keep it warm after the sun has set and through the night, and possibly for a day or two of sunless weather.

The location of the storage mass in a passive building depends upon which passive method is used. In the popular "direct gain" approach, the room collects sunlight through large south-facing windows. Solar radiation strikes the darkened surface of the thick masonry walls and floor, where it is absorbed as heat. The heat remains in the masonry until the air in the room cools. The heat then radiates into the room, drawn, as heat always is, from a warmer to a cooler spot until both places are of equal temperature. The surface area of storage exposed to direct sunlight in a "direct gain" building should be at least three times the area of the south-facing window. Storage walls and floors should be between 4 and 8 in. thick. An adequately sized storage mass prevents extensive overheating during the day and provides warmth through the night. It is important to remember that storage floors cannot be carpeted and that walls cannot be covered. These surfaces must be exposed if they are to absorb and radiate heat effectively.

There is also what is called a Trombe wall (originally Trombe-Michel, after the two French scientists who developed it). The outside surface of the wall is painted black to absorb heat. Gradually migrating through the wall, the heat reaches the room later in the day and continues to radiate into the room through the night. For a Trombe wall, the area of the window and the area of the storage wall's surface should be the same. The wall should be from 8 to 16 in. thick.

Most Trombe walls also have vents at the floor and ceiling. While the sun shines, the heated air in the space between the glazing and the wall rises and enters the room through the upper vents, drawing cooler air into the space through the lower vents. This natural circulation of heat continues as long as the sun shines.

Water-filled containers (fiberglass tubes, 55-gal. drums, etc.) can be used to create a "water wall"; water has about twice the heat storage capacity of masonry of equivalent volume. A variation on the water wall is the roof pond. Roof ponds are essentially waterbeds made of sturdy plastic. They rest on a special ceiling structure and are covered and insulated on winter nights by a movable roof. During the day the mechanically operated roof is opened exposing the roof pond to the sun so that it can absorb heat and later radiate it to the living area.

When an attached greenhouse (Fig. 32-1) is used to collect solar heat, the storage may include the masonry floor of the greenhouse, the wall separating the greenhouse from the main building and water containers, such as the 55-gal. drums shown in the illustration. The wall may be either masonry or water-filled containers, and it may include windows for heating with direct sunlight.

Fig. 32-1. This greenhouse is one type of solar collector. The basic green-house concept is used with some systems.

ACTIVE SYSTEMS

In active systems, air, water, or other liquid passes through the collector (Fig. 32-2) which is warmed by the sun's rays. It then is routed by tubing of some sort to a storage tank or bin (in the case of air) where it is stored until use.

The components of an active system are generally housed in the basement or in a special utility area. Active-system storage is linked to both rooftop flat-plate collectors and a mechanical heat distribution system that makes use of pumps or air-handling units.

Ideally, storage bins or tanks for active systems should be located next to or under those rooms that require the greatest

697

Fig. 32-2. A solar collector is small, but it can gather a great deal of solar energy.

amount of heat. The storage area should be accessible for maintenance and repairs.

Air Systems

Solar heating systems that circulate air through rooftop collectors commonly use a bin filled with rocks for heat storage. The rock
bin stores heat when hot air from the collectors is blown through
the rocks. The air first enters a plenum at the top of the bin. It is
distributed so that it flows through the bin evenly. As the air
reaches the bottom of the bin, the rocks have absorbed most of its
heat. The air enters another plenum at the bottom and is returned
to the collectors for reheating.

To retrieve heat from the storage bin, house air is drawn into the
lower plenum and blown upward through the rocks. The warmed
air is drawn off at the top of the bin and distributed through the

house. If the air is not warm enough, an auxiliary heater boosts its temperature before it is distributed.

It is not practical to store and withdraw heat simultaneously. If the hot air from the collector is needed to heat the house directly, it bypasses the rock bin and flows directly into the heat distribution system. Similarly, an air system that heats both the house and the household water should bypass the bin in the summer to heat the water directly. If a system heats the rock bin in order to preheat hot water, it will add to the house's cooling load in the summer.

Rock bins can be made from cinderblock, concrete or wood. When treated plywood is used for bin construction, it should be lined with drywall and a vapor barrier. This will protect the rocks and the entire system from any gases released by the plywood. The greatest problem of air systems is air leakage from the storage bin, as well as from ducts and dampers. Because leaks drastically reduce the efficiency of the system, the bin must be tightly constructed and sealed. Sealing also prevents vermin and insects from entering.

Air should flow through the system at the rate of 2 to 5 cu. ft. per minute for each square foot of collector. Ducts should be sized for a velocity of 5 to 10 ft. per second, and the rock bin should provide $\frac{1}{2}$ to 1 cu. ft. of storage for every square foot of collector. This is roughly two and one-half to three times the volume of a water tank that would provide an equivalent storage capacity in a liquid system.

The plenums at both the top and bottom of the bin ensure uniform airflow through the rock. Air filters are needed between both of the plenums and in the ducts to and from the bin. These filters prevent dust from blowing into either the collectors or the house. The filters should be inspected periodically and replaced when necessary.

Dense rock, such as river rock (which is predominantly quartz), performs best. The rocks should be of uniform size, roughly $\frac{3}{4}$ in. to $1\frac{1}{2}$ in. in diameter. Before the rocks are put into the bin, they must be washed to remove dirt and insect eggs. Problems with mould, mildew, and insects can be prevented by keeping the rocks inside the bin dry.

Because of the weight of the rocks, the bin is usually located in the basement of the house. (Outside underground bins are another option, but they are difficult to insulate and waterproof and should be avoided. Ground water can ruin the insulation and may corrode the walls of the bin if the waterproofing materials fail.) To reduce heat loss, an indoor bin should be insulated to a value of not less than R-19 (R-30 or greater if it is outside). The ducts to and from the bin should be insulated to a value of R-16.

Liquid Systems

Water is the common storage medium in active systems that circulate liquid through collectors. From 1 to 2½ gal. of water are needed for every square foot of collector. The tanks used to store water can be made of concrete, steel, or fiberglass-reinforced plastic. All should be insulated to a value of R-19 or better, and all piping to a value of R-4. The cost and availability of suitable tanks vary widely across the country.

Steel tanks should be lined to prevent corrosion. Glass (domestic heating only) can all be used as linings. When applying a liner to a tank, the manufacturer's recommendations should be followed carefully. It is especially important to consider the type of system, its volume and maximum temperature and the wet vs. dry temperature limitations of the liner material. A rust inhibitor, such as disodium phosphate (NA_2HPO_4), can be used instead of a lining, but inhibitors degenerate and must be monitored regularly for pH balance. If the pH is not what is specified by the manufacturer, the inhibitor must be replaced. Concrete tanks, such as the ones used for septic systems are sometimes used outdoors where steel is more likely to corrode. Since concrete is porous, the tanks should have either vinyl or other liner to prevent leaks.

All tanks should have drains, as well as a means of access for routine maintenance, such as cleaning heat exchangers. Tanks should also be inspected for leaks before they are installed. And they should not be built into the structure of the house because they may someday need to be replaced. Temperature sensors for water storage tanks can and should be positioned for easy replacement.
commercial water softener should be used to prevent problems with "scaling" (i.e., carbonate residue).

The water in the storage tank is heated in one of two ways. It is either circulated directly through the collectors or indirectly by a heat-transfer fluid that circulates from storage to the collectors within a "closed loop." In the latter instance, the transfer fluid absorbs heat from the collector and then passes through a heat exchanger, which transfers the heat to the water inside the storage tank. The heat exchanger is usually a coiled copper tube that conducts heat from the transfer fluid flowing through it to the storage water in which it is immersed. Because exposure to oxygen encourages corrosion, the heat exchanger should be placed completely below the water level inside the tank.

If the system is also providing hot water for domestic use, the heat exchanger needs to be double-walled so that drinking water cannot be contaminated by the heat-transfer fluid.

HYBRID SYSTEMS

Elements of passive and active solar heating systems can be combined to create hybrid systems. The most common example uses a greenhouse collector (passive) linked to a rock storage bin (active) by way of fans and ducts. Hybrid systems offer more options to solar designers and builders and increase the overall versatility of solar heating technology. For example, collectors and/or storage bins in hybrid systems do not need to be located directly in the main living spaces and interior walls, doors, and stairwells can be designed with less regard for free air movement than is required with a strictly passive approach. It is important not to mistake a situation where two separate systems are present— one active, the other passive—for a hybrid system. A system is considered hybrid only when components are mixed within a single system.

PHASE-CHANGE MATERIALS

Another storage medium, one that could be of use for both passive and active systems in the future, is phase-change materials. These are substances that store energy as latent heat. Latent heat is the heat absorbed by a material as it changes phase, such as from a solid to a liquid at constant temperature—for example, the melting of ice or wax. This heat is given up again as the liquid returns to its solid form. Two to five times more heat is stored per unit volume by this process than by raising the temperature of an equal volume of water, rock, or concrete. Researchers are investigating several types of materials, including salts and paraffin waxes. At present, storage in phase-change materials is more expensive than in water or rocks. Further testing and field applications will be necessary to prove the reliability of phase-change storage.

CHAPTER 33

Fastening to Masonry or Drywall

How to fasten something to concrete, concrete block, or brick, and how to get a screw to hold in plaster or drywall are tasks completed by certain basic devices that have become standbys for the purpose.

FASTENERS FOR MASONRY

These items include bolts cast into wet concrete, expansion anchors for lag screws, wood screws, and machine screws, drive anchors, and masonry nails. A number of adhesives are also available, but their holding power is generally less than that of mechanical fasteners and their set-up time can make them inconvenient where time is a factor.

Sill Bolts

Whenever possible, the need to fasten something extremely large or heavy to concrete should be anticipated, and suitable bolts set in place on the forms before the concrete is poured. A sill bolt (Fig. 33-1), sometimes called a foundation bolt, hook bolt, or anchor bolt, is made of black iron ½-in. in diameter and 12 in. long. The hooked end is buried in wet concrete far enough to allow the threaded end to protrude through a hole bored in a sill, or whatever is to be anchored. A heavy washer and nut completes the installation. Sill bolts can be suspended in concrete forms and have concrete poured around them, such as is done in poured foundations, or they can be inserted in a concrete-block core opening and the opening then packed with concrete. The nut should not be torqued heavily until the fresh concrete has had a chance to cure, for heavy torque applied prematurely can draw the bolt and ruin its holding power. On a building foundation, tighten the nuts firmly onto the sills and proceed with construction. Tighten them again before the cavities are closed.

Fig. 33-1. A sill bolt, also called a foundation bolt, hook bolt, or anchor bolt.

Expansion Anchors

The least expensive way to do occasional, small-quantity fastening to masonry is with expansion anchors made from soft metal alloy or plastic (Fig. 33-2). These will provide fastenings not as strong as cast-in-place bolts but stronger than masonry nails, and have the advantage of easy disassembly when necessary. They are generally available in enough sizes to accommodate any threaded

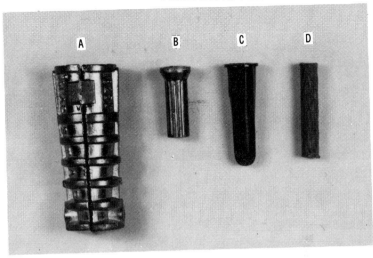

Fig. 33-2. Expansion anchors: (A) made from lead alloy for use with lag screws; (B) made from a softer lead alloy for use with wood screws; (C) made from plastic and best used with sheet-metal screws; (D) made from fiber-jacketed lead—a plug-type anchor sized here for small wood screws.

fastener normally found in nonindustrial applications and are also available in large sizes from specialty suppliers for industrial use.

An expansion anchor must be installed in a hole of a specified diameter and depth, bored into the masonry or its mortar with a carbide-tipped masonry bit in a power or hand drill. A star drill and a hammer can also be used (and sometimes *must* be used when there is no power available), but this is hard, slow, dirty work when done by hand, and accuracy is none too good. Masonry bits (Fig. 33-3) range in size from ⅛ in. to ⅞ in. and should be used at drilling speeds no higher than 900 r/min. Keep constant pressure on the drill and sharpen it often with a silicon-carbide grinding wheel. The bits will drill mortar, concrete, brick, slate, cement block, ceramic tile, and other masonry products, including natural stone.

The principle of expansion anchors designed for use with lag or wood screws is to have the screw expand the soft material of the anchor tightly against the sides of the hole, where it grips firmly and provides a surface for the screw threads to bite into. The grip

Fig. 33-3. A carbide-tipped masonry bit.

of such an anchor against the sides of the hole is improved by texturing the outside surface of the anchor during manufacture, and the grip of a lag screw on the inside surface of the anchor is also improved by molding female threads on each of the split halves.

Since a screw will expand an anchor only a limited amount, the diameter of the hole is a critical factor in the holding power of the finished job. Anchor manufacturers therefore specify not only the necessary diameter of the hole, but also the diameter of the screw to be used, and this information is cast or stamped into the outside surface of the larger units. All anchors have the information printed on their packaging as it comes from the factory, and it is therefore easily obtainable from the dealer.

The hole should be drilled as deep as the anchor is long, plus whatever depth is necessary to clear the pilot end of the screw when it is driven home. Failure to drill a blind hole deep enough to clear the length of the screw will result in the anchor being drawn out of the hole as the screw is tightened (obviously, this problem will not occur when the hole extends through the masonry and into a cavity behind it, as a hole through the side of a concrete block).

Tap the anchor gently into the hole until it is flush with the masonry surface and then insert the screw. Most expansion anchors have a collar at their roomside end which prevents the anchor from being driven too deeply into the hole and provides a neat finish with the hole's edge. Plug-type anchors made from plastic or from fiber-jacketed lead are straight cylinders, however, and are best used in blind holes the depth of the anchor to avoid losing them. They can be used in deep holes if carefully fitted.

Load capacities for expansion anchors are generally greater for the metal versions than the plastic ones. Both types are excellent

706

for loads which occur at right angles to the axis of the anchor, but this rating becomes sharply poorer as the direction of the load shifts to one parallel to the axis of the anchor. Their efficiency is poorest when the direction of force is a straight outward pull, which can strip the screw out of the anchor or, more likely, yank both screw and anchor out of the hole. Consideration of the direction and amount of exerted force is therefore necessary for success with expansion anchors used with lag or wood screws.

Machine-Screw Anchors

Machine screws and bolts, carriage bolts, studs, and threaded steel rod must be used with yet another type of anchor which provides probably the greatest holding power of all—up to 12,000 pounds in the large sizes. It consists of a hard, cone-shaped steel core which is longitudinally drilled and threaded to accept various standard sizes of machine screws and bolts. Surrounding the core is a separate cylinder of soft lead alloy (Fig. 33-4).

Fig. 33-4. A machine-screw anchor made to be set in a blind hole with a special setting tool.

Any debris is removed from the hole and the anchor is inserted with the large end of the cone to the bottom of the hole. A specially designed setting tool is then placed over the end of the anchor and struck a few blows with a hammer. The tool forces the lead cylinder down onto the tapered base of the steel core, squashing the lead outward and into the sides of the hole under heavy pressure; as the tool continues to be struck, it packs the lead tighter and tighter into the hole until it can be packed no further. Once set, the anchor is virtually permanently in place.

Care should be used in setting this type of anchor in the shell of a hollow masonry unit because it can be driven through the shell while being set; a similar possibility exists if the anchor is set in the mortar joints of concrete block which has been laid with face shell bedding. With block, it is best to drill the shell at a point over one of the core webs to assure solid material behind the hole. If this is impossible, an expansion anchor or toggle bolt should be used instead of the machine-screw anchor.

Masonry Nails

There are two basic masonry nails which are driven "free-hand" into masonry, using a hammer of up to 2 lb. Cut nails (Fig. 33-5) are blunt-pointed nails of rectangular cross-section having a continuous taper from head to tip. They are driven primarily into mortar and act to wedge themselves into place by friction of the masonry particles on the steel of the nail. They drive quite easily into mortar which is air-entrained, but are not so good with portland cement mortar.

Fig. 33-5. Cut nail, used primarily to drive into mortar.

In use, a blind hole the diameter of the lead cylinder is bored into the masonry the exact depth of the length of the steel core.

Hardened masonry nails (Fig. 33-6) can be driven into concrete, concrete block, or mortar, but attempting to drive them into brick can snap the nail or crack the brick. Driving these nails should only be done while wearing shatterproof safety goggles, because a heavy hammer-blow delivered any way but square with the nail-head can break the nail or whip it out of the hole and send pieces of it flying with vicious force. To preclude this, and for other advantages, some manufacturers recommend predrilling a hole in the masonry ¾ in. deep and slightly smaller than the root diameter of the nail. In all cases, whether predrilled or not, the proper length nail is one which does not penetrate the masonry more than ¾ in. Hardened masonry nails are manufactured with spiral striations along their length to aid holding-power.

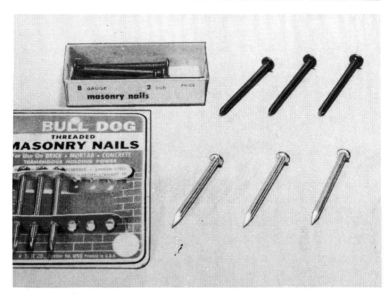

Fig. 33-6. Hardened steel masonry nails with striated shanks.

Drive Pins and Studs

These are actually a sophisticated form of masonry nail applied with a driving tool and a small, two-to-four-pound sledge. The nail-type drive pins are available in several lengths, while the drive studs come in one length for concrete and another for concrete block (Fig. 33-7). Both have smoothly shaped, sharp points reminiscent of an armor-piercing shell, and a ⅜-in.-diameter washer friction-fitted on the shank, behind the point. As the pin is driven into the masonry, the washer bears on the surface of the object being fastened while the pin slides through it; when the pin is driven home, the washer acts as a ⅜-inch nailhead and prevents the fastened object from tearing loose. On drive studs, the washer acts to prevent and/or cover up chipping as the stud is driven. When driven home, the studs provide at least ½ in. of male machine threads above the surface of the masonry.

The driving tool provides several advantages over freehand masonry nailing, but is effective only with drive pins or studs and *not* with ordinary masonry nails.

Fig. 33-7. Threaded studs (left), drive pins, and driving tool (top) with pin in place, ready to be driven.

The tool is capable of holding the drive pin absolutely perpendicular to the masonry surface for superior holding power and ease of driving; the pin is held firmly and safely (it is fully enclosed by the tool) with no danger to the eyes or body from flying nails, metal splinters, or masonry particles. A large-diameter vinyl shield deflects missed hammer blows away from the hand which grips the tool, and the anvil surface of the drive piston is easy to hit squarely. The piston transfers maximum impact to the head of the drive pin, which should be completely set after five or six hammer blows.

When fastening to well-cured concrete, the length of the drive pin used should equal the thickness of the material being fastened, plus ½ to 1 in. to allow for proper penetration into the concrete. If a threaded stud is to be driven, the one recommended for concrete should be used.

Pins driven into concrete-block shells, mortar, or relatively new concrete should equal the thickness of the fastened material, plus ¾ to 1 ¼ in. Threaded studs should be of the length recommended for such masonry.

710

Fig. 33-8 shows the tool in use with threaded studs to fasten the back of a workbench to a concrete basement wall.

Powder-Driven Hammers

Some see these as the ultimate in fast, effective masonry fastening; others see them as dangerous and expensive. They are all these things and more, depending on the point of view.

There is probably no faster or more labor-saving way to anchor something to concrete, concrete block, or mortar than by the use of a powder hammer, and there are circumstances where a fastening could be made no other way, except by extensive tearing-out and rebuilding. They are more expensive than hand-driven nails, but can be less expensive than screws-and-anchors; on really tough jobs, they can save as much as 20 minutes per fastening, which, if time is counted as money, can amount to enormous savings. They can be dangerous in almost the same way that a loaded gun can be dangerous; they can also be dangerous in other ways, but safety devices, precautions, and common sense can minimize this aspect.

A powder hammer is essentially a refinement of the driving tool used with the drive pins and studs described earlier, except that the force used to drive the fastener comes from a gunpowder charge instead of a hammer blow. The power of this charge, which looks like a blank cartridge, is precisely measured according to the size fastener to be used and the type masonry into which it is to be driven. In use, the fastener desired is muzzle-loaded into the barrel of the hammer, the cartridge (called a "power load") is placed in the chamber, and the hammer placed perpendicular to the masonry surface with the muzzle pressed firmly against the material to be fastened. A trigger pull or a tap with a hammer on the firing pin drives the fastener home.

The fasteners are specially designed and hardened for use in powder hammers, and the proper power load will drive the fastener home through a 2 × 4 or fairly thick, solid steel. No other type of fastener should ever be used with these devices.

Although they were formerly used only by contractors and industry maintenance people, professional-quality powder hammers have been rental items for some time (they are too expensive to buy only for occasional use), and a light-duty unit should be within the cost range of the do-it-yourselfer. In the interest of time,

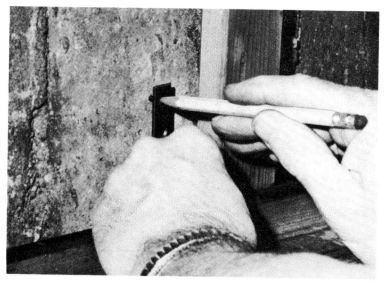

(A) Mark the spot.

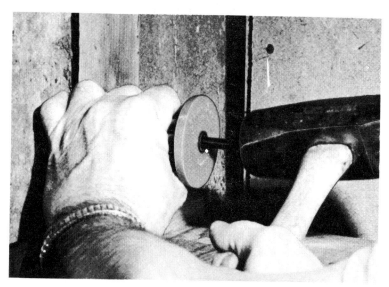

(B) Set the stud according to directions.

Fig. 33-8. Driving a threaded stud into a concrete

(C) The stud held securely in the concrete.

(D) A bracket secures the bench to the wall.

wall, using the equipment shown in Fig. 33-7.

safety, and money, the operating instructions should be followed to the letter.

FASTENERS FOR PLASTER OR DRYWALL

Because of the relatively fragile nature of plaster and drywall in comparison to brick, stone, and concrete, fasteners used with the former must necessarily be different from those used with the latter. Whenever weight of any consequence is involved or a direct outward pull is to be exerted, a fastening is best accomplished with standard wood screws or lag screws inserted through the object to be fastened and driven through the plaster or drywall directly into the studs, rafters, or other framing material beneath. When this is impossible, anchor directly to the plaster or drywall with one or more of the fastening devices discussed in the following paragraphs.

Expansion Anchors

Metal expansion anchors are unsuitable for use with plaster or gypsum board because they tend to crush the walls of the hole into which they are inserted and then fall or pull out easily, assuming they can be tightened in place to begin with. Plastic expansion anchors (Fig. 33-2) are better in this regard, perform their best with radial loads, and poorest of any anchor on axial loads. This poor axial-load performance can be countered, to a degree, by using more than one anchor to support the load, as in a ceiling-mounted traverse rod, for example.

Holes for plastic expansion anchors are best bored with a twist or push drill in both plaster and gypsum board in order to get an accurate fit. Holes jabbed with an ice pick, screwdriver, or similar tool are seldom sized correctly for the best friction fit and may have considerable material knocked away from the edge of the hole, inside the wall, making the site useless for an anchor. Bore the hole the diameter specified on the anchor package and use the screw-size specified there, also. The length of the screw should be equal to the length of the anchor plus the thickness of the object to be fastened, as a minimum. As a general rule, sheet-metal screws work better in plastic anchors than do wood screws, possibly because their comparative lack of body taper causes a more effective expansion of the anchor.

Hollow Wall Screw Anchors

These devices are manufactured by a number of different companies and consist of a metal tube having a large flange at one end and an internally threaded collar at the other. A machine screw is inserted through a hole in the flange, extended the length of the tube, and screwed into the threaded collar (Fig. 33-9).

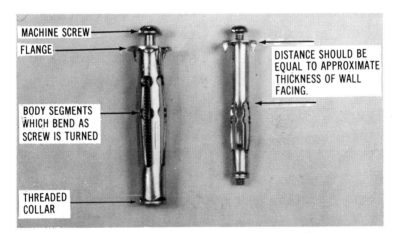

Fig. 33-9. Two sizes of hollow wall screw anchors.

In use, a hole of specified diameter is bored through the gypsum board or plaster and an anchor (Table 33-1) of the proper grip range (depends on the thickness of the drywall or plaster) and screw size (depends on the weight of the object to be anchored) is inserted so that its length is inside the wall and its flange rests against the wall's surface; the anchor is then lightly tapped with the butt of a screwdriver to seat it and prevent it from turning in the hole. The screw is then turned clockwise with a screwdriver.

As the screw is turned, it draws the collar end of the anchor toward the flange end; four slots cut lengthwise into the tube allow the sections of the tube between the slots to bend outward in

715

Table 1. Allowable Carrying Loads for Anchor Bolts.

TYPE FASTENER	SIZE	ALLOWABLE LOAD	
		½" WALLBOARD	⅝" WALLBOARD
HOLLOW WALL	⅛" dia. SHORT	50 LBS.	——
SCREW ANCHORS	³⁄₁₆" dia. SHORT	65 LBS.	——
	¼", ⁵⁄₁₆", ⅜" dia. SHORT	65 LBS.	——
	³⁄₁₆" dia. LONG	——	90 LBS.
	¼", ⁵⁄₁₆", ⅜" dia. LONG	——	95 LBS.
COMMON	⅛" dia.	30 LBS.	90 LBS.
TOGGLE BOLTS	³⁄₁₆" dia.	60 LBS.	120 LBS.
	¼", ⁵⁄₁₆", ⅜" dia.	80 LBS.	120 LBS.

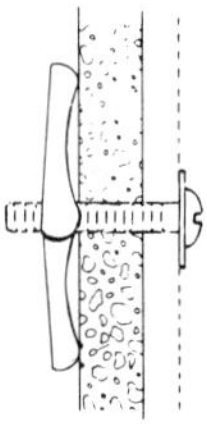

Courtesy National Gypsum Co.

response to pressure from the collar until they lie flat against the inside surface of the wall, drawing the flange tightly against the outside surface and locking the anchor securely in place. The screw is then removed, inserted through whatever object is to be fastened, and replaced in the anchor body, an action which can be performed repeatedly without loosening the anchor body.

Hollow wall screw anchors are also manufactured with pointed screws and tapered threaded collars; these can be driven into the wall without drilling a hole first. Very short anchors are also available for use in thin wood paneling and hollow-core flush doors.

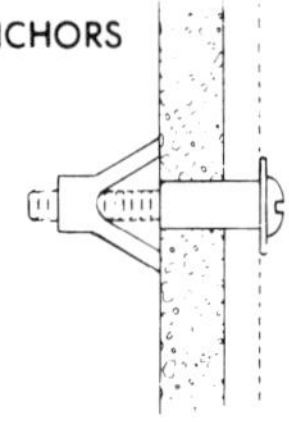

Once they are in place, the anchors are removable only with some ingenuity if a large hole in the wall is to be avoided. One method that works, but needs to be done gently, is to replace the screw in the anchor with another one of the same size and thread, but longer. This replacement screw needs to be threaded into the anchor only one turn, if at all. Once it is in place, smack the head of the screw with a hammer. This action may straighten out the bent legs of the anchor so that it can be withdrawn from the wall intact; more frequently, it will either break off the legs or break off the flange. Remove the screw, and in the former case, pull the flanged section of the anchor out of the wall with the fingers (if it won't come out all the way, pull it out as far as possible, cut it in two with diagonal cutters, and let the stubborn half drop down inside the wall cavity). If the flange breaks off, push the remainder of the anchor back inside the wall. The only hole to be patched will be the one originally bored for the anchor, although a too-vigorous hammer blow can produce an additional dimple in the wall surface. The method is a little risky with very soft or very thin wall facings because their relative lack of substance may allow the flange to be driven backward through the facing; let good judgment be the guide—it may be better to leave the anchor alone and learn to live with it.

Toggle Bolts

These differ from hollow wall anchors in that they must be attached to the object to be fastened before they are inserted into the wall. In their larger sizes, they will carry a heavier load (Table 33-1); in all sizes, they are good for axial as well as radial loads, need a larger hole for mounting, and once mounted, cannot be reused if the screw is removed from the toggle. A longer screw is necessary in order to allow the wings of the toggle to unfold in the wall cavity.

Toggle bolts (Fig. 33-10) are simple devices consisting of a center-hinged crosspiece pierced in the middle by a long machine screw. The crosspiece (called the toggle) is composed of two halves (called the wings) hinged around a threaded center through which the screw runs. The wings are normally held at almost a right angle to the screw by spring pressure, but can be folded flat along the screw to allow insertion into the wall. Once inside the wall cavity, they automatically snap upright again (they fold only one way—toward the head of the screw) and prevent removal of the unit. Tightening the screw squeezes the toggle firmly against the inner wall surface and the object to be fastened against the outer wall surface. Removing the screw allows the toggle to drop into the wall cavity; hence, the unit is easily removed, but in most

717

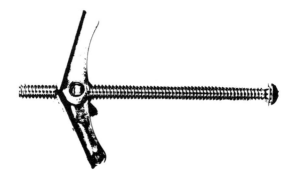

Fig. 33-10. Common toggle bolt.

cases, is not reusable because the toggle cannot be recovered.

Toggle bolts are commonly available with screw diameters from ⅛-in. to ⅜-in. and screw lengths to 6 inches, although they can be fitted with screws of any maximum practical length by using threaded rod. Minimum screw length should equal the thickness of the object to be fastened, plus the thickness of the wall facing, plus the length of the wings when folded, plus ¼ in. The maximum length should not exceed the minimum by much, if at all, or the screw can bottom against the opposite wall facing and be impossible to tighten.

Hanger Bolts, Dowel Screws, and Toggle Studs

Whenever a hook is to be installed to a plaster or gypsum board ceiling, it should be driven through the ceiling material and into a joist if at all possible. Since most common ornamental hooks are supplied with female machine threads, a device called a hanger bolt (Fig. 33-11) is necessary to accomplish this.

Hanger bolts are relatively short lengths of steel rod having a machine-screw thread on one end and a wood-screw thread on the other. The machine-screw thread is turned into the ornamental hook until it bottoms, and then the whole unit is turned to sink the wood-screw threads into the ceiling and ceiling joist. A pilot hole should be bored into the joist to make turning easier and to prevent possible breakage of the hook if too much twisting force is applied.

Fig. 33-11. Hanger bolt with wood screw threads at top and machine-screw threads at bottom.

Although ornamental hooks are mentioned here as an example, almost any device can be mounted to the ceiling or wall by using a hanger bolt in a similar manner or by using a standard nut on the machine-screw threads protruding from the wall or ceiling.

Dowel screws, which are identical to hanger bolts except that they have wood-screw thread on both ends, can be used to fasten something of wood to a ceiling or wall in the same way.

If an object having female machine threads cannot be fastened to a ceiling joist or wall stud by a hanger bolt, a toggle stud is used instead. All this amounts to is a toggle bolt without a head on the screw, so that the screw can be turned into the female machine threads in the device to be mounted. A toggle stud will not bear as much weight as a hanger bolt, but should be adequate for a small-to-medium-size flower pot and hanger. It is used the same as a toggle bolt.

Glossary of Building Terms

Anchor Bolts—Bolts to secure a wooden sill to a concrete or masonry floor, foundation, or wall.

Astragal—A moulding, attached to one of a pair of swinging doors, against which the other door strikes.

Balusters—Small spindles or members forming the main part of a railing for a stairway or balcony, fastened between a bottom and top rail.

Base Shoe—Moulding used next to the floor on interior baseboard. Sometimes called a carpet strip.

Batter Board—One of a pair of horizontal boards nailed to posts set at the corners of an excavation, used to indicate the desired level. Also a fastening for stretched strings to indicate the outlines of foundation walls.

Bearing Wall—A wall that supports any vertical load in addition to its own weight.

Blind-Nailing—Nailing in such a way that the nailheads are not visible on the face of the work.

Casing—Wide moulding of various widths and thicknesses used to trim door and window openings.

Casement Frames and Sash—Frames of wood or metal enclosing part or all of the sash which may be opened by means of hinges affixed to the vertical edges.

Collar Beam—A beam connecting pairs of opposite rafters above the attic floor.

Corner Bead—A strip of formed iron, sometimes combined with a strip of metal lath, placed on the corners before plastering, to reinforce them. Also, a strip of wood finish three-quarters-round or angular placed over a plastered corner for protection.

Corner Boards—Used as trim for the external corner of a house or other frame structure against which the ends of the siding are finished.

Corner Braces—Diagonal brackets let into studs to reinforce corners of frame structures.

Cornice—A decorative element made up of moulded members usually placed at or near the top of an exterior or interior wall.

Cornice Return—That portion of the cornice that returns on the gable end of a house.

Counterflashing—A flashing usually used on chimneys at the roof-line to cover shingle flashing and to prevent moisture entry.

Cove Moulding—A three-sided moulding with concave face used wherever small angles are to be covered.

Crawl Space—A shallow space below the living quarters of a house. It is generally not excavated or paved and is often enclosed for appearance by a skirting or facing material.

Cricket—A small drainage diverting roof structure of single or double slope placed at the junction of larger surfaces that meet at an angle.

Crown Moulding—A moulding used on a cornice or wherever a large angle is to be covered.

Dado—A rectangular groove in a board or plank. In interior decoration, a special type of wall treatment.

Direct Nailing—To nail perpendicular to the initial surface or to the junction of the pieces joined. Also called *face nailing*.

Dressed and Matched (Tongue and Groove)—Boards or planks machined in such a manner that there is a groove on one edge and a corresponding tongue on the other.

Eaves—The margin or lower part of a roof projecting over the wall.

Expansion Joint—A fiber, rubber, or plastic strip to separate blocks or units of concrete to prevent cracking due to expansion as a result of temperature changes.

Facia or Fascia—A flat board, band, or face, used sometimes by itself but usually in combination with mouldings, often located at the outer face or the cornice.

Fire Stop—a solid, tight closure of a concealed space, placed to prevent the spread of fire and smoke through such a space.

Flue Lining—Fire clay or terra-cotta pipe, round or square, usually made in all of the ordinary flue sizes and in 2-ft. lengths, used for the inner lining of chimneys with the brick or masonry work around the outside.

Footing—The spreading course or courses at the base or bottom of a foundation wall, pier, or column.

Foundation—The supporting portion of a structure below the first-floor construction, or below grade, including the footings.

Frieze—Any sculptured or ornamental band in a building. Also the horizontal member of a cornice set vertically against the wall.

Frostline—The depth of frost penetration in soil. This depth varies in different parts of the country. Footings should be placed below this depth to prevent movement.

Gable—That portion of a wall contained between the slope of a single-sloped roof and a line projected horizontally through the lowest elevation of the roof construction.

Girder—A large or principal beam used to support concentrated loads at isolated points along its length.

Grain, Edge (Vertical)—Edge-grain lumber has been sawed parallel to the pith of the log and approximately at right angles to the growth rings; i.e., the rings form an angle of 45° or more with the surface of the piece.

Grain, Flat—Flat-grain lumber has been sawed parallel to the pith of the log and approximately at right angles to the growth rings; i.e., the rings form an angle of less than 45° with the surface of the piece.

Grounds—Strips of wood, of the same thickness at the lath and plaster, that are attached to walls before plastering. Used around windows, doors, and other openings as a plaster stop and in other places for the purpose of attaching baseboards or other trim.

Header—(a) A beam placed perpendicular to joists and to which joists are nailed in framing for chimney, stairway, or other opening. (b) A wood lintel.

Hip—The external angle formed by the meeting of two sloping sides of a roof.

Insulating Board or Fiberboard—A low-density board made of wood, sugarcane, cornstalks, or similar materials, usually formed by a felting process, dried and usually pressed to thicknesses of $\frac{1}{2}$ and $\frac{5}{32}$ in.

Jack rafter—A rafter that spans the distance from the wallplate to a hip, or from a valley to a ridge.

Jamb—The side post or lining of a doorway, window, or other opening.

Joint Cement—A powder that is usually mixed with water and used for joint treatment in gypsum-wallboard finish. Often called "spracle."

Joist—One of a series of parallel beams used to support floor and ceiling loads, and supported in turn by larger beams, girders, or bearing walls.

Landing—A platform between flights of stairs or at the termination of a flight of stairs.

Ledger Strip—A strip of lumber nailed along the bottom of the side of a girder on which joists rest.

Lintel—A horizontal structural member that supports the load over an opening such as a door or window.

Lumber, Boards—Wood which is less than 2 in. thick and 2 or more in. wide.

Lumber, Dimension—Boards ranging from 2 in. to 5 in. thick, and 2 or more inches wide. Includes joists, rafters, studding, planks, and small timbers.

Mantel—The shelf above a fireplace. Originally referred to the beam or lintel supporting the arch above the fireplace opening.

Mullion—A slender bar or pier forming a division between panels or units of windows, screens, or similar frames.

Muntin—The members dividing the glass or openings of sash, doors, and the like.

Newel—Any post to which a stair railing or balustrade is fastened.

Paper, Building—A general term for papers, felts, and similar sheet material used in buildings without reference to their properties or uses.

Penny—As applied to nails, it originally indicated the price per hundred. The term now serves as a measure of nail length and is abbreviated by the letter d.

Pitch—The incline or rise of a roof. Pitch is expressed in inches or rise per foot of run, or by the ratio of the rise to the span.

Rabbet—A rectangular longitudinal groove cut in the corner of a board or other piece of material.

Rafter, Hip—A rafter that forms the intersection of an external roof angle.

Rafter, Jack—A rafter that spans the distance from a wallplate to a hip or from a valley to a ridge.

Rafter, Valley—A rafter that forms the intersection of an internal roof angle.

Ridge Board—The board placed on edge at the ridge of the roof to support the upper ends of the rafters.

Rise—The height a roof rising in horizontal distance (run) from the outside face of a wall supporting the rafters or trusses to the ridge of the roof. In stairs, the perpendicular height of a step or flight of steps.

Riser—Each of the vertical boards closing the spaces between the treads of the stairways.

Run—In reference to roofs, the horizontal distance from the face of a wall to the ridge of the roof. Referring to stairways, the net width of a step; also the horizontal distance covered by a flight of steps.

Sash—A single frame containing one or more panes of glass.

Siding, Bevel (Lap Siding)—Used as the finish siding on the exterior of a house or other structure. It is usually manufactured by resawing dry square-surfaced boards diagonally to produce two wedge-shaped pieces. These pieces commonly run from $13/16$ in. thick on the thin edge to $\frac{1}{2}$ to $\frac{3}{4}$ in. thick on the other edge, depending on the width of the siding.

Siding, Drop—Usually $\frac{3}{4}$ in. thick and 6 in. wide, machined into various patterns. Drop siding has tounge-and-groove joints, is heavier, has more structural strength, and is frequently used on buildings that require no sheathing, such as garages and barns.

Sill—the lowest member of the frame of a structure, resting on the foundation and supporting the uprights of the frame.

Tail Beam—A relatively short beam or joist supported in a wall on one end and by a header on the other.

726

Trimmer—A beam or joist to which a header is nailed in framing for a chimney, stairway, or other opening.

Toenailing—To drive a nail at a slant with the initial surface in order to permit it to penetrates into a second member.

Truss—A frame or jointed structure designed to act as a beam of long span, while each member is usually subjected to longitudinal stress only, either tension or compression.

Index

**Over a Century of Excellence
for the Professional
and
Vocational Trades and the Crafts**

Order now from your local bookstore
or use the convenient order form
at the back of this book.

AUDEL

These fully illustrated, up-to-date guides and manuals mean a better job done for mechanics, engineers, electricians, plumbers, carpenters, and all skilled workers.

CONTENTS

ELECTRICAL

HOUSE WIRING (Sixth Edition)
ROLAND E. PALMQUIST
5 1/2 x 8 1/4   Hardcover   256 pp.   150 Illus.
ISBN: 0-672-23404-1 $14.95
The rules and regulations of the National Electrical Code as they apply to residential wiring fully detailed with examples and illustrations.

PRACTICAL ELECTRICITY
(Fifth Edition)
ROBERT G. MIDDLETON;
revised by L. DONALD MEYERS
5 1/2 x 8 1/4   Hardcover   512 pp.   335 Illus.
ISBN: 0-02-584561-6 $19.95
The fundamentals of electricity for electrical workers, apprentices, and others requiring concise information about electric principles and their practical applications.

GUIDE TO THE 1987 NATIONAL ELECTRICAL CODE
ROLAND E. PALMQUIST
5 1/2 x 8 1/4   Hardcover   664 pp.   225 Illus.
ISBN: 0-02-594560-2 $22.50
The most authoritative guide available to interpreting the National Electrical Code for electricians, contractors, electrical inspectors, and homeowners. Examples and illustrations.

MATHEMATICS FOR ELECTRICIANS AND ELECTRONICS TECHNICIANS
REX MILLER
5 1/2 x 8 1/4   Hardcover   312 pp.   115 Illus.
ISBN: 0-8161-1700-4 $14.95
Mathematical concepts, formulas, and problem-solving techniques utilized on-the-job by electricians and those in electronics and related fields.

FRACTIONAL-HORSEPOWER ELECTRIC MOTORS
REX MILLER and
MARK RICHARD MILLER
5 1/2 x 8 1/4   Hardcover   436 pp.   285 Illus.
ISBN: 0-672-23410-6 $15.95
The installation, operation, maintenance, repair, and replacement of the small-to-moderate-size electric motors that power home appliances and industrial equipment.

ELECTRIC MOTORS (Fourth Edition)
EDWIN P. ANDERSON;
revised by REX MILLER
5 1/2 x 8 1/4   Hardcover   656 pp.   405 Illus.
ISBN: 0-672-23376-2 $14.95
Installation, maintenance, and repair of all types of electric motors.

HOME APPLIANCE SERVICING (Fourth Edition)
EDWIN P. ANDERSON;
revised by REX MILLER
5 1/2 x 8 1/4   Hardcover   640 pp.   345 Illus.
ISBN: 0-672-23379-7 $22.50
The essentials of testing, maintaining, and repairing all types of home appliances.

TELEVISION SERVICE MANUAL (Fifth Edition)

ROBERT G. MIDDLETON;
revised by JOSEPH G. BARRILE

5 1/2 x 8 1/4   Hardcover   512 pp.   395 Illus.
ISBN: 0-672-23395-9 $16.95

A guide to all aspects of television transmission and reception, including the operating principles of black and white and color receivers. Step-by-step maintenance and repair procedures.

ELECTRICAL COURSE FOR APPRENTICES AND JOURNEYMEN (Third Edition)

ROLAND E. PALMQUIST

5 1/2 x 8 1/4   Hardcover   478 pp.   290 Illus.
ISBN: 0-02-594550-5 $19.95

This practical course in electricity for those in formal training programs or learning on their own provides a thorough understanding of operational theory and its applications on the job.

QUESTIONS AND ANSWERS FOR ELECTRICIANS EXAMINATIONS (Ninth Edition)

ROLAND E. PALMQUIST

5 1/2 x 8 1/4   Hardcover   316 pp.   110 Illus.
ISBN: 0-02-594691-9 $18.95

Based on the 1987 National Electrical Code, this book reviews the subjects included in the various electricians examinations—apprentice, journeyman, and master. Question and Answer format.

MACHINE SHOP AND
MECHANICAL TRADES

MACHINISTS LIBRARY
(Fourth Edition, 3 Vols.)

REX MILLER

5 1/2 x 8 1/4   Hardcover   1352 pp.   1120 Illus.
ISBN: 0-672-23380-0 $52.95

An indispensable three-volume reference set for machinists, tool and die makers, machine operators, metal workers, and those with home workshops. The principles and methods of the entire field are covered in an up-to-date text, photographs, diagrams, and tables.

Volume I: Basic Machine Shop
REX MILLER

5 1/2 x 8 1/4   Hardcover   392 pp.   375 Illus.
ISBN: 0-672-23381-9 $17.95

Volume II: Machine Shop
REX MILLER

5 1/2 x 8 1/4   Hardcover   528 pp.   445 Illus.
ISBN: 0-672-23382-7 $19.95

Volume III: Toolmakers Handy Book
REX MILLER

5 1/2 x 8 1/4   Hardcover   432 pp.   300 Illus.
ISBN: 0-672-23383-5 $14.95

MATHEMATICS FOR MECHANICAL TECHNICIANS AND TECHNOLOGISTS

JOHN D. BIES

5 1/2 x 8 1/4   Hardcover   342 pp.   190 Illus.
ISBN: 0-02-510620-1 $17.95

The mathematical concepts, formulas, and problem-solving techniques utilized on the job by engineers, technicians, and other workers in industrial and mechanical technology and related fields.

MILLWRIGHTS AND MECHANICS GUIDE
(Fourth Edition)

CARL A. NELSON

5 1/2 x 8 1/4   Hardcover   1,040 pp.   880 Illus.
ISBN: 0-02-588591-x $29.95

The most comprehensive and authoritative guide available for millwrights, mechanics, maintenance workers, riggers, shop workers, foremen, inspectors, and superintendents on plant installation, operation, and maintenance.

WELDERS GUIDE (Third Edition)

JAMES E. BRUMBAUGH

5 1/2 x 8 1/4   Hardcover   960 pp.   615 Illus.
ISBN: 0-672-23374-6 $23.95

The theory, operation, and maintenance of all welding machines. Covers gas welding equipment, supplies, and process; arc welding equipment, supplies, and process; TIG and MIG welding; and much more.

WELDERS/FITTERS GUIDE

JOHN P. STEWART

8 1/2 x 11   Paperback   160 pp.   195 Illus.
ISBN: 0-672-23325-8 $7.95

Step-by-step instruction for those training to become welders/fitters who have some knowledge of welding and the ability to read blueprints.

SHEET METAL WORK

JOHN D. BIES

5 1/2 x 8 1/4   Hardcover   456 pp.   215 Illus.
ISBN: 0-8161-1706-3 $19.95

An on-the-job guide for workers in the manufacturing and construction industries and for those with home workshops. All facets of sheet metal work detailed and illustrated by drawings, photographs, and tables.

POWER PLANT ENGINEERS GUIDE (Third Edition)

FRANK D. GRAHAM;
revised by CHARLIE BUFFINGTON

5 1/2 x 8 1/4   Hardcover   960 pp.   530 Illus.
ISBN: 0-672-23329-0 $27.50

This all-inclusive, one-volume guide is perfect for engineers, firemen, water tenders, oilers, operators of steam and diesel-power engines, and those applying for engineer's and firemen's licenses.

MECHANICAL TRADES POCKET MANUAL (Second Edition)

CARL A. NELSON

4 x 6   Paperback   364 pp.   255 Illus.
ISBN: 0-672-23378-9   10.95

A handbook for workers in the industrial and mechanical trades on methods, tools, equipment, and procedures. Pocket-sized for easy reference and fully illustrated.

PLUMBING

PLUMBERS AND PIPE FITTERS LIBRARY (Fourth Edition, 3 Vols.)

CHARLES N. McCONNELL

5 1/2 x 8 1/4   Hardcover   952 pp.   560 Illus.
ISBN: 0-02-582914-9 $68.45

This comprehensive three-volume set contains the most up-to-date information available for master plumbers, journeymen, apprentices, engineers, and those in the building trades. A detailed text and clear diagrams, photographs, and charts and tables treat all aspects of the plumbing, heating, and air conditioning trades.

Volume I: Materials, Tools, Roughing-In

CHARLES N. McCONNELL;
revised by TOM PHILBIN

5 1/2 x 8 1/4   Hardcover   304 pp.   240 Illus.
ISBN: 0-02-582911-4 $20.95

Volume II: Welding, Heating, Air Conditioning

CHARLES N. McCONNELL;
revised by TOM PHILBIN

5 1/2 x 8 1/4   Hardcover   384 pp.   220 Illus.
ISBN: 0-02-582912-2 $22.95

Volume III: Water Supply, Drainage, Calculations

CHARLES N. McCONNELL;
revised by TOM PHILBIN

5 1/2 x 8 1/4   Hardcover   264 pp.   100 Illus.
ISBN: 0-02-582913-0 $20.95

HOME PLUMBING HANDBOOK (Third Edition)

CHARLES N. McCONNELL

8 1/2 x 11   Paperback   200 pp.   100 Illus.
ISBN: 0-672-23413-0 $13.95

An up-to-date guide to home plumbing installation and repair.

THE PLUMBERS HANDBOOK (Seventh Edition)

JOSEPH P. ALMOND, SR.

4 x 6   Paperback   352 pp.   170 Illus.
ISBN: 0-672-23419-x $11.95

A handy sourcebook for plumbers, pipe fitters, and apprentices in both trades. It has a rugged binding suited for use on the job, and fits in the tool box or conveniently in the pocket.

QUESTIONS AND ANSWERS FOR PLUMBERS EXAMINATIONS (Second Edition)

JULES ORAVITZ

5 1/2 x 8 1/4   Paperback   256 pp.   145 Illus.
ISBN: 0-8161-1703-9 $9.95

A study guide for those preparing to take a licensing examination for apprentice, journeyman, or master plumber. Question and answer format.

HVAC

AIR CONDITIONING: HOME AND COMMERCIAL (Second Edition)

EDWIN P. ANDERSON;
revised by REX MILLER

5 1/2 x 8 1/4   Hardcover   528 pp.   180 Illus.
ISBN: 0-672-23397-5 $15.95

A guide to the construction, installation, operation, maintenance, and repair of home, commercial, and industrial air conditioning systems.

HEATING, VENTILATING, AND AIR CONDITIONING LIBRARY
(Second Edition, 3 Vols.)
JAMES E. BRUMBAUGH
*5 1/2 x 8 1/4 Hardcover 1,840 pp. 1,275 Illus.
ISBN: 0-672-23388-6 $53.95*
An authoritative three-volume reference library for those who install, operate, maintain, and repair HVAC equipment commercially, industrially, or at home.

Volume I: Heating Fundamentals, Furnaces, Boilers, Boiler Conversions
JAMES E. BRUMBAUGH
*5 1/2 x 8 1/4 Hardcover 656 pp. 405 Illus.
ISBN: 0-672-23389-4 $17.95*

Volume II: Oil, Gas and Coal Burners, Controls, Ducts, Piping, Valves
JAMES E. BRUMBAUGH
*5 1/2 x 8 1/4 Hardcover 592 pp. 455 Illus.
ISBN: 0-672-23390-8 $17.95*

Volume III: Radiant Heating, Water Heaters, Ventilation, Air Conditioning, Heat Pumps, Air Cleaners
JAMES E. BRUMBAUGH
*5 1/2 x 8 1/4 Hardcover 592 pp. 415 Illus.
ISBN: 0-672-23391-6 $17.95*

OIL BURNERS (Fourth Edition)
EDWIN M. FIELD
*5 1/2 x 8 1/4 Hardcover 360 pp. 170 Illus.
ISBN: 0-672-23394-0 $16.95*
An up-to-date sourcebook on the construction, installation, operation, testing, servicing, and repair of all types of oil burners, both industrial and domestic.

REFRIGERATION: HOME AND COMMERCIAL (Second Edition)
EDWIN P. ANDERSON;
revised by REX MILLER
*5 1/2 x 8 1/4 Hardcover 768 pp. 285 Illus.
ISBN: 0-672-23396-7 $19.95*
A reference for technicians, plant engineers, and the home owner on the installation, operation, servicing, and repair of everything from single refrigeration units to commercial and industrial systems.

PNEUMATICS AND HYDRAULICS

HYDRAULICS FOR OFF-THE-ROAD EQUIPMENT (Second Edition)
HARRY L. STEWART;
revised by TOM PHILBIN
*5 1/2 x 8 1/4 Hardcover 256 pp. 175 Illus.
ISBN: 0-8161-1701-2 $13.95*

This complete reference manual on heavy equipment covers hydraulic pumps, accumulators, and motors; force components; hydraulic control components; filters and filtration, lines and fittings, and fluids; hydrostatic transmissions; maintenance; and troubleshooting.

PNEUMATICS AND HYDRAULICS (Fourth Edition)
HARRY L. STEWART;
revised by TOM STEWART
*5 1/2 x 8 1/4 Hardcover 512 pp. 315 Illus.
ISBN: 0-672-23412-2 $19.95*
The principles and applications of fluid power. Covers pressure, work, and power; general features of machines; hydraulic and pneumatic symbols; pressure boosters; air compressors and accessories; and much more.

PUMPS (Fourth Edition)
HARRY STEWART;
revised by TOM PHILBIN
*5 1/2 x 8 1/4 Hardcover 508 pp. 360 Illus.
ISBN: 0-672-23400-9 $17.95*
The principles and day-to-day operation of pumps, pump controls, and hydraulics are thoroughly detailed and illustrated.

CARPENTRY AND CONSTRUCTION

CARPENTERS AND BUILDERS LIBRARY (Fifth Edition, 4 Vols.)
JOHN E. BALL; revised by TOM PHILBIN
*5 1/2 x 8 1/4 Hardcover 1,224 pp. 1,010 Illus.
ISBN: 0-672-23369-x $43.95*

Also available as a boxed set at no extra cost:
ISBN: 0-02-506450-9 $43.95

This comprehensive four-volume library has set the professional standard for decades for carpenters, joiners, and woodworkers.

Volume I: Tools, Steel Square, Joinery
JOHN E. BALL; revised by TOM PHILBIN
*5 1/2 x 8 1/4 Hardcover 384 pp. 345 Illus.
ISBN: 0-672-23365-7 $10.95*

Volume II: Builders Math, Plans, Specifications
JOHN E. BALL; revised by TOM PHILBIN
*5 1/2 x 8 1/4 Hardcover 304 pp. 205 Illus.
ISBN: 0-672-23366-5 $10.95*

Volume III: Layouts, Foundations, Framing
JOHN E. BALL; revised by TOM PHILBIN
*5 1/2 x 8 1/4 Hardcover 272 pp. 215 Illus.
ISBN: 0-672-23367-3 $10.95*

Volume IV: Millwork, Power Tools, Painting

JOHN E. BALL; revised by TOM PHILBIN

5 1/2 x 8 1/4   Hardcover   344 pp.   245 Illus.
ISBN: 0-672-23368-1 $10.95

COMPLETE BUILDING CONSTRUCTION (Second Edition)

JOHN PHELPS; revised by TOM PHILBIN

5 1/2 x 8 1/4   Hardcover   744 pp.   645 Illus.
ISBN: 0-672-23377-0 $22.50

Constructing a frame or brick building from the footings to the ridge. Whether the building project is a tool shed, garage, or a complete home, this single fully illustrated volume provides all the necessary information.

COMPLETE ROOFING HANDBOOK

JAMES E. BRUMBAUGH

5 1/2 x 8 1/4   Hardcover   536 pp.   510 Illus.
ISBN: 0-02-517850-4 $29.95

Covers types of roofs; roofing and reroofing; roof and attic insulation and ventilation; skylights and roof openings; dormer construction; roof flashing details; and much more.

COMPLETE SIDING HANDBOOK

JAMES E. BRUMBAUGH

5 1/2 x 8 1/4   Hardcover   512 pp.   450 Illus.
ISBN: 0-02-517880-6 $24.95

This companion volume to the *Complete Roofing Handbook* includes comprehensive step-by-step instructions and accompanying line drawings on every aspect of siding a building.

MASONS AND BUILDERS LIBRARY (Second Edition, 2 Vols.)

LOUIS M. DEZETTEL;
revised by TOM PHILBIN

5 1/2 x 8 1/4   Hardcover   688 pp.   500 Illus.
ISBN: 0-672-23401-7 $27.95

This two-volume set provides practical instruction in bricklaying and masonry. Covers brick; mortar; tools; bonding; corners, openings, and arches; chimneys and fireplaces; structural clay tile and glass block; brick walls; and much more.

Volume I: Concrete, Block, Tile, Terrazzo

LOUIS M. DEZETTEL;
revised by TOM PHILBIN

5 1/2 x 8 1/4   Hardcover   304 pp.   190 Illus.
ISBN: 0-672-23402-5 $13.95

Volume 2: Bricklaying, Plastering, Rock Masonry, Clay Tile

LOUIS M. DEZETTEL;
revised by TOM PHILBIN

5 1/2 x 8 1/4   Hardcover   384 pp.   310 Illus.
ISBN: 0-672-23403-3 $13.95

WOODWORKING

WOOD FURNITURE: FINISHING, REFINISHING, REPAIRING (Second Edition)

JAMES E. BRUMBAUGH

5 1/2 x 8 1/4   Hardcover   352 pp.   185 Illus.
ISBN: 0-672-23409-2 $12.95

A fully illustrated guide to repairing furniture and finishing and refinishing wood surfaces. Covers tools and supplies; types of wood; veneering; inlaying; repairing, restoring, and stripping; wood preparation; and much more.

WOODWORKING AND CABINETMAKING

F. RICHARD BOLLER

5 1/2 x 8 1/4   Hardcover   360 pp.   455 Illus.
ISBN: 0-02-512800-0 $18.95

Essential information on all aspects of working with wood. Step-by-step procedures for woodworking projects are accompanied by detailed drawings and photographs.

MAINTENANCE AND REPAIR

BUILDING MAINTENANCE (Second Edition)

JULES ORAVETZ

5 1/2 x 8 1/4   Hardcover   384 pp.   210 Illus.
ISBN: 0-672-23278-2 $11.95

Professional maintenance procedures used in office, educational, and commercial buildings. Covers painting and decorating; plumbing and pipe fitting; concrete and masonry; and much more.

GARDENING, LANDSCAPING AND GROUNDS MAINTENANCE (Third Edition)

JULES ORAVETZ

5 1/2 x 8 1/4   Hardcover   424 pp.   340 Illus.
ISBN: 0-672-23417-3 $15.95

Maintaining lawns and gardens as well as industrial, municipal, and estate grounds.

HOME MAINTENANCE AND REPAIR: WALLS, CEILINGS AND FLOORS

GARY D. BRANSON

8 1/2 x 11   Paperback   80 pp.   80 Illus.
ISBN: 0-672-23281-2 $6.95

The do-it-yourselfer's guide to interior remodeling with professional results.

PAINTING AND DECORATING

REX MILLER and GLEN E. BAKER

5 1/2 x 8 1/4   Hardcover   464 pp.   325 Illus.
ISBN: 0-672-23405-x $18.95

A practical guide for painters, decorators, and homeowners to the most up-to-date materials and techniques in the field.

TREE CARE (Second Edition)

JOHN M. HALLER

8 1/2 x 11   Paperback   224 pp.   305 Illus.
ISBN: 0-02-062870-6 $9.95

The standard in the field. A comprehensive guide for growers, nursery owners, foresters, landscapers, and homeowners to planting, nurturing and protecting trees.

UPHOLSTERING (Updated)

JAMES E. BRUMBAUGH

5 1/2 x 8 1/4   Hardcover   400 pp.   380 Illus.
ISBN: 0-672-23372-x $15.95

The essentials of upholstering fully explained and illustrated for the professional, the apprentice, and the hobbyist.

AUTOMOTIVE AND ENGINES

DIESEL ENGINE MANUAL
(Fourth Edition)

PERRY O. BLACK;
revised by WILLIAM E. SCAHILL

5 1/2 x 8 1/4   Hardcover   512 pp.   255 Illus.
ISBN: 0-672-23371-1 $15.95

The principles, design, operation, and maintenance of today's diesel engines. All aspects of typical two- and four-cycle engines are thoroughly explained and illustrated by photographs, line drawings, and charts and tables.

GAS ENGINE MANUAL
(Third Edition)

EDWIN P. ANDERSON;
revised by CHARLES G. FACKLAM

5 1/2 x 8 1/4   Hardcover   424 pp.   225 Illus.
ISBN: 0-8161-1707-1 $12.95

How to operate, maintain, and repair gas engines of all types and sizes. All engine parts and step-by-step procedures are illustrated by photographs, diagrams, and troubleshooting charts.

SMALL GASOLINE ENGINES

REX MILLER and MARK RICHARD MILLER

5 1/2 x 8 1/4   Hardcover   640 pp.   525 Illus.
ISBN: 0-672-23414-9 $16.95

Practical information for those who repair, maintain, and overhaul two- and four-cycle engines—including lawn mowers, edgers, grass sweepers, snowblowers, emergency electrical generators, outboard motors, and other equipment with engines of up to ten horsepower.

TRUCK GUIDE LIBRARY (3 Vols.)

JAMES E. BRUMBAUGH

5 1/2 x 8 1/4   2,144 pp.   1,715 Illus.
ISBN: 0-672-23392-4 $45.95

This three-volume set provides the most comprehensive, profusely illustrated collection of information available on truck operation and maintenance.

Volume 1: Engines

JAMES E. BRUMBAUGH

5 1/2 x 8 1/4   Hardcover   416 pp.   290 Illus.
ISBN: 0-672-23356-8 $16.95

Volume 2: Engine Auxiliary Systems

JAMES E. BRUMBAUGH

5 1/2 x 8 1/4   Hardcover   704 pp.   520 Illus.
ISBN: 0-672-23357-6 $16.95

Volume 3: Transmissions, Steering, and Brakes

JAMES E. BRUMBAUGH

5 1/2 x 8 1/4   Hardcover   1,024 pp.   905 Illus.
ISBN: 0-672-23406-8 $16.95

DRAFTING

INDUSTRIAL DRAFTING

JOHN D. BIES

5 1/2 x 8 1/4   Hardcover   544 pp.   Illus.
ISBN: 0-02-510610-4 $24.95

Professional-level introductory guide for practicing drafters, engineers, managers, and technical workers in all industries who use or prepare working drawings.

ANSWERS ON BLUEPRINT
READING (Fourth Edition)
ROLAND PALMQUIST;
revised by THOMAS J. MORRISEY

5 1/2 x 8 1/4   Hardcover   320 pp.   275 Illus.
ISBN: 0-8161-1704-7 $12.95

Understanding blueprints of machines and tools, electrical systems, and architecture. Question and answer format.

HOBBIES

COMPLETE COURSE IN STAINED GLASS
PEPE MENDEZ

8 1/2 x 11   Paperback   80 pp.   50 Illus.
ISBN: 0-672-23287-1 $8.95

The tools, materials, and techniques of the art of working with stained glass.

.